Springer Advanced Texts in Chemistry

Springer
New York
Berlin
Heidelberg
Barcelona
Hong Kong
London
Milan
Paris
Singapore
Tokyo

Springer Advanced Texts in Chemistry

Series Editor: Charles R. Cantor

Principles of Protein Structure
G.E. Schulz and R.H. Schirmer

Bioorganic Chemistry: A Chemical Approach to Enzyme Action
(Third Edition)
H. Dugas

Protein Purification: Principles and Practice (Third Edition)
R.K. Scopes

Principles of Nucleic Acid Structure
W. Saenger

Biomembranes: Molecular Structure and Function
R.B. Gennis

Basic Principles and Techniques of Molecular Quantum Mechanics
R.E. Christoffersen

Energy Transduction in Biological Membranes:
A Textbook of Bioenergetics
W.A. Cramer and D.B. Knaff

Principles of Protein X-Ray Crystallography (Second Edition)
J. Drenth

Essentials of Carbohydrate Chemistry
J.F. Robyt

Jan Drenth

Principles of Protein X-Ray Crystallography

Second Edition

With 154 Illustrations

 Springer

Jan Drenth
Laboratory of Biophysical Chemistry
Nijenborgh 4
9747 AG Groningen
The Netherlands
j.drenth@chem.rug.nl

Series Editor:
Charles R. Cantor
Boston University
Center for Advanced Biotechnology
Boston, MA 02215
USA

Cover illustration: Courtesy of Adrian R. Ferré-D'Amaré and Stephen K. Burley, The Rockefeller University. The four-α-helix bundle moiety of transcription factor Max. Reproduced with permission from *Nature* 363:38–45 (1993).

Library of Congress Cataloging-in-Publication Data
Drenth, Jan.
    Principles of protein x-ray crystallography / Jan Drenth.—[2nd.
  ed.]
      p. cm.—(Springer advanced texts in chemistry)
    Includes bibliographical references and index.
    ISBN 0-387-98587-5 (hardcover : alk. paper)
    1. Proteins—Analysis.   2. X-ray crystallography.   I. Title.
  II. Series.
  QD431.D84   1999
  547′.75046—dc21   98-26970

Printed on acid-free paper.

Production coordinated by Chernow Editorial Services, Inc., and managed by Francine McNeill; manufacturing supervised by Nancy Wu.
Typeset by Best-set Typesetter Ltd., Hong Kong.
Printed and bound by Sheridan Books, Inc., Ann Arbor, MI.
Printed in the United States of America.

9 8 7 6 5 4 3 2 (Corrected second printing, 2002)

ISBN 0-387-98587-5   SPIN 10860622

Springer-Verlag   New York Berlin Heidelberg
A member of BertelsmannSpringer Science+Business Media GmbH

Series Preface

New textbooks at all levels of chemistry appear with great regularity. Some fields such as basic biochemistry, organic reaction mechanisms, and chemical thermodynamics are well represented by many excellent texts, and new or revised editions are published sufficiently often to keep up with progress in research. However, some areas of chemistry, especially many of those taught at the graduate level, suffer from a real lack of up-to-date textbooks. The most serious needs occur in fields that are rapidly changing. Textbooks in these subjects usually have to be written by scientists actually involved in the research that is advancing the field. It is not often easy to persuade such individuals to set time aside to help spread the knowledge they have accumulated. Our goal, in this series, is to pinpoint areas of chemistry where recent progress has outpaced what is covered in any available textbooks, and then seek out and persuade experts in these fields to produce relatively concise but instructive introductions to their fields. These should serve the needs of one-semester or one-quarter graduate courses in chemistry and biochemistry. In some cases, the availability of texts in active research areas should help stimulate the creation of new courses.

Charles R. Cantor

Preface to the Second Edition

Since the publication of the previous edition in 1994, X-ray crystallography of proteins has advanced by improvements in existing techniques and by addition of new techniques. Examples are, for instance, MAD, which has developed into an important method for phase determination. Least squares as a technique for refinement is gradually being replaced by the formalism of maximum likelihood. With several new sections, the book has been updated, and I hope it will be as well received as the previous edition.

In the preparation of this second edition, I was greatly assisted by experts who commented on relevant subjects. I acknowledge the contributions of Jan Pieter Abrahams, Eleanor Dodson, Elspeth Garman, Eric de La Fortelle, Keith Moffat, Garib Murshudov, Jorge Navaza, Randy Read, Willem Schaafsma, George Sheldrick, Johan Turkenburg, Gert Vriend, Charles Weeks, and my colleagues in the Groningen Laboratory.

I am especially grateful to Bauke Dijkstra for the generous hospitality in his laboratory.

Jan Drenth

Preface to the First Edition

Macromolecules are the principal nonaqueous components of living cells. Among the macromolecules (proteins, nucleic acids, and carbohydrates), proteins are the largest group. Enzymes are the most diverse class of proteins because nearly every chemical reaction in a cell requires a specific enzyme. To understand cellular processes, knowledge of the three-dimensional structure of enzymes and other macromolecules is vital. Two techniques are widely used for the structural determination of macromolecules at atomic resolution: X-ray diffraction of crystals and nuclear magnetic resonance (NMR). While NMR does not require crystals and provides more detailed information on the dynamics of the molecule in question, it can be used only for biopolymers with a molecular weight of less than 30,000. X-ray crystallography can be applied to compounds with molecular weight up to at least 10^6. For many proteins, the difference is decisive in favor of X-ray diffraction.

The pioneering work by Perutz and Kendrew on the structure of hemoglobin and myoglobin in the 1950s led to a slow but steady increase in the number of proteins whose structure was determined using X-ray diffraction. The introduction of sophisticated computer hardware and software dramatically reduced the time required to determine a structure while increasing the accuracy of the results. In recent years, recombinant DNA technology has further stimulated interest in protein structure determination. A protein that was difficult to isolate in sufficient quantities from its natural source can often be produced in arbitrarily large amounts using expression of its cloned gene in a microorganism. Also, a protein modified by site-directed mutagenesis of its gene can be created for scientific investigation and industrial application. Here, X-ray diffraction plays a crucial role in guiding the molecular biologist to the best amino acid positions for modification. Moreover, it is often important to learn what effect a change in a protein's sequence will have on its three-dimensional structure. Chemical and pharmaceutical companies have become very active in the field of protein structure determination because of their interest in protein and drug design.

This book presents the principles of the X-ray diffraction method. Although I will discuss *protein* X-ray crystallography exclusively, the same techniques can in principle be applied to other types of macro-molecules and macromolecular complexes. The book is intended to serve both as a textbook for the student learning crystallography, and as a reference for the practicing scientist. It presupposes a familiarity with mathematics at the level of upper level undergraduates in chemistry and biology, and is designed for the researcher in cell and molecular biology, biochemistry, or biophysics who has a need to understand the basis for crystallographic determination of a protein structure.

I would like to thank the many colleagues who have read the manuscript and have given valuable comments, especially Aafje Vos, Shekhar and Sharmila Mande, Boris Strokopytov, and Risto Lapatto.

Jan Drenth

Contents

Chapter 3
Crystals   50

Chapter 4
The Theory of X-ray Diffraction by a Crystal   70

Chapter 5
Average Reflection Intensity and Distribution of Structure
Factor Data   117

Chapter 6
Special Forms of the Structure Factor   125

Chapter 1
Crystallizing a Protein

1.1. Introduction

Students new to the protein X-ray crystallography laboratory may under-standably be confused when colleagues discuss Fouriers and Pattersons or molecular replacement and molecular dynamics refinement. However, they understand immediately that the first requirement for protein struc-ture determination is to grow suitable crystals. Without crystals there can be no X-ray structure determination of a protein! In this chapter we discuss the principles of protein crystal growth and as an exercise give the recipe for crystallizing the enzyme lysozyme. We shall also generate an X-ray diffraction picture of a lysozyme crystal. This will provide an introduction to X-ray diffraction. The chapter concludes with a discussion of the problems encountered.

1.2. Principles of Protein Crystallization

Obtaining suitable single crystals is the least understood step in the X-ray structural analysis of a protein. The science of protein crystallization is an underdeveloped area, although interest is growing, spurred especially by microgravity experiments in space flights (McPherson et al., 1995). Protein crystallization is mainly a trial-and-error procedure in which the protein is slowly precipitated from its solution. The presence of impurities, crystalliza-tion nuclei, and other unknown factors plays a role in this process. As a general rule, however, the purer the protein, the better the chances to grow crystals. The purity requirements of the protein crystallographer are differ-ent and more stringent than the requirements of the biochemist, who would be satisfied if, for example, the catalytic activity of an enzyme is sufficiently high. On the other hand, to achieve protein crystallization not only should other compounds be absent, but all molecules of the protein should have the same surface properties, especially the same charge distribution on their surface, since this influences the packing of the molecules in the crystal. Mass spectrometry is a valuable tool in protein crystallization, for example, in checking the expression of recombinant protein, the purity of a preparation, heavy atom derivatives, and the nature of protein constructs (Cohen, 1996).

The crystallization of proteins involves four important steps:

1. The purity of the protein is determined. If it is not extremely pure, further purification will generally be necessary to achieve crystallization.

2. The protein is dissolved in a suitable solvent from which it must be precipitated in crystalline form. The solvent is usually a water–buffer solution, sometimes with an organic solvent such as 2-methyl-2,4-pentanediol (MPD) added. Normally, the precipitant solution is also added, but only to such a concentration that a precipitate does not develop. Membrane proteins, which are insoluble in a water–buffer or a water–organic solvent, require in addition a detergent.

3. The solution is brought to supersaturation. In this step small aggregates are formed, which are the nuclei for crystal growth. For the crystallization of small molecules, which is much better understood than the crystallization of proteins, the spontaneous formation of nuclei requires a supply of surface tension energy. Once the energy barrier has been passed, crystal growth begins. The energy barrier is easier to overcome at a higher level of supersaturation. Therefore, spontaneous formation of nuclei is best achieved at a high supersaturation. We assume that this is also true for the crystallization of proteins. Formation of nuclei can be studied as a function of supersaturation and other parameters by a number of techniques, including light scattering, fluorescence depolarization, and electron microscopy.

4. Once nuclei have formed, actual crystal growth can begin. As for low-molecular-weight compounds, the attachment of new molecules to the surface of a growing crystal occurs at steps on the surface. This is because the binding energy is larger at such positions than if the molecule attaches to a flat surface. These steps are either created by defects in the crystalline order or occur at nuclei formed randomly on the surface.

To achieve crystal growth, supersaturation must be reduced to a lower level; maintaining a high supersaturation would result in the formation of too many nuclei and therefore too many small crystals (Figure 1.1). Also, crystals should grow slowly to reach a maximum degree of order in their structure. In practice, however, this fundamental rule is not always obeyed. The easiest way to change the degree of supersaturation is by changing the temperature.

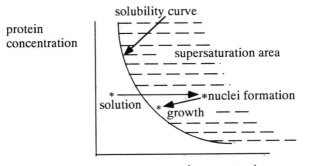

Increasing effective conc. of protein drives precipitation

Figure 1.1. A typical solubility curve for a protein, as a function of the salt concentration or another parameter. For best results the crystals should be grown at a lower level of supersaturation than is required for formation of nuclei.

Precipitation of the protein can be achieved in more than one way. A common precipitation method involves increasing the effective concentration of the protein, usually by adding a salt (salting-out) or polyethyleneglycol (PEG). Either serves to immobilize water, thereby increasing the effective protein concentration. The most popular salt is ammonium sulfate because of its high solubility.

Some proteins are poorly soluble in pure water but do dissolve if a small amount of salt is added. By removing the salt, the protein precipitates. This "salting-in" effect is explained by regarding the protein as an ionic compound. According to the Debye–Hückel theory for ionic solutions, an increase in the ionic strength lowers the activity of the ions in the solution and increases the solubility of ionic compounds. Alternatively, one can regard salting-in as the result of a competition between charged groups on the surface of the protein molecule and the ions in the solution. In the absence of solvent ions the protein precipitates by Coulomb attraction between opposite charges on different protein molecules. If ions are added they screen the charged groups on the protein and increase the solubility. A second method of protein precipitation is to diminish the repulsive forces between the protein molecules or to increase the attractive forces. These forces are of different types: electrostatic, hydrophobic, and hydrogen bonding. Electrostatic forces are influenced by an organic solvent such as alcohol, or by a change in pH. The strength of hydrophobic interactions increases with temperature.

To summarize, the usual procedure for crystallizing a protein is to:

1. check the purity carefully;
2a. slowly increase the concentration of the precipitant such as PEG, salt, or an organic solvent; or
2b. change the pH or the temperature.

In practice the amount of protein available for crystallization experiments is often very small. To determine the best crystallization conditions it is usually necessary to carry out a great number of experiments; hence a minimum amount of protein should be used per experiment. A single protein crystal of reasonable size ($0.3 \times 0.3 \times 0.3$ mm = 0.027 mm^3) weighs approximately $15\,\mu g$. Therefore, 1 mg of purified protein is sufficient to perform about 65 crystallization experiments.

Membrane proteins are insoluble in water and notoriously difficult to crystallize. The conventional strategy is to use detergents to solubilize them in an aqueous solution and then follow one of the procedures for water-soluble proteins (Michel, 1990; Sowadski, 1994). Landau and Rosenbusch (1996) have introduced a new method which could be promising for the crystallization of membrane proteins. They apply lipidic cubic phases as crystallization matrices for a variety of compounds. Lipidic cubic phases are liquid crystals with a cubic symmetry. They can be formed in a mixture of lipid and water (Lindblom and Rilfors, 1989). Several types of lipidic cubic phases exist. Landau et al. (1997) succeeded in growing well-ordered crystals of the membrane protein bacteriorhodopsin in a certain type of lipidic cubic phase (see also Gouaux, 1998). It remains to be seen whether this method will be successful in the crystallization of more membrane proteins.

1.3. Crystallization Techniques

1.3.1. Batch Crystallization

This is the oldest and simplest method for protein crystallization. The principle is that the precipitating reagent is instantaneously added to a protein solution, suddenly bringing the solution to a state of high supersaturation. With luck crystals grow gradually from the supersaturated solution without further processing. An automated system for microbatch crystallization has been designed by Chayen et al. (1990, 1992).

In their microbatch method, they grow protein crystals in 1–$2\,\mu l$ drops containing the protein and the precipitant. The drops are suspended in an oil, for example, paraffin oil. The oil acts as a sealant to prevent evaporation. It does not interfere with the common precipitants, but it does with organic compounds that dissolve in the oil (Chayen, 1997; see also Chayen, 1998).

1.3.2. Liquid–Liquid Diffusion

In this method the protein solution and the solution containing the precipitant are layered on top of each other in a small-bore capillary; a melting point capillary can conveniently be used (Figure 1.2). The lower

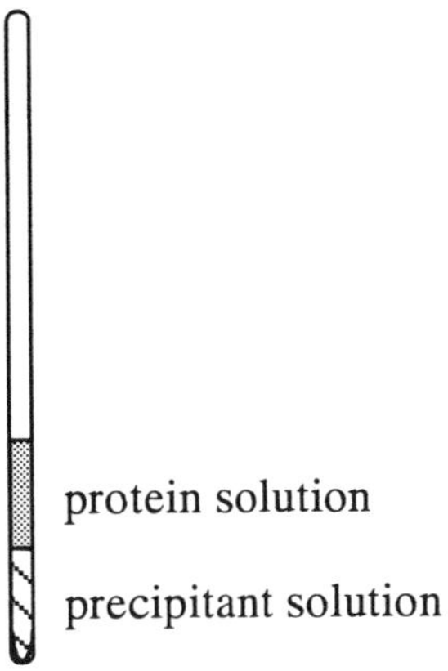

Figure 1.2. Liquid–liquid diffusion in a melting point capillary. If the precipitant solution is the denser one, it forms the lower layer.

protein solution

precipitant solution

Org. precipitant solution will form upper layer while inorganic precipitant will form org lower layer as shown.

layer is the solution with higher density (for example, a concentrated ammonium sulfate or PEG solution). If an organic solvent such as MPD is used as precipitant, it forms the upper layer. For a 1:1 mixture the concentration of the precipitant should be two times its desired final concentration. The two solutions (approximately 5 µl of each) are introduced into the capillary with a syringe needle, beginning with the lower one. Spinning in a simple swing-out centrifuge removes air bubbles. The upper layer is added and a sharp boundary is formed between the two layers. They gradually diffuse into each other.

García-Ruiz and Moreno (1994) have developed the technique of liquid/liquid diffusion further to the acupuncture method. The protein solution is sucked-up into narrow tubes by capillary force; one end of each tube is closed. Next, the open end is pushed into a gel contained in a small vessel. The gel keeps the tubes in a vertical position, and the protein solution is in contact with the gel. The solution with the precipitating agent is then poured over the gel, and the whole setup is kept in a closed box to avoid evaporation. The diffusion time of the precipitating agent through the gel and the capillary can be controlled by the penetration depth of the capillary in the gel. A range of supersaturations is created in the protein solution, high at the bottom and low at the top. This can be used as additional information in screening the best crystallization conditions.

1.3.3. Vapor Diffusion

1.3.3.1. By the Hanging Drop Method

In this method drops are prepared on a siliconized microscope glass cover slip by mixing 3–10 µl of protein solution with the same volume of precipitant solution. The slip is placed upside down over a depression in a tray; the depression is partly filled with the required precipitant solution

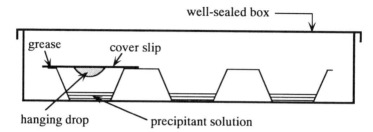

Figure 1.3. The hanging drop method of protein crystallization. Using a tray with depressions, the protein solution is suspended as a drop from a glass cover slip above the precipitant solution in a sealed depression. The glass slip is siliconized to prevent spreading of the drop. Equilibrium is reached by diffusion of vapor from the drop to the precipitating solution or vice versa. All of the depressions in the tray can of course be used.

(approximately 1 ml). The chamber is sealed by applying oil or grease to the circumference of the depression before the cover slip is put into place (Figure 1.3).

1.3.3.2. Sitting Drop

If the protein solution has a low surface tension, it tends to spread out over the cover slip in the hanging drop method. In such cases the sitting drop method is preferable. A schematic diagram of a sitting drop vessel is shown in Figure 1.4.

1.3.4. Dialysis

As with the other methods for achieving protein crystallization, many variations of dialysis techniques exist. The advantage of dialysis is that the precipitating solution can be easily changed. For moderate amounts of protein solution (more than 0.1 ml), dialysis tubes can be used, as shown in Figure 1.5a. The dialysis membrane is attached to a tube by means of a rubber ring. The membrane should be rinsed extensively with water

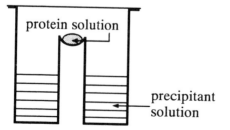

Figure 1.4. The sitting drop method for performing protein crystallization.

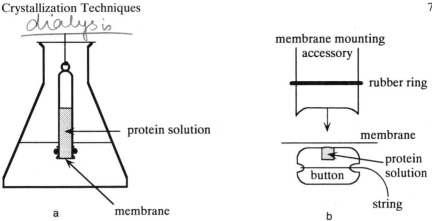

Figure 1.5. Protein crystallization by dialysis. If a relatively large amount of protein is available, dialysis can be performed as in (a). Smaller amounts can be crystallized in a button (b).

before use or, preferably, boiled in water for about 10 mins. For microliter amounts of protein solution, one may use either a thick-walled micro-capillary (Zeppezauer method) or a plexiglass "button" covered with a dialysis membrane (Figure 1.5b). The disadvantage of the button is that any protein crystals in the button cannot be observed with a polarizing microscope.

Another microdialysis procedure is illustrated in Figure 1.6. Five microliters of protein solution is injected into a capillary, which is covered by a dialysis membrane. The membrane may be fastened with a piece of tubing. The protein solution is spun down in a simple centrifuge and the capillary closed with modeling clay. The capillary is then placed in an Eppendorf tube containing the dialysis solution.

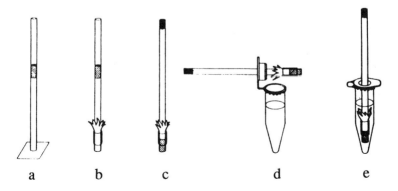

Figure 1.6. Protein crystallization by microdialysis procedure. In (a) and (b) the protein solution is still somewhere in the capillary. In (c) it has been spun to the bottom of the tube and is in contact with the membrane. In (d) and (e) the capillary is mounted in an Eppendorf tube containing dialysis solution.

1.3.5. Final Remarks

Crystallization of a protein is a multiparameter problem in which the parameters are varied in the search for optimal crystallization conditions. The most common parameters that are changed include protein concentration, the nature and concentration of the precipitant, pH, and temperature. Specific additives that affect the crystallization can also be added in low concentration. An appropriate statistical design to use for this problem is the factorial or incomplete factorial method, in which a table is constructed with a number of values entered for each parameter (e.g., pH set at 5, 6, and 7). The table is constructed to take into account reasonable boundary conditions. After approximate crystallization conditions have been found, the conditions can be optimized (Carter, 1990).

An organized screening technique is called sparse matrix sampling (Jancarik and Kim, 1991): a standard set of a very large number of conditions is listed for screening purposes. McPherson (1992) described a method that is in between pure trial and error and the factorial design approach for identifying initial crystallization conditions. However, in his method a large number of conditions must still be tried (Table 1.1). Other strategies for crystal growth are given in Cudney et al. (1994). For a review see Gilliland and Ladner (1996). Crystal and crystallization data of all crystal forms of biological macromolecules have been compiled in the Biological Macromolecule Crystallization Database, Version 3.0 (Gilliland, 1997).

Robotic workstations can be highly effective in conducting such multifactorial experiments (an example is found in Chayen et al., 1990).[1] Hampton Research is a supplier of a large variety of solutions, reagents and gadgets for protein crystallization.[2]

1.4. Crystallization of Lysozyme

After learning the principles of protein crystallization, it is time to do a crystallization experiment. The most convenient protein to start with is hen egg white lysozyme. It can be obtained commercially in pure form, is relatively cheap, and can be used immediately for a crystallization experiment. In addition to native lysozyme, a heavy atom derivative of the protein will be crystallized, because the determination of a protein structure sometimes requires X-ray diffraction patterns of crystals of the native protein as well as of one or more heavy-atom-containing derivatives. In

---

[1] Protein crystallization is extensively discussed in: Ducruix, A. and Giegé, R. (1992). For the crystallization of membrane proteins, see Michel, H. (1990) and Sowadski (1994).
[2] Hampton Research, 27632 El Lazo Road, Suite 100, Laguna Niguel, CA 92677 USA. Tel.: 1-714-425-1321; Fax: 1-714-425-1611, http://www.hamptonresearch.com; e-mail: xtalrox@aol. com

Table 1.1. Screening Reagents[a]

Box 1 and 2 buffers	Box 4 salt solutions
1. pH 4.0 0.1 M citrate or acetate	1. 50% sat. sodium phosphate, pH 7.0
2. pH 5.0 0.1 M acetate	2. 50% sat. magnesium sulfate
3. pH 5.5 0.1 M MES or cacodylate	3. 50% sat. sodium citrate pH, 7.0
4. pH 6.0 0.1 M MES or cacodylate or phosphate	4. 50% sat. lithium sulfate
5. pH 6.5 0.1 M cacodylate or phosphate	5. 50% sat. ammonium phosphate, pH 7.0
6. pH 7.0 0.1 M phosphate	6. 50% sat. sodium acetate
7. pH 7.5 0.1 M Tris-HCl	7. 50% sat. sodium chloride
8. pH 8.0 0.1 M HEPES or Tris	8. 50% sat. calcium sulfate
9. pH 8.5 0.1 M Tris or glycine	9. 50% sat. ammonium formate
Droplet = 5 μl protein + 5 μl buffer + 5 μl reservoir Reservoir Box 1 = 14% PEG 3350 Reservoir Box 2 = 45% sat. ammonium sulfate	Droplet = 5 μl protein + 5 μl salt + 5 μl water Reservoir = 45% sat. ammonium sulfate

Box 3 nonvolatile organics	Cryschem or Linbro tray
1. 16.5% polyeneamine	1. n-Propanol
2. 50% v/v PEG 400	2. Isopropanol
3. 15% w/v PEG 20,000	3. Dioxane
4. 30% v/v Jeffamine ED 4000	4. Ethanol
5. 50% sat. PEG 1000 monostearate	5. tert-Butanol
6. 30% PEG 1000	6. DMSO
7. 30% Jeffamine ED 2001	
8. 50% propylene glycol	
9. 60% hexanediol	

Droplet = 5 μl protein + 5 μl organic + 5 μl water
Reservoir = 45% sat. ammonium sulfate
Alternates: Methyl pentanediol
    30% Jeffamine ED 900
    30% Jeffamine ED-600
	Droplet = 5 μl protein + 5 μl water
	Reservoirs 0.75 ml of 35% organic solvent
	Alternates: Methanol
	    Acetone
	    n Butanol

[a] From McPherson (1992), with permission.

this experiment the mercury-containing reagent p-chloromercuriphenyl sulfonate is used.

1. Prepare a sodium acetate buffer solution of pH 4.7 by dissolving
   a. 1.361 g of sodium acetate in 50 ml water (purified by reverse osmosis or double distillation);

  b. 0.572 ml glacial acetic acid in 50 ml water.
  The acetic acid solution is added to the salt solution until the pH
  reaches 4.7.
2. Prepare a precipitant solution of sodium chloride
  a. for the native crystals by preparing 30 ml of 10% (w/v) NaCl in the
     sodium acetate buffer;
  b. for the derivative crystals by dissolving 0.041 g of the sodium salt of
     p-chloromercuriphenyl sulfonate (PCMS) in 5 ml of the precipitant
     solution prepared in 2a.

Warning! Heavy atom reagents can be very toxic or radioactive. Handle
them with care and wear gloves. Discard excess reagent and
solution in a special container.

---

3. Dissolve the native lysozyme in the acetate buffer to a concentration
   of 50 mg/ml (e.g., 10 mg in 200 μl). The heavy-atom derivative is less
   soluble and its concentration should be only 15 mg/ml in the same
   buffer. To remove insoluble particles the solutions are centrifuged at
   10,000 rpm and 4°C for 10 min prior to setting up the crystallization
   experiments.
4. Take one or two trays for the native and for the derivative protein,
   with at least 5 × 3 depressions in each (Figure 1.7). The crystallization
   experiments are performed with five concentrations of the precipitant
   solution: 2, 3, 4, 5, and 6% NaCl and two replicates per condition.
   Use the extra row of depressions to label the experimental conditions.
   Fill each depression with 1 ml of the precipitant solution at the required
   concentration. Prepare the hanging drops on the siliconized side of
   microscope cover slips: pipette 10 μl of the enzyme solution onto a

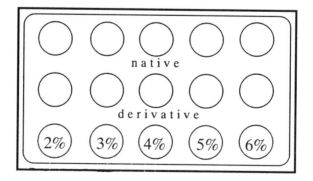

Figure 1.7. A tray for the crystallization of lysozyme. Five experiments each for
the native protein and for the heavy atom derivative can be performed with
increasing concentrations of the precipitant solution: 2–6% NaCl.

cover slip and mix with 10 μl of the precipitant solution. Brush a film of immersion oil around each depression on the surface of the tray to seal (Figure 1.3).

The lysozyme crystals will grow in a day or so and you will probably be anxious to see how your crystals will diffract X-rays. However, we will first consider some general features about crystals.

1.5. A Preliminary Note on Crystals

Crystals occur in a great variety of shapes and colors and naturally grown crystals have been used as gems since prehistoric times. Their flat faces reflect the regular packing of the molecules, atoms, or ions in the crystal. Crystals are distinguished from amorphous substances by their flat faces and by their anisotropy: some of their physical properties are dependent on the direction of measurement in the crystal. For some properties crystals may be isotropic, but usually not for all of them. You can easily observe the anisotropy of the lysozyme crystals you have just grown by examining them between crossed polarizers in a polarizing microscope. They display beautiful colors because the crystals are birefringent: they have two different refraction indices. If you rotate the microscope platform holding the crystals, they will become completely dark at a certain position and every 90° away from it; extinction thus occurs four times in a complete revolution.

The flat faces and anisotropy of crystals reflect their regular packing of molecules, atoms, or ions. This regular arrangement cannot be observed with the naked eye or with a light microscope because the particles are too small, but can be visualized with an electron microscope (Figure 1.8). If the resolution of the microscope would be high enough, the atomic structure inside the large protein molecules could be observed. Unfortunately this is not possible because biological substances can be observed in an electron microscope with a resolution of only 10–20 Å, due to limitations in specimen preparation. Although atomic resolution cannot be reached in this way, electron micrographs of crystals show convincingly the regular packing of molecules in the crystals. It is because of this regular arrangement that the crystals diffract X-rays.

Before starting the X-ray diffraction experiment it can be illuminating to demonstrate diffraction with visible light and a grating. The principle is the same, the only difference being that the grating is two-dimensional and the crystal is three-dimensional. Take a laser pointer as light source and ask an electron microscopist for one of the copper grids on which specimens are mounted. In a dark room you will see a beautiful pattern of diffraction spots. Rotation of the grid causes the pattern to rotate with it. If you can borrow grids with different spacings, the pattern from the

Figure 1.8. An electron micrograph of a crystal face of the oxygen transporting protein hemocyanin from *Panulirus interruptus*. The molecular weight of this protein is 450,000; magnification: 250,000×. Courtesy of E.F.J. van Bruggen.

grid with larger spacing will have the diffraction spots closer together: here you observe "reciprocity" between the diffraction pattern and the grid. Alternatively, you can perform the experiment with transparent woven fabric. Note that stretching the fabric horizontally causes the diffraction pattern to shrink horizontally.

1.6. Preparation for an X-ray Diffraction Experiment

You should now perform the real X-ray diffraction experiment using one of the crystals you have just grown. However, you should be aware of an important difference between crystals of small molecules and crystals of a protein. In protein crystals, the spherical or egg-shaped molecules are loosely packed with large solvent-filled holes and channels, which normally occupy 40–60% of the crystal volume. This is an advantage for the reaction of the protein with small reagent molecules; they can diffuse through these channels and reach reactive sites on all protein molecules in the crystal. However, the high solvent content causes problems in handling the crystals because loss of solvent destabilizes the crystals. Therefore, protein crystals should always be kept in their mother liquor, in the saturated vapor of their mother liquor, or at a sufficiently low temperature to prevent evaporation of the solvent.

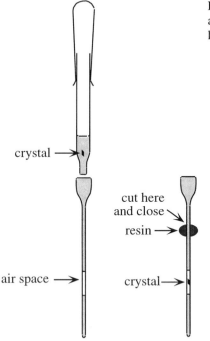

Figure 1.9. Mounting a protein crystal in a glass capillary between layers of mother liquor.

crystal →

cut here
and close ↘

resin →

air space →   crystal→

For the collection of X-ray data at or near room temperature, the crystals are mounted in thin-walled glass capillaries of borosilicate glass or quartz (Figure 1.9). It has several advantages in keeping the crystals at a very low temperature (100–120°K) during data collection (Section 1.7). A capillary is no longer required because at that temperature the water vapor pressure is so low that the crystals do not loose water.

For your first data collection the crystal with be mounted in a capillary. The procedure is as follows. Fill the X-ray capillary with the mother liquor in which the crystals were grown or with the precipitation solution in the bottom of the tray, but leave an air space where the crystal should be. Suck up one of the crystals together with mother liquor, let it descend to the tip of the pipet, and then touch the pipet tip to the liquid in the capillary. The crystal continues its way down until it reaches the air space. Push it carefully into the open space with a thin glass fiber. Excess liquid must be removed and the capillary closed. This can be done with a resin that can be melted using a soldering iron. Finally, for handling and for mounting on the X-ray camera, a piece of modeling clay is wrapped around the resin. Lysozyme crystals have a 4-fold symmetry axis and for the diffraction experiment they should be mounted with this axis perpendicular to the axis of the capillary, as seen in Figure 1.10. This has the advantage that in the diffraction experiment the symmetry axis can be aligned parallel to the X-ray beam.

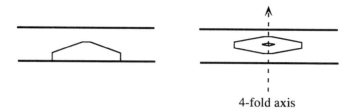

4-fold axis

Figure 1.10. Two perpendicular views of a lysozyme crystal mounted in a capillary. The size of the crystal is 0.2–0.5 mm.

Warning! The human body should not be exposed to X-rays because of their damaging effect on tissues

---

The crystal in its capillary is attached to a goniometerhead (Figure 1.11). This gadget has two perpendicular arcs, allowing rotation of the crystal along two perpendicular axes. For further adjustment and centering, the upper part of the goniometerhead can be moved along two perpendicular sledges. After preadjustment under a microscope, the

Figure 1.11. A goniometerhead.

goniometerhead is screwed on to an area detector. At this point you need the assistance of a colleague with experience in handling the area detector.

Be sure that the crystal is in the X-ray beam path and adjust it with its 4-fold symmetry axis along the direction of the X-ray beam. Observe the X-ray pattern with the crystal oscillating over a tiny angle (e.g., 0.15°). You will see reflection spots arranged in circles (Figure 1.12a). These circles can be regarded as the intersection of a series of parallel planes with a sphere (Figure 1.13). Try to adjust the orientation of the crystal such that the planes are perpendicular to the X-ray beam. The inner circle will then disappear and the others will be concentric around the beam, which is hidden by the beamstop. From this position oscillate the crystal over a 3° range; many more reflections appear on the screen. The inner circle appears again and you observe that the plane to which it corresponds has spots nicely arranged in perpendicular rows (Figure 1.12b). This is true for all planes and your crystal produces thousands of diffraction spots in a well-ordered lattice. From the position of these spots the repeating distances in the crystal can be derived and from their intensities the structure of the lysozyme molecules.

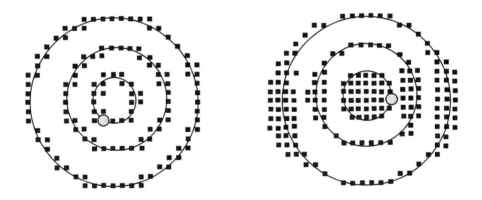

⊚ shadow of the beamstop

a   b

Figure 1.12. (a) A schematic representation of the diffraction pattern of a stationary lysozyme crystal. The diffraction spots are arranged in circles. The innermost circle passes through the origin, which is behind the beamstop. The latter prevents the strong primary beam to reach the detector. (b) A three degree oscillation picture of the same lysozyme crystal.

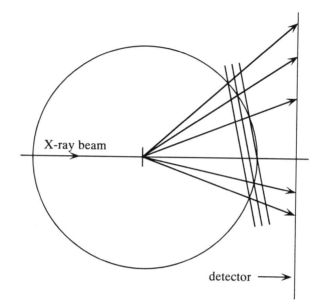

Figure 1.13. Diagrammatic representation of the generation of an X-ray diffraction pattern. The circles shown in Figure 1.12a can be regarded as the intersection of a series of parallel planes with a sphere. The planes are part of a lattice composed of diffraction spots. However, diffraction occurs only if the spots are on, or pass through, the surface of the sphere. The planes are determined by the lysozyme crystal and the sphere by the wavelength of the X-ray radiation (Chapter 4).

The mercury derivative crystals will produce the same pattern as crystals of the native enzyme. However, close inspection will show that although the position of the spots is the same, the intensity of the diffracted beams is slightly different.

1.7. Cryocooling

Complete data sets are often collected with the protein crystal in a stream of cold nitrogen gas at a temperature in the range of 100–120 °K with a stability of 1 °K. The main reason for cooling protein crystals is that they are liable to radiation damage if exposed to X-rays, and this damage can be so serious that the X-ray pattern dies away after a few hours of exposure at room temperature. Cooling the crystals slows the destructive process appreciably, and radiation damage virtually disappears (Section 3.8). The method and its development is extensively discussed by Garman and Schneider (1997). Cooling the crystals from room temperature to cryogenic temperature must occur suddenly, because the water in the mother liquor and in the crystal must freeze to a vitreous structure. The water should not crystallize because ice formation would damage the protein crystal structure by expanding the water structure in its transition to crystalline ice. Therefore, the method is called flash freezing or shock cooling. To prevent ice formation in and around the crystal completely, it is carefully transferred to a solvent containing an antifreeze that acts as a cryoprotectant. Glycerol, MPD, ethyleneglycol, and low molecular weight PEG are popular cryoprotectants. It is an advantage if the crystal was already grown in the presence of a cryoprotectant, for examples an alcohol. A high salt concentration in the original mother liquor can cause problems by precipitation of the salt in the cryoprotectant containing solution and cracking the crystals. The salt must then be replaced by a more soluble salt or by additional cryoprotectant.

The effect of cooling is often small changes in cell dimensions and an increased divergence of the diffracted beams caused by strain developed in the crystal (increased mosaicity, Figure 4.29). This must be kept within acceptable limits.

Cooling is accomplished by nitrogen gas, boiled-off from liquid nitrogen (boiling point 77 °K at atmospheric pressure) or cooled in a heat exchanger. The accumulation of ice on the crystal and on the diffraction instrument is a problem. To prevent its growth on the protein crystal, the cold gas stream is coaxially surrounded by a warm and dry stream of air or nitrogen. Turbulence between the two streams must be avoided by adjusting their flow speed. Ice formation can further be prevented by enclosing the entire diffraction apparatus in a box.

When working with liquid nitrogen take the necessary precautions. For instance, always wear goggles and gloves when filling Dewars. Moreover, the room where liquid nitrogen evaporates should be well ventilated. Volatile organic cryogens, for instance propane, require additional precautions, because there is danger of explosion and fire.

For data collection, the crystals are lifted from the cryoprotectant solution by a loop made from a thin fiber (Teng, 1990). Surface tension in the liquid keeps the crystal in the thin liquid layer spanning the loop. The size of the loop should approximate the size of the crystal or a little larger ("lasso technique"). If the liquid has a low surface tension it may not hold the crystal in the loop. Under these conditions, a loop slightly smaller than the crystal should be chosen ("spatula technique"). Rayon or nylon fibers are preferred, but glass fibers—although somewhat brittle—are also suitable. Both ends of the loop are cemented in a capillary (Figure 1.14). The

Figure 1.14. A cryoloop, approximately the size of the crystal, is mounted in a metal capillary that can be fixed to a base. This base has a magnetic plate at the bottom to attach it quickly to a goniometer head. The base can also be screwed to a cap for transport and storage.

thin capillary is fixed into a base that can accurately and quickly be positioned on the top of a goniometerhead. It can also be screwed into a cap for transport and storage. Data collection with flash freezing always starts by determining the optimal conditions as to kind and concentration of cryoprotectant. Salts must be replaced if required. A crystal is picked up from the solution with the loop and—to prevent loss of solvent by evaporation—as quickly as possible transferred to the cold nitrogen stream close to the cryonozzle. Alternatively, the crystal is dropped into liquid propane or immersed into liquid nitrogen for fast cooling and from there transferred to the cold nitrogen stream. The crystal is now ready for data collection. The diffraction pattern must be checked for strain in the crystal causing increased mosaicity. If the flash-cooled crystal shows unacceptably high mosaicity, it is sometimes helpful to "anneal" it: the crystal is quickly transferred from cryogenic conditions to the original cyrosolvent, at the temperature the crystal was grown, and incubated for at least three minutes before reflash-cooling (Harp et al., 1997, 1998)

Additional advantages of cryocrystallography are:

- The crystals can be stored and transported at liquid nitrogen temperature for data collection at a later stage.
- There is less absorption by surrounding nonprotein material than in the capillary technique.

1.8. Notes

1. Crystallization: The first essential step in determining the X-ray structure of a protein is to grow crystals of sufficient size and quality. It is always amazing to see that these big molecules can arrange themselves so neatly that crystals with flat faces and sharp edges grow. There is great excitement in the laboratory when the first crystals of a protein appear

and the excitement is even greater if those crystals produce a high quality X-ray diffraction pattern. Although more steps are required to complete a structure determination, the growth of good quality crystals indicates a reasonable chance of success and suggests that the protein structure can indeed be determined.

What are the best conditions for crystal growth? Can they be predicted? These questions are difficult to answer for proteins because so many parameters play a role. In general, the chances of growing good-quality crystals are higher if the protein solution is monodisperse (all molecules of the same size) (Ferré-d'Amaré and Burley, 1994). George and Wilson (1994) proposed to use the second virial coefficient as a predictor. They found that this coefficient is, for a protein solution, within a restricted range for conditions under which good-quality crystals can be grown. If it is outside that range, the chance of growing good crystals is low. The second virial coefficient is a measure of the interaction between the protein molecules in the solution and can be derived, for instance, from static light scattering experiments.

However, choosing a crystallization technique and crystallization conditions is largely a matter of taste. The hanging or sitting drop methods are the most popular. Liquid–liquid diffusion in a small-bore capillary has the advantage of being somewhat slower in reaching equilibrium and this is sometimes an advantage. Dialysis is not frequently used but it must be applied for salting-in precipitation.

After preparing the crystallization experiments, leave them in a quiet place with a minimum of vibration and keep the temperature fairly constant.

Protein crystallization is in essence a trial-and-error method, and the results are usually unpredictable. Serendipity also plays a role. In the author's laboratory a failure in the air-conditioning system, causing an unexpected rise in temperature, led to the growth of perfect crystals of an enzyme, which could not be obtained under more normal circumstances.

Commercially available robots make it possible to perform more experiments in the same time and to determine the optimum crystallization conditions more quickly. However, human intelligence is still required to tell the robot what to do.

If crystallization does not occur, even after many experiments, the following can be tried:

    a. Crystallization of a homologous protein from another source.

    b. Crystallization of one or more proteolytic fragments of the protein. The polypeptide chain of the native protein is split by a proteolytic enzyme at a limited number of positions, or, alternatively, the required fragment is expressed and purified from a bacterial or eucariotic expression system.

2. Mounting the crystals: X-ray capillaries are very fragile because of their thin glass wall. Why not mount them in a stronger glass tube? Although you can see the crystal just as well through a much thicker glass

Table 1.2. Transmission through Pyrex Glass

	Glass thickness (mm)	Transmission (%)
$\lambda = 1.54$ Å	0.01	93
(X-ray tube with	0.1	50
Cu-anode)	1.0	0.01
$\lambda = 0.71$ Å	0.1	93
(X-ray tube with	1.0	40
Mo-anode)		

wall, X-rays cannot. A glass plate of 0.01 mm thickness has 93% transmission for the common X-ray wavelength of 1.54 Å, but this diminishes exponentially with increasing thickness of the glass (Table 1.2). The data show that transmission is highly dependent on the X-ray wavelength and is much higher for a shorter wavelength. This is one of the reasons that X-ray data collection at a synchrotron is preferably done around 1 Å instead of 1.5 Å, the usual wavelength in the home laboratory. The disadvantage of a lower diffraction intensity at shorter wavelength is compensated by the strong intensity of the synchrotron X-ray beam. Other advantages of a shorter wavelength are the more favorable crystal lifetime and better response for a detector with a fluorescent screen.

X-ray wavelength and atomic distances are usually expressed in Ångström units: 1 Å $= 10^{-10}$ m $= 10^{-1}$ nm. Although the Ångström unit is not an SI unit it has the advantage of giving simpler numbers. For example, the C–C distance in ethane is 1.54 Å, and a frequently used wavelength in the X-ray diffraction of proteins is 1.5418 Å. Sometimes the photon energy (E) is given instead of the wavelength λ. The relationship is λ (in Å) $= 12.398/E$ (in keV).

3. The X-ray pattern: You have observed X-ray diffraction spots nicely arranged in rows. You should compare this with the laser + electron microscope grid experiment. There you saw a two-dimensional diffraction pattern with rows of spots produced by a two-dimensional grid and the diffraction pattern was reciprocal to the actual electron microscope grid. A crystal can be regarded as a three-dimensional grid and you can imagine that this will produce a three-dimensional X-ray diffraction pattern. As with the electron microscope grid, the pattern is reciprocal to the crystal lattice.

The planes that intersect the sphere in Figure 1.13 are layers in a three-dimensional lattice, which is not the crystal lattice, but its reciprocal lattice. The unit distances in this lattice are related reciprocally to the unit distances in the crystal and that is why the lattice is called a reciprocal lattice. Each reciprocal lattice point corresponds to one diffracted beam. The reciprocal lattice is an imaginary but extremely convenient concept for determining the direction of the diffracted beams. If the crystal rotates, the reciprocal lattice rotates with it. In an X-ray diffraction experiment the direction of the diffracted beams depends on two factors: the unit cell distances in the crystal from which the unit cell distances in the reciprocal lattice are derived and the X-ray wavelength.

As indicated in Figure 1.13 diffraction conditions are determined not only by the reciprocal lattice, but also by the radius of the sphere, which is called the sphere of reflection or "Ewald sphere."[3] Its radius is reciprocal to the wavelength λ; it is equal to $1/\lambda$. From the diffraction experiment with lysozyme you can determine that not all diffracted beams occur at the same time. Only the ones corresponding to reciprocal lattice points on the Ewald sphere in Figure 1.13 are actually observed. Other points can be brought to diffraction by rotating the crystal and with it the reciprocal lattice, to bring these new lattice points on the sphere.

Diffracted beams are often called "reflections." This is due to the fact that each of them can be regarded as a reflection of the primary beam against planes in the crystal. Why this is so will be explained in Chapter 4.

Summary

Protein crystal growth is mainly a trial-and-error process but the primary rule is that the protein should be as pure as possible. The higher the purity, the better the chance of growing crystals by batch crystallization, by liquid–liquid or vapor diffusion, or by dialysis. The experiment with hen egg white lysozyme, the growth of nicely shaped crystals, and the huge number of reflections in the X-ray diffraction pattern have provided an introduction to protein X-ray crystallography. From the X-ray diffraction experiment it became clear that the X-ray pattern can be regarded as derived from the intersection of a lattice and a sphere. This lattice is not the crystal lattice but the reciprocal lattice, which is an imaginary lattice related to the crystal lattice in a reciprocal way. The sphere is called the Ewald sphere and has a radius of $1/\lambda$.

---

[3] Paul P. Ewald, 1888–1985, professor of physics at several universities in Europe and the United States, was the first to apply the reciprocal lattice and the sphere named after him to the interpretation of an X-ray diffraction pattern.

Chapter 2
X-ray Sources and Detectors

2.1. Introduction

In Chapter 1 you learned how crystals of a protein can be grown and you observed a diffraction pattern. The crystalline form of a protein is required to determine the protein's structure by X-ray diffraction, but equally necessary are the tools for recording the diffraction pattern. These will be described in this chapter on hardware. The various X-ray sources and their special properties are discussed, followed by a description of cameras and detectors for quantitative and qualitative X-ray data collection.

2.2. X-ray Sources

The main pieces of hardware needed for the collection of X-ray diffraction data are an X-ray source and an X-ray detector. X-rays are electromagnetic radiation with wavelengths of $10^{-7}-10^{-11}$ m (1000–0.1 Å). Such radiation was discovered by Roentgen[1] in 1895, but as the nature of the radiation was not yet understood, Roentgen called them X-rays. Von Laue's[2] diffraction theory, which he developed around 1910, inspired his assistants, Friedrich and Knipping, to use a crystal as a diffraction grating. Their results, published in 1912, were direct

---

[1] Wilhelm Conrad Roentgen, 1845–1925, discovered X-rays on November 8, 1895 in Würzburg, Germany.
[2] Max von Laue, 1879–1960, German physicist, developed the theory of X-ray diffraction by a three-dimensional lattice.

proof for the existence of lattices in crystals and for the wave nature of X-rays.

For the X-ray diffraction experiment with lysozyme crystals in Chapter 1 an X-ray generator in the home laboratory was used. These are typically either a generator with a sealed tube and fixed target or a more powerful system with a rotating anode. Both instruments are push-button operated but require some maintenance:

1. a sealed tube must be replaced if the filament burns out and
2. a rotating anode tube occasionally needs a new filament (cathode) and new seals.

Particle accelerators as synchrotrons and storage rings are the most powerful X-ray sources, but they are so complicated technically that the X-ray crystallographer is just a user at the front end. Because protein molecules are very large, their crystals diffract X-ray beams much less than do crystals of small molecules. The reason is that diffraction is a cooperative effect between the molecules in the crystal; for larger molecules there are fewer in a crystal of same size and therefore the diffracted intensity is lower. Moreover, proteins consist mainly of C, N, and O. These are light elements with only a few electrons (6–8) per atom. Since the electrons are responsible for the diffraction, atoms of these light elements scatter X-rays much more weakly than do heavier elements. Because of this phenomenon of relatively low scattering power, protein crystallographers prefer a high intensity source: a rotating anode tube rather than a sealed tube. For crystals of a very small size (<0.1 mm) or with extremely large molecules, synchrotron radiation is required for data collection.

Although X-ray generators with a sealed tube produce a relatively low beam intensity, they have not disappeared from protein crystallography laboratories because of their ease of operation and simple maintenance. For preliminary qualitative work they are fine, especially for relatively small proteins. Sealed tubes have a guaranteed life time of 1000 hr, but in practice they last much longer: 1 year of continuous operation is not unusual. Quantitative measurements should be collected with a rotating anode tube as X-ray source; they have a radiation intensity approximately four times higher than that of a sealed X-ray tube. Although your part in operating an X-ray source mainly consists of pushing the correct buttons, you should nevertheless be aware of the properties of radiation from various sources. Therefore, we shall discuss them in more detail (Phillips, 1985).

2.2.1. Sealed X-ray Tubes

In a sealed X-ray tube a cathode emits electrons (Figure 2.1). Because the tube is under vacuum and the cathode is at a high negative potential with respect to the metal anode, the electrons are accelerated and reach the anode at high speed. For protein X-ray diffraction the anode is usually a copper plate onto which the electron beam is focused, to a focal spot, normally of 0.4 × 8mm. Most of the electron energy is converted to heat, which is removed by cooling the anode, usually with water. However, a small part of the energy is emitted as X-rays in two different ways: as a smooth function of the wavelength and as sharp peaks at specific wavelengths (Figure 2.2). The continuous region is due to the physical phenomenon that decelerated (or accelerated) charged particles emit electromagnetic radiation called "Bremsstrahlung." This region has a sharp cut-off at the short wavelength side. At this edge the X-ray photons obtain their full energy from the electrons when they reach the anode. The electron energy is e × (accelerating voltage V), where e is the electron charge. The photon energy is hv = h × (c/λ), where h is Planck's constant, v is the frequency of the radiation, c is the speed of light, and λ is the wavelength. Therefore,

$$\lambda_{min} = \frac{h \times c}{e \times V} = \frac{12.4}{V}$$

where V is in kilovolts. At V = 40 kV the cut-off edge is at 0.31 Å.

The sharp peaks in the spectrum are due to electron transitions between inner orbitals in the atoms of the anode material. The high energy electrons reaching the anode shoot electrons out of low lying orbitals in the anode atoms. Electrons from higher orbitals occupy the empty positions and the energy released in this process is emitted as X-ray radiation of specific wavelength: K_α radiation if it comes from a transition from the L-shell to the K-shell and K_β for a transition from the M- to the K-shell. Because of the fine structure in the L-shell, K_α is split up in $K_{\alpha 1}$ and $K_{\alpha 2}$. The energy levels in the M-shell are so close that for K_β one wavelength value is given (Figure 2.3).

When copper is the anode material the following values for λ are given:

λ (Å)		
$K_{\alpha 1}$	1.54051	The weight average value for $K_{\alpha 1}$ and $K_{\alpha 2}$ is taken as 1.54178 Å
$K_{\alpha 2}$	1.54433	because the intensity of $K_{\alpha 1}$ is twice that of $K_{\alpha 2}$
K_β	1.39217	

Figure 2.1. (a) A schematic drawing of a sealed X-ray tube. The windows are made of thin berylium foil that has a low X-ray absorption. (b) Cross section of an X-ray tube: Courtesy Philips, Eindhoven, The Netherlands.

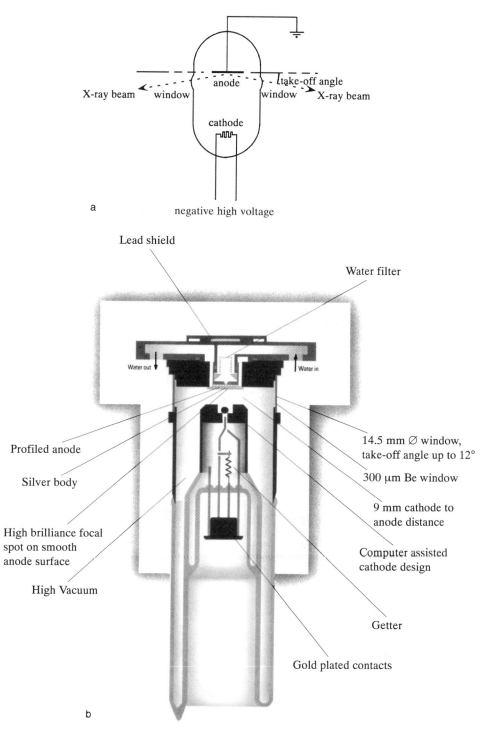

Figure 2.1. *Continued*

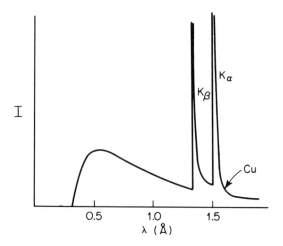

Figure 2.2. The spectrum from an X-ray tube with a copper anode. It shows a continuous spectrum and in addition two sharp peaks due to quantized electrons in the copper. I is the energy of the emitted radiation on an arbitrary scale.

For emission of the characteristic lines in the spectrum a minimum excitation voltage is required. For example, for the emission of the CuK_α line, V should be at least $8\,kV$. If a higher voltage is applied the intensity of the line is stronger with respect to the continuous radiation, up to about $V/V_{min} = 4$. The intensity of the line is also proportional to the tube current, at least as long as the anode is not overloaded. A normal setting is $V = 40\,kV$ with a tube current of $37\,mA$ for a 1.5-kW tube.

2.2.2. Rotating Anode Tubes

The heating of the anode caused by the electron beam at the focal spot limits the maximum power of the tube. Too much power would ruin the

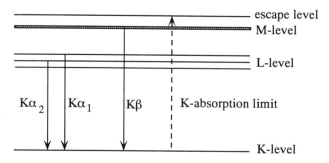

Figure 2.3. Schematic representation of the atomic energy levels and transitions causing characteristic X-ray wavelengths.

anode. This limit can be moved to a higher power loading if the anode is a rotating cylinder instead of a fixed piece of metal. With a rotating anode tube small source widths (0.1–0.2 mm) with a high brilliance[3] are possible. The advantage over the sealed tube is the higher radiation intensity, but a disadvantage is that it requires continuous pumping to keep the vacuum at the required level.

The take-off angle (Figure 2.1a) is usually chosen near 4°. Then the observer "sees" the focal spot with dimensions of 0.4×0.5 mm for a sealed tube or smaller for a rotating anode tube. Higher brilliance could be obtained with a smaller take-off angle, but a smaller angle results in a longer X-ray path in the anode material and, therefore, higher absorption and lower beam intensity.

2.2.3. Synchrotron Radiation

Synchrotrons are devices for circulating electrically charged particles (negatively charged electrons or positively charged positrons) at nearly the speed of light [a more detailed discussion is given in Helliwell (1992)]. The particles are injected into the storage ring directly from a linear accelerator or through a booster synchrotron (Figure 2.4a). Originally these machines were designed for use in high energy physics as particle colliders. When the particle beam changes direction the electrons or positrons are accelerated toward the center of the ring and therefore emit electromagnetic radiation, and consequently lose energy. This energy loss is compensated for by a radiofrequency input at each cycle. The physicists' main aim was to study colliding particles; they were not interested in the radiation, which they regarded as an annoying byproduct and wasted energy. However, chemists and molecular biologists discovered (Rosenbaum et al., 1971) that the radiation was a useful and extremely powerful tool for their studies to the extent that radiation-dedicated synchrotrons have been constructed.

Synchrotrons are extremely large and expensive facilities, the ring having a diameter of 10 to a few hundred m. The trajectory of the particles is determined by their energy and by the magnetic field, which causes the charged particles to change their direction. There are four types of magnetic devices in storage rings:

1. *bending magnets* needed to guide the electrons in their orbit.
   The other three devices extend the spectrum to shorter wavelength. These devices can be inserted into straight sections and give no net displacement of the particle trajectory:
2. a *wavelength shifter*, with a stronger local magnetic field and a sharper curvature than the bending magnets,

---

[3] Brilliance is defined as number of photons/sec/mrad^2/mm^2/0.1% relative bandwidth.

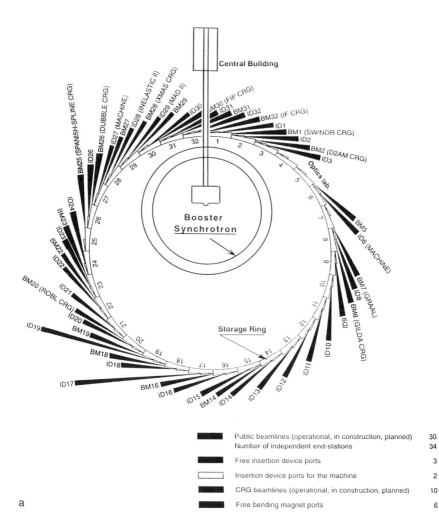

Central Building

BM25 (SPANISH-SPLINE CRG)
ID26
BM26 (DUBBLE CRG)
ID27 (MACHINE)
BM27
ID28 (INELASTIC II)
BM28 (XMAS CRG)
ID29 (MAD II)
BM29
ID30
BM30 (FIP CRG)
ID31
BM31
ID32
BM32 (IF CRG)
ID1
BM1 (SW/NOR CRG)
ID2
BM2 (D2AM CRG)
ID3

Optics lab.

BM5
ID6 (MACHINE)
BM7 (GRAAL)
ID8
BM8 (GILDA CRG)
6GI
ID10
ID11
ID12
ID13
ID14
BM14
ID15
BM16
ID16
ID17
BM18
ID18
BM19
ID19
BM20 (ROBL CRG)
ID20
BM22
ID21
BM22
ID23
BM23
ID24

Booster
Synchrotron

Storage Ring

▬	Public beamlines (operational, in construction, planned)	30
	Number of independent end-stations	34
▬	Free insertion device ports	3
▭	Insertion device ports for the machine	2
▬	CRG beamlines (operational, in construction, planned)	10
▬	Free bending magnet ports	6

a

b

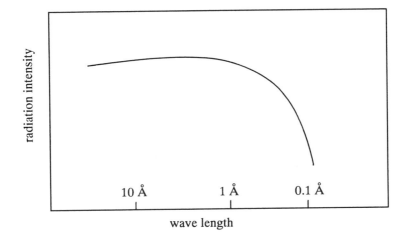

Figure 2.5. A typical curve of radiation intensity as a function of wavelength for a synchrotron radiation source.

3. a *wiggler*, producing a number of sharp extra bends in the electron trajectory, and
4. an *undulator*, similar to a wiggler but with the difference that interference effects cause the emission of radiation at more specific wavelengths.

Because freely traveling electrons (and positrons) are not quantized, the radiation ranges over a wide wavelength region depending on the energy of the charged particles and on the strength of the magnetic field. A useful quantity is the median of the distribution of power over the spectral region, called the critical photon energy E_c; it divides the power spectrum into two equal parts.

$$E_c = 0.665 E^2 B \quad \text{or} \quad \lambda_c = \frac{18.64}{E^2 B} \tag{2.1}$$

E_c is in keV and E is the circulating power in GeV:

$$E = \frac{\text{particle energy} \times \text{current}}{\text{revolution frequency}} \tag{2.2}$$

B is the magnetic field strength in Tesla and λ_c is in Å. The main photon flux is close to E_c, but above E_c it drops exponentially as a function of the photon energy (Figure 2.5). The European Synchrotron Radiation Facility (ESRF) in Grenoble (Figure 2.4b) has a circumference of 844.39 m, is

Figure 2.4. The European Synchrotron Radiation Facility (ESRF) in Grenoble, France. (a) The booster synchrotron and the storage ring are drawn with the large number of beamlines. (b) An artist picture of the Facility. Reproduced with permission from ESRF.

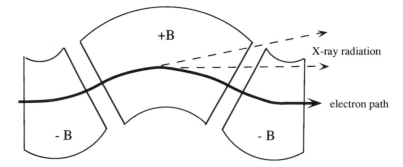

Figure 2.6. A wave length shifter. B is the magnetic field strength. The emission of radiation comes essentially from the top of the bump in the electron path.

operated with an energy of 6 GeV, and has bending magnets with 0.86 T field strength. Therefore λ_c is 0.6 Å.

The radius of a storage ring depends on E and B: it is proportional to E and inversely proportional to B. If one wants to design a synchrotron for the home laboratory, relatively small dimensions of the instrument are required and therefore E should be low and B high. The most powerful superconducting magnets for laboratory use are found in NMR instruments; they have a field strength of $B = 15$ T. If λ_c has to be 0.6 Å for the home instrument, as it is for the Grenoble facility, and assume that suitable magnets of 15 T can be constructed, we calculate for E:

$$E = \sqrt{E^2} = \sqrt{\frac{18.64}{\lambda_c \times B}} = 1.44 \text{ GeV}$$

The diameter of the instrument would then be approximately 4 m. However, the feasibility of such a project also depends on other factors, such as the injection system, which must be at least 50 m in length. Other problems are the spreading of the beam and the slow rate at which the magnetic field can be increased after beam injection.

1. The wavelength shifter: According to Eq. (2.1) the critical photon energy E_c is determined by E and B. E is everywhere constant in the ring, but the magnetic field strength B can be increased locally resulting in an increase of E_c and the production of higher intensity X-rays at shorter wavelength. A schematic picture of a wavelength shifter is shown in Figure 2.6.

2. The multipole wiggler: A series of wavelength shifters constitutes a multipole wiggler (Figure 2.7). The radiation from consecutive magnets is independent and adds up incoherently in the general direction of propagation of the electron beam. The total flux is simply $2N$ times the flux generated by a single period, where N is the number of periods. It is easily tunable to the desired wavelength.

3. The undulator: Undulators are multipole wigglers but with moderate

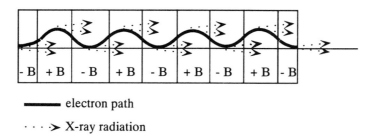

——— electron path

· · · ⟶ X-ray radiation

Figure 2.7. A multipole wiggler. B is the magnetic field strength.

magnetic fields and a large number of poles close together. The effect of this difference is that in the undulator strong interference occurs between the radiation from the consecutive magnets, which results in a spectral profile with a peak at a specific wavelength and a few harmonics. The radiation is tunable by adjustment of the magnetic pole piece distance. Its emitted intensity can be N^2 times that of a single period, with N the number of poles. Moreover, the beam divergence of the radiation from an undulator is extremely small.

2.2.3.1. Properties of Synchrotron Radiation

Intensity

The main advantage of synchrotron radiation for X-ray diffraction is its high intensity, which is two orders of magnitude stronger than for a conventional X-ray tube, at least for radiation from a bending magnet. For radiation from a multipole wiggler or an undulator it is again a few orders of magnitude stronger. This high intensity is profitably used by protein X-ray crystallographers for data collection on weakly diffracting specimens, such as very tiny crystals or crystals with extremely large unit cells. Another advantage is the low divergence of the beam resulting in sharper diffraction spots.

Tunability

Synchrotron radiation also differs from tube radiation in its tunability. Any suitable wavelength in the spectral range can be selected with a monochromator. This property is used in multiple wavelength anomalous dispersion (Section 9.5) and for Laue diffraction studies (Chapter 12). In the latter method a wide spectral range is used. Several types of modern detectors for X-ray radiation have fluorescent material as their X-ray-sensitive component. This is more sensitive for radiation with shorter wavelength, e.g., 1 Å instead of the conventional 1.5 Å from a copper target. Therefore, in protein X-ray diffraction experiments, a synchrotron is tuned to 1 Å or even a shorter wavelength. The additional advantage of this shorter wavelength is lower absorption along its path and in the crystal. Moreover, radiation damage to the protein crystal is appreciably reduced and often all required diffraction data can be collected from just one crystal, especially if the crystal is flash frozen and held at 100 °K.

Lifetime

The lifetime of a storage ring filling is limited. Typical lifetimes are between a few and several hours. When the intensity of the radiation has fallen to a certain minimum value, a new injection is required. The factors determining the lifetime are many and of a complicated nature, e.g., collision with residual gas atoms. The positively charged ions that form accumulate in the electron beam, leading to instabilities in the beam due to the continuous loss of particles. A beam composed of positrons has a longer lifetime than an electron beam, because positrons repel positive ions created in the residual gas, thus avoiding ion trapping in the beam. The beam lifetime in the European Synchrotron Radiation Facility is between 50 and 80 hours, depending on the current.

Time Structure

Synchrotron radiation, in contrast to X-ray tube radiation, is produced in flashes by the circulating bunches of charged particles. The ESRF operates in single-bunch or multibunch mode with a bunch length in the psec. range. This allows structural changes in the nsec. time scale to be observed.

Polarization

The X-ray beam from an X-ray tube is not polarized; synchrotron radiation is highly polarized. If the radiation from a bending magnet is observed in the plane of the orbit, it is fully polarized with the electric vector in the orbit plane (parallel polarization). If the observer moves away from the plane, a small perpendicular component is added. The polarization of the X-ray beam from a synchrotron has not yet found extensive application in X-ray diffraction. However, it must be considered when applying the correct polarization factor (Section 4.14.1) Moreover the polarization of the beam has an effect on the so-called anomalous X-ray scattering of atoms, which occurs when the X-ray wavelength approaches an absorption edge wavelength (Section 7.8).

2.3. Monochromators

Except for Laue diffraction, where the crystal is exposed to a spectrum of wavelengths (Chapter 12), monochromatic X-rays are used in all other diffraction methods. Therefore, a narrow wavelength band must be selected from the spectrum supplied by the source. If a sealed X-ray tube with copper anode is the radiation source, the wavelength selected for protein X-ray diffraction is the high intensity K_α doublet ($\lambda = 1.5418\,\text{Å}$).

The K_β radiation can be removed with a nickel filter. A nickel foil of 0.013 mm thickness reduces the K_β radiation to 2% and K_α to 66% of the original intensity. The continuous spectrum is also reduced but certainly not eliminated.

Much cleaner radiation can be obtained with a monochromator. For X-ray radiation from a tube the monochromator is a piece of graphite that reflects X-rays of 1.5418 Å against its layer structure at a scanning angle of 13.1°. For synchrotron radiation the preferable monochromators are made of germanium or silicium because they select a wavelength band two orders of magnitude narrower ($\delta\lambda/\lambda = 10^{-4}$–$10^{-5}$, whereas for a graphite monochromator it is 2.6×10^{-3}). Monochromators for synchrotron radiation are of the single or double type. Single type monochromators can be either flat or bent. The advantage of the bent monochromators is that they focus the divergent beam from the synchrotron, preferably onto the specimen. The focusing is in one direction only, producing a line focus. Focusing in the other direction is obtained with toroidal mirrors, made from highly polished quartz or glass. The mirrors are often coated with a thin layer of gold or platinum. The X-ray beam is focused by total reflection. Short wavelengths are absorbed by the mirror instead of reflected. By adjusting the reflection angle, mirrors can be effective as filters to remove harmonics or K_β. Thin diamond plates form excellent mirrors because they can withstand an extremely high heat flux without deformation. Because of their high energy resolution, they are often applied for splitting a synchrotron X-ray beam into a number of beams for different purposes: A first diamond crystal selects and reflects a very narrow wavelength region from the primary beam. The rest of the beam passes through the diamond, still with high intensity, and at least two other reflected beams can be generated with other diamond crystals, selecting slightly different wavelengths. A disadvantage of diamond crystals is their limited size which restricts their use to the rather narrow undulator beams (Freund, 1996).

The single type monochromators have a disadvantage: if they are tuned to another wavelength, the scanning angle of the monochromator changes and the entire X-ray diffraction equipment must be moved. This is not necessary for a double monochromator, where the direction of the beam is independent of the wavelength (Figure 2.8).

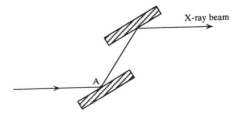

Figure 2.8. A double monochromator. The rotation axis for changing the wavelength is at A, perpendicular to the plane of the page. This gives a small vertical shift to the exit beam after rotation, but the incoming and exit beam are parallel.

2.4. Introduction to Cameras and Detectors

In the X-ray diffraction experiment with lysozyme in Chapter 1 a diffraction pattern was observed that could be regarded as corresponding to a three-dimensional lattice, reciprocal to the actual crystal lattice. For a crystal structure determination the intensities of all (or a great many) diffracted beams must be measured. To do so, all corresponding reciprocal lattice points must be brought to diffracting condition by rotating the lattice (that is, by rotating the crystal) until the required reciprocal lattice points are on the sphere with radius $1/\lambda$. It follows that an X-ray diffraction instrument consists of two main parts:

1. A mechanical part for rotating the crystal.
2. A detecting device to measure the position and the intensity of the diffracted beams; it should be noted that this intensity is always the result of two measurements: the total intensity in the direction of the diffracted beam and, subtracted from it, the background scattering in that same direction.[4]

In the last decade data collection in protein crystallography has changed tremendously (Pflugrath, 1992). Formerly a well-equipped protein crystallography laboratory had some precession cameras, a rotation camera, and a diffractometer. In both types of camera, X-ray film was the detector. The precession camera has the advantage of giving an undistorted image of the reciprocal lattice see Figure 4.27b. Unit cell dimensions and symmetry in the crystal can easily be derived from such an undistorted image as can the quality of the crystal. For full three-dimensional X-ray data collection the precession camera is not suitable, because it registers one reciprocal layer per exposure.

A rotation camera registers the data more efficiently, but the recognition of the diffraction spots is more complicated. Moreover, for each exposure the crystal is oscillated over a small angle (e.g., 1–2°) to avoid overlap of spots. Depending on the symmetry in the crystal the total oscillating range for a complete data collection may be 60°, 90°, or 180°.

The third classical detector is a computer-controlled diffractometer, which has a single counter, normally a scintillation counter. This has been the workhorse in protein X-ray crystallography and still is so for the X-ray diffraction of small molecules. It measures the diffracted beams with high accuracy but only one at a time. Although it is computer-controlled and automatically finds the diffracted beams, it is extremely slow.

---

[4]Background scattering is mainly caused by the air through which the X-ray beam passes from the collimator to the beamstop. If the airpath is long and absorption serious, it can appreciably be reduced if a cone filled with helium is put between the crystal and the plate.

The classical picture has been changed completely by the introduction of much faster image plates and electronic area detectors. Rotation "cameras" are now equipped with such a detector, either image plate or electronic area detector, though the principle of the "camera" has not changed. With an image plate, which must be read after each exposure, the data collection is still in the "film mode," that is contiguous oscillations over a small angle and reading (analogous to film processing) after each exposure. If the instrument has an electronic area detector the oscillations are much smaller (e.g., 0.1°), and the data are immediately processed by an on-line data acquisition system.

Before describing these instruments in some detail, we shall discuss the various types of X-ray-detecting systems including their advantages and disadvantages (Helliwell, 1992; Gruner, 1994).

2.5. Detectors

2.5.1. Single Photon Counters

Single photon counters have been used since the early days of X-ray diffraction and are now usually of the scintillation type. They give very accurate results, but because they measure X-ray reflections sequentially it takes several weeks to collect a complete data set from a protein crystal (of the order of 10,000 to 100,000 reflections).

Although the instrument with a single photon counter is no longer used very often for data collection in protein X-ray crystallography, we shall present its mechanical construction, because the angle nomenclature is standard in X-ray crystallography (Figure 2.9a). In this diffractometer the X-ray beam, the counter, and the crystal are in a horizontal plane. To measure the intensity of a diffracted beam the crystal must be oriented such that this beam will also be in the horizontal plane. This orientation is achieved by the rotation of the crystal around three axes: the ϕ-, the ω-, and the χ-axis (see Figure 2.9a). The counter can be rotated in the horizontal plane around the 2θ-axis, which is coincident with, but independent of the ω-axis. Data collection is done either with the ω- and the 2θ-axis coupled or with the 2θ-axis fixed and the crystal scanned by rotation around the ω-axis.

An alternative construction is found in the CAD4 diffractometer produced by Delft Instruments. It has two advantages over the classical design. The rather bulky χ circle is absent; instead the instrument has another axis, the oblique κ-axis (Figure 2.9b). The χ rotation is mimicked by a combined rotation around the κ-, ϕ-, and ω-axes. The other advantage is that rotation around an axis is mechanically more accurate than sliding along an arc.

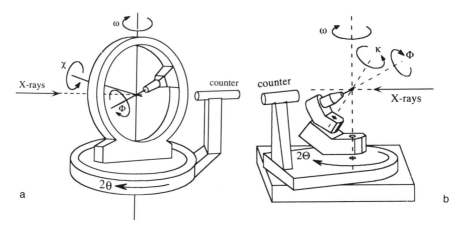

Figure 2.9. (a) In this four circle diffractometer of classical design, the crystal is located in the center of the large circle. It can be rotated around three axes: by ϕ around the axis of the goniometer head, by ω around the vertical axis, and by χ through sliding of the block that holds the goniometerhead, along the large circle. The counter can be rotated around a fourth axis by the angle 2θ; this axis is coincident with the ω-axis. (b) A four circle diffractometer with the κ construction. The κ-axis is at 50° with respect to the ω-axis. The ϕ rotation is again around the goniometerhead axis. The ω- and the 2θ-axes are as in the classical design. The crystal is at the intersection of all axes.

2.5.2. Photographic Film

Photographic film is a classical detector for X-ray radiation, but it is not used much anymore because of the availability of far more sensitive image plates and area detectors. The single advantage of film over other present day area detectors is its superior resolution resulting from its fine grain. For historical reasons we shall devote one paragraph to this classical detector.

X-ray film is double-coated photographic film. The double coating of the film base prevents curling and provides a thick layer of sensitive silver-base material, which absorbs approximately two-thirds of the incoming photons. Processing of X-ray film is—although not difficult—somewhat cumbersome and time consuming because of the developing process, the labeling of the films, as well as the handling of chemicals. Furthermore, for quantitative work, the density of the spots on the film must be measured with a densitometer. The density is defined as $\log(I_0/I)$, where I_0 is the intensity of the light beam in the densitometer before, and I after it has passed the film. The density of the spots is (up to a certain level) proportional to the number of X-ray photons that are absorbed by the film. Density, plotted as a function of the exposure time, is a straight line over a long exposure range but levels off at very high density values.

These higher density values should be used only with careful calibration. This limited dynamic range (1:200) requires that for quantitative measurements, in which the full range of X-ray intensities must be measured, a pack of three consecutive films is used. The weakest spots are measured on the first film and the stronger ones on the last film. Altogether, photographic film is a simple but rather slow and in the long run even expensive detector.

2.5.3. Image Plates

Image plates are used in the same manner as X-ray film but have several advantages. They are therefore replacing conventional X-ray film in most laboratories. Image plates are made by depositing a thin layer of an inorganic storage phosphor on a flat base. X-ray photons excite electrons in the material to higher energy levels. Part of this energy is emitted very soon as normal fluorescent light in the visible wavelength region. However, an appreciable amount of energy is retained in the material by electrons trapped in color centers; it is dissipated only slowly over a period of several days. This stored energy is released on illumination with light. In practical applications a red laser is used for scanning the plate and blue light is emitted. The red light is filtered away and the blue light is measured with a photomultiplier (Figure 2.10a). With certain precautions the light emitted is proportional to the number of photons to which that particular position of the plate was exposed. The pixel size depends mainly on the reading system and is between 100×100 and $200 \times 200 \mu m^2$ (Amemiya, 1997).

Image plates are at least 10 times more sensitive than X-ray film and their dynamic range is much wider $(1:10^4-10^5)$. The entire range from strong to weak reflections can therefore be collected with one exposure on a single plate. The plates can be erased by exposure to intense white light and used repeatedly. Another advantage for application with synchrotron radiation is their high sensitivity at shorter wavelengths (e.g., 0.65Å). A further advantage of short wavelengths is that the absorption of the X-ray beam in the protein crystal becomes negligible and no absorption corrections are required. But image plates are similar to photographic film in the sense that they require a multistep process: exposure as the first step and processing (in this case reading) as the second step. Reading takes only a few minutes.

The first commercially available instruments had a rather small size plate but in new models this has been increased, for instance in the Mar Research instrument (Figure 2.10b) from 18 to 30 cm. If synchrotron radiation is used, the smaller size is not a problem because with the shorter wavelength the diffraction pattern is more compressed. A disadvantage of image plates is that the stored image fades away gradually. This decay is

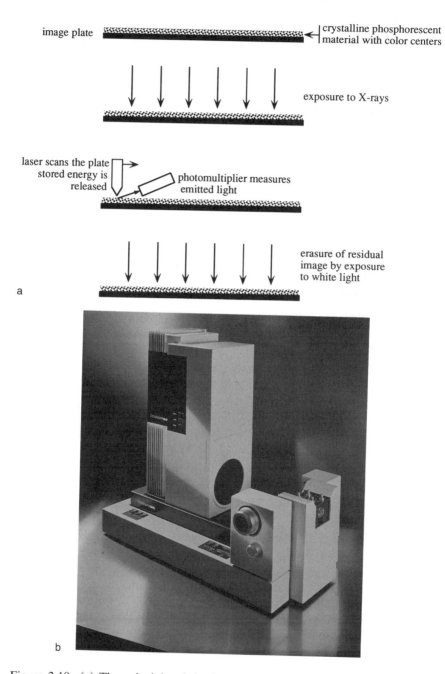

Figure 2.10. (a) The principle of the image plate as an X-ray detector. (b) The image plate instrument produced by Mar Research, Norderstedt Germany. It has a circular plate; for scanning the plate rotates and the laser scans along a radial line. Courtesy Mar Research.

rather rapid in the first few minutes but then slows down: it takes a few hundred hours for a 50% decrease in the stored energy.

2.5.4. Area Detectors

Although photographic film and image plates are area detectors, the use of this term is restricted to electronic devices that detect X-ray photons on a two-dimensional surface and process the signal immediately after photon detection. They are also called position sensitive detectors, because both the intensity of a diffracted beam and the position where it hits the detector are determined. A basic difference with image plate and photographic film is that area detectors scan through a diffraction spot every $0.1°$ or so, giving a three-dimensional picture of the spot. In contrast, for film and image plates a much larger oscillation angle (e.g., $2°$) is used for each exposure and therefore no profile is obtained of the diffraction spot in the oscillation direction.

Area detectors are currently based on either a gas-filled ionization chamber or an image intensifier coupled to a video system or a CCD. The gas-filled chambers are essentially single photon counting X-ray detectors. Their construction is shown schematically in Figure 2.11a.

In the absorption gap X-ray photons cause ionization of gas atoms with the formation of ions and electrons. The liberated electrons ionize neighboring gas atoms by collision with the result that about 300 ions and electron pairs are formed by the absorption of a single 8-keV ($\lambda =$ 1.55 Å) X-ray photon. This is not enough for a measurable signal and, therefore, amplification is applied by accelerating the electrons in an electric field between a cathode and an anode. The anode consists of many parallel wires, $1-2$ mm apart. The accelerated electrons cause secondary ionization, mainly at the anode wires. The electrons hit the anode and the ions the second cathode, which—like the anode—consists of many parallel wires that are perpendicular to the anode wires. The event is now registered as one count and the position of the incident photon is electronically determined. A disadvantage of this type of detector is the limitation in counting rate due to the build-up of charges in the chamber and to limitations in the processing electronics. The peak counting rate is about 10^5 Hz. This prevents the use of area detectors of the multiple wire proportional counter type in X-ray crystallography with high-intensity synchrotron radiation. Another disadvantage of this type of detector, in combination with synchrotron radiation, is the lower sensitivity at shorter wavelength due to poor absorption of the X-ray photons in the gas chamber.

In video-based area detectors the diffraction pattern is collected on a fluorescent screen. The resultant light is amplified with an image intensifier, stored in the target of a video camera tube, read out, and fed

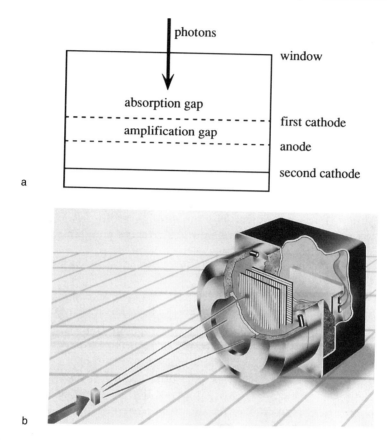

photons

window

absorption gap

first cathode

amplification gap

anode

second cathode

a

b

Figure 2.11. (a) The principle of the design of a multiple wire proportional counter. The gas-filled chamber has two cathodes and an anode, each consisting of parallel wires, either horizontal or vertical. (b) The Siemens area detector, which is filled with xenon gas for maximum absorption of the X-ray beam.

into a computer. The system is schematically represented in Figure 2.12. Because of the storage in the video tube the diffraction pattern is integrated over short time periods. Overloading is a less serious problem than for the gas-filled detectors. Therefore, they are better suited for synchrotron radiation than are the gas-filled multiple wire detectors. They have a spatial resolution of about 0.1 mm. Electronic noise is rather high but can be kept under control, e.g., by maintaining the components at a constant temperature. This noise level limits the dynamic range to $1:10^3$. By remeasuring the strong intensities with a lower high voltage setting of the video-camera the dynamic range is increased by a factor of 10.

   The performance of the two types of area detectors, the gas-filled and the TV systems, is approximately the same. They are both very sensitive, compared with X-ray film. This, and the rapid processing of the data,

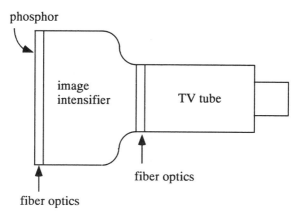

phosphor

image intensifier

TV tube

fiber optics

fiber optics

Figure 2.12. An area detector of the television type. X-ray photons enter on the left and are transformed into light photons by the phosphor screen. The light is amplified in the image intensifier and registered in the video tube.

allows a data collection speed that is about 50 times the speed obtained with X-ray film. Calibration of area detectors is required because of nonuniform sensitivity over the area and geometric distortion of the diffraction pattern in the system.

In another kind of area detector the video tube is replaced by a charge coupled device (CCD). Much effort is being invested to improve them and to make CCDs of the relatively large size and high sensitivity required for X-ray detection. They have a high dynamic range, combined with excellent spatial resolution, low noise, and high maximum count rate (Walter et al., 1995).

2.6. The Rotation (Oscillation) Instrument

In protein X-ray crystallography efficient data collection is always screenless, whether photographic film, image plate, or electronic area detector is used. The crystal (and the reciprocal lattice) is rotated in small oscillation steps through the Ewald sphere (Chapter 1). Rotation around any arbitrary axis is not the correct procedure. An efficient strategy for data collection requires careful consideration of the crystal properties and the available hardware, such as X-ray source and detector. It is important to minimize systematic errors and maximize completeness; high redundancy is also an advantage. In Section 2.6.1 a few of the problems will be discussed. For a complete discussion of data collection strategy with a rotation camera one should consult Dauter (1997). The mechanical part of the first rotation instruments was very simple: just one rotation axis perpendicular to the direction of the X-ray beam. The detector, for instance a fluorescent plate, is flat. The consequence of this simple design is that reciprocal lattice points near the rotation axis never pass through the Ewald sphere (Figure 2.13a). Therefore, it requires—in principle—mounting of another crystal in a different orientation to measure the diffracted beams in the blind region. However, this is not always necessary. Suppose the crystal has three perpendicular axes and it is oriented with one of the crystal axes along the rotation axis; then the blind region is indeed a problem: assuming sufficient symmetry in the crystal, the four shaded areas in Figure 2.13a contain reciprocal lattice points belonging to reflections with the same intensity in each region and none of the four regions is recorded. However, if the crystal is slightly misset, the problem is solved because now all reflections in at least one of the four regions can be recorded (Figure 2.13b).

Modern oscillation camera's have the same sophisticated system for rotating the crystal as electronic area detectors like the Siemens X-100 (Figure 2.14) and the Fast area detector made by Delft Instruments (Figure 2.15) the Siemens instrument has the classical χ circle and the Delft instrument the κ construction. They allow the crystal to rotate around three axes. Therefore, the same crystal can easily be reoriented with respect to the X-ray beam. The disadvantage of screenless data collection is the seemingly disordered arrangement of the spots on the film or plate. The spot positions are determined by

1. the crystal orientation,
2. the unit cell parameters in the crystal,
3. the crystal-to-film distance and the wavelength, and
4. the film center.

However, this problem can be solved with intelligent software, which can recognize the spots, apply correction factors, and supply the crystallographer with a final data set.

Because of basic differences between instruments equipped with an image plate and ones equipped with an electronic area detector, we shall treat them separately.

2.6.1. Rotation Instruments with an Image Plate

This is the modern version of the rotation camera with film, originally designed by U.W. Arndt.[6] Except for the detector, there are no differences in principle between the old film and the new image plate instruments. Data are collected in contiguous oscillation ranges, each of

---

[6] This instrument is discussed in detail in Arndt and Wonacott (1977).

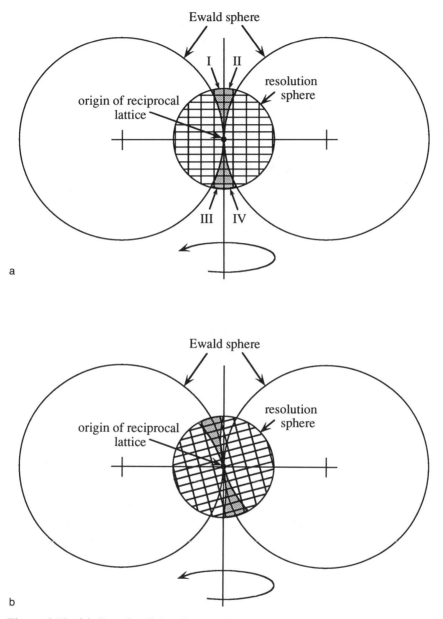

Figure 2.13. (a) For simplicity the Ewald sphere is rotated instead of the reciprocal lattice. The vertical line in the center is the rotation axis. In its rotation
around the axis the Ewald sphere never passes through the reciprocal lattice
regions I–IV. This is the "blind region." (b) Here the reciprocal lattice is in a
skew position. In this example the reciprocal lattice has three perpendicular axes
and, contrary to the situation in (a), all reflections within one octant formed by
the three reciprocal lattice axes can be measured.

Figure 2.14. The Siemens X100 area detector. This instrument is equipped with a multiple wire proportional counter. The crystal can be oriented around three axes. The counter can swing out around another axis for collecting reflections with large diffraction angle. New models, X1000 and Hi-star, have an improved detector system. Courtesy Siemens AG, Analytical Systems, Karlsruhe, Germany.

approximately 2°. The exact value is determined by the distance between the reciprocal lattice points, the maximum angle of reflection (resolution), and the width of the spots. In performing the oscillation, the maximum displacement is found for reciprocal lattice points in the plane through the origin perpendicular to the rotation axis at the edge of maximum diffraction angle (the resolution limit) (Figure 2.16). If the distance between adjacent points at this edge is d^* and their distance from the origin d_{lim}, the angle Δ_1 has $\tan \Delta_1 = d^*/d_{lim}$. Because $d^* \ll d_{lim}$ the angle Δ_1 is approximately equal to d^*/d_{lim} (in radians[7]). However, the reflections have a certain angular width Δ_2, determined by the crystal size, its mosaicity, and the divergence of the X-ray beam. Therefore, the maximum oscillation angle is $\Delta_1 - \Delta_2$, because a reflection spot is allowed to move for the next exposure until it reaches the region of its neighbor in the current exposure.

To minimize the number of exposures the oscillation range should be as large as possible and therefore the crystal should be oriented with its

---

[7] A full circle is 2π radians.

shortest reciprocal distance along the rotation axis. (This is not required for area detectors.) Because of the reciprocity between the crystal lattice and the reciprocal lattice, this orientation corresponds with the longest unit distance in the crystal lattice along the rotation axis. Another feature to consider for minimizing the number of exposures is the symmetry in the crystal. This symmetry is also present in the reciprocal lattice. Lattice points related by this symmetry belong to diffracted beams with the same intensity and, therefore, fewer diffracted beams need to be measured.

Two disadvantages of the screenless rotation method with image plates are immediately apparent: (1) the background is relatively high and (2) some spots appear partly on one and partly on the next or previous exposure. This second problem is especially severe for large unit cells in the crystal because of the close spacing of reciprocal lattice points and a very small oscillation range. For the solution to this problem the two "partials" are treated as individual reflections and their intensities are added in a later stage. This requires extreme precision in the mechanical part of the instrument. Nevertheless, the accuracy in the intensity of

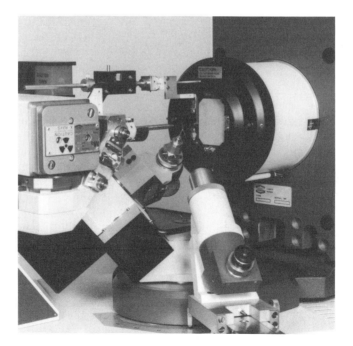

Figure 2.15. The Fast area detector produced by Delft Instruments, Delft, The Netherlands. This is a video type instrument. It has the κ construction for the crystal orientation, derived from the successful CAD-4 four circle diffractometer of this company. The counter can swing out around the 2θ-axis. Courtesy Delft Instruments.

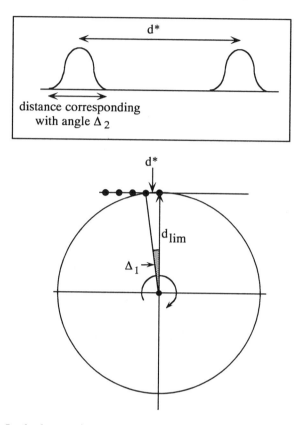

Figure 2.16. In the lower picture the rotation axis is perpendicular to the plane of the paper. The black dots are reciprocal lattice points at the edge of the limiting resolution sphere, which has a radius d_{lim}. The angular distance between adjacent reflections is Δ_1 with $\tan \Delta_1 = d^*/d_{lim}$. In the box two adjacent reflections are drawn. Suppose the right reflection has just passed through the Ewald sphere and has caused a spot on the image plate. In the ongoing oscillation the left reflection should not be allowed to reach the Ewald sphere before the right reflection has passed. If it does, then it would overlap the image of the right reflection. Therefore, the maximum oscillation angle permitted is $\Delta_1 - \Delta_2$.

reflections composed of partials is somewhat lower than for fully recorded reflections and for this reason they are sometimes completely neglected. The first problem is a signal-to-noise ratio problem. The background noise is proportional to the exposure time and thus to the oscillation angle. The intensity of a fully recorded reflection is independent of the oscillation angle. The signal-to-noise ratio is more favourable if the X-ray beam has a smaller divergence and the mosaic spread of the crystal is low.

A smaller oscillation angle also improves the ratio but causes many more reflections to be recorded as partials.

After exposing the crystal in a few oscillation ranges, the software is able to find an approximate orientation of the crystal with rough unit cell dimensions (if these are not already known). It is assumed that the crystal-to-plate distance and the position of the center of the plate are fixed instrument parameters. The approximate orientation and cell parameters can then be refined during the processing of all the data. In this process a rectangular box is defined around each spot. The integrated intensity in this box is the spot intensity; the background is estimated in the surrounding region. The measurement of weak reflections can be appreciably improved by profile fitting. A reliable two-dimensional spot profile is derived from the strong spots and then applied to the weak ones, of course assuming that the profiles are equal for strong and weak spots.

Image plates are also the detectors in Weissenberg cameras for protein X-ray crystallography. This very classical instrument, in which the rotation of the crystal is coupled to a translation of the plate, has been adapted to protein work by Sakabe (1983, 1991) at the Photon Factory (Tskuba, Japan). See also Stuart and Jones (1993).

2.7. Electronic Area Detectors

It was pointed out before that basic differences exist between rotation instruments equipped with an image plate and electronic area detectors. The main difference is that the oscillation angle for the area detectors is much smaller; frames with an oscillation angle of 0.1° are contiguously measured. This is possible due to the immediate processing of the data from each frame. The advantages are that the background is low and that a three-dimensional profile of the reflection spots can be constructed, which is even more favorable for measuring the weak intensities than the two-dimensional profile.

With the more sophisticated mechanical part of the instrument a crystal can be easily adjusted to nearly every orientation. The search for the initial orientation is also easier with these instruments than with a rotation instrument because the smaller oscillation angle defines more precisely the position of a spot.

Radiation Protection

The high energy photons of X-rays have a harmful effect on living tissue. Therefore, they must be used with great care, taking all necessary precautions. Local regulations for the protection of personnel should be obeyed and unauthorized use of X-ray equipment must be forbidden. A somewhat confusing number of radiation units are in use and, therefore, a definition of them will be given.

The Curie (Ci) is the old unit for radioactivity:

$$1\,Ci = 37 \times 10^9 \text{ disintegrations/sec}$$

1 Ci is approximately equal to the activity of 1 g of radium. The Ci is not an SI unit and has therefore been replaced by the Becquerel (Bq), which stands for 1 disintegration/sec:

$$1\,Ci = 37 \times 10^9\,Bq$$

The radiation absorbed dose (rad) is a measure for the amount of radiation that corresponds to the energy absorption in a certain medium, such as a tissue.

1 rad is the dose of radiation corresponding to an energy absorption of 0.01 J/kg medium.

The rad has been replaced by an SI unit: the Gray (Gy), corresponding to 1 J/kg or $1\,m^2/sec^2$

$$1\,Gy = 100\,rad$$

The relative biological effect (RBE) was introduced because it has been found that the same absorbed dose from different types of radiation does not always have the same harmful effect in biological systems. Therefore, a quality factor Q has been introduced. In the rem (X-ray equivalent man) this quality factor has been taken into account:

$$1\,rem = 1\,rad \times Q$$

The rem has now been replaced by the Sievert (Sv):

$$1\,Sv = 100\,rem$$

A summary of the names and symbols is given in Table 2.1.

Table 2.1. Radiation Units and Symbols

Physical or biological property			SI unit	
Name	Symbol	Value	Symbol	Value
Radioactivity	A	Becquerel	Bq	1 disintegration/sec
		Old: Curie	Ci	$1\,Ci = 37 \times 10^9\,Bq$
Absorbed dose	D	Gray	Gy	$1\,J\,kg^{-1} = 1\,m^2/sec^2$
		Old: rad	rad	$1\,rad = 10^{-2}\,Gy$
Dose equivalent	H	Sievert	Sv	$1\,J\,kg^{-1} = 1\,m^2/sec^2$
		Old: rem	rem	$1\,rem = 10^{-2}\,Sv$

Summary

In this chapter X-ray sources were discussed as well as instruments for the registration and measurement of the diffracted beams. Conventional X-ray sources in the laboratory are sealed tubes and tubes with a rotating anode, the latter being preferred for protein X-ray crystallography because of their higher intensity. The most commonly used radiation from a tube has a wavelength of 1.5418 Å. This is the characteristic K_α wavelength emitted by a copper anode and is selected from the spectral distribution by a filter or, preferably, with a graphite monochromator. The extremely high intensity X-ray radiation from a synchrotron is of great value for collecting data from weakly diffracting specimens. The beam is not only strong, it is also highly parallel, causing smaller but more brilliant spots on the detector. Therefore, with synchrotron radiation the resolution is somewhat better than in the home laboratory (more diffracted beams at high diffraction angle with greater details in the resulting protein structure).

Another advantage of synchrotron radiation is its tunability, which allows the selection of radiation with a wavelength at or below 1 Å. Although the specimen diffracts this radiation more weakly than the 1.5418 Å copper radiation, the fluorescent type detectors (image plate and Fast area detector) are more sensitive at this shorter wavelength. Another important advantage is the lower absorption of radiation with shorter wavelength and, consequently, less radiation damage to the crystals. For a protein structure determination the number of diffracted beams to be recorded is extremely high, of the order of 10^4–10^5. To achieve this within a reasonable time requires highly efficient hardware, either an electronic area detector or an image plate. These instruments have completely changed protein X-ray crystallography by considerably reducing the time for exposure and data processing, and in this way have solved the previously time-consuming data collection problem.

Chapter 3
Crystals

3.1. Introduction

The beauty and regularity of crystals impressed people to such an extent that in the past crystals were regarded as products of nature with mysterious properties. Scientific investigation of crystals started in 1669, when Nicolaus Steno, a Dane working as a court physician in Tuscan, proposed, that during crystal growth *the angles between the faces remained constant.*

For a given crystal form individual crystals may differ in shape, that is, in the development of their faces, but they always have identical angles between the same faces (Figure 3.1). The specific morphology may depend on factors such as the supply of material during growth, on the presence of certain substances in the mother liquor, or on the mother liquor itself. For a single crystal form the angles between the faces are constant, but this is not true if the crystals belong to different crystal forms. Figure 3.2 shows four different crystal forms of deoxyhemoglobin from the sea lamprey *Petromyzon marinus*. Their appearance depends on the buffer and on the precipitating agent, although occasionally two different forms appear under the same conditions.

Before the famous first X-ray crystallographic diffraction experiment by von Laue, Friedrich, and Knipping in 1912, the internal regularity of a crystal was suggested but never proven. X-ray crystallography has dramatically changed this situation.

Determining the atomic structure of a molecule, particularly one as complex as a protein molecule, is greatly facilitated if a large number of identical molecules can be aggregated in a regular arrangement. The highest order is present in crystals of the material, although structural

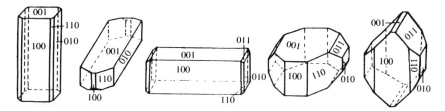

Figure 3.1. Crystals of trimethylammonium bromide belonging to the same crystal form but exhibiting a range of morphologies.

information can also be obtained from fibers. Not much can be done with an amorphous solid or a solution, which give weak and diffuse X-ray diffraction patterns and from which little structural information can be derived. In this book we restrict ourselves to crystals.

A crystal of organic material is a three-dimensional periodic arrangement of molecules. When the material precipitates from a solution, its molecules attempt to reach the lowest free energy state. This is often accomplished by packing them in a regular way; in other words, a crystal grows. It is surprising to observe that even large protein molecules follow this principle, although occasionally they unfortunately do not crystallize. Flat planes at the surface of a well-developed crystal reflect the regular packing of the molecules in the crystal. In this regular packing three repeating vectors $\mathbf{a}^1$, $\mathbf{b}$, and $\mathbf{c}$ can be recognized with the angles α, β, and γ between them. These three vectors define a unit cell in the crystal lattice (Figure 3.3).

If the content of the unit cells is neglected for the moment, the crystal can be regarded as a three-dimensional stack of unit cells with their edges forming a grid or lattice (Figure 3.4). The line in the $\mathbf{a}$ direction is called

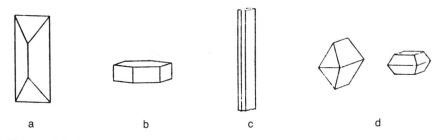

a   b   c   d

Figure 3.2. The crystals a–d belong to four different crystal forms of deoxyhemoglobin from the sea lamprey *Petromyzon marinus*. Reproduced with permission from Hendrickson et al. (1968).

---

[1] Vectors are in boldface type.

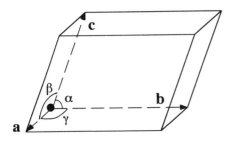

Figure 3.3. One unit cell in the crystal lattice.

● Origin

the x-axis of the lattice; the y-axis is in the **b** direction and the z-axis is in the **c** direction. The x-, y- and z-axes together form a coordinate system that by convention is right-handed. We shall later see (Section 4.7.2) that diffraction of X-rays by a crystal can be regarded as reflection against planes in the lattice. These planes are constructed through the lattice points and a great many sets of these planes can be drawn (Figure 3.5).

Within a set, the planes are parallel and equidistant with perpendicular distance d. As can be derived from Figure 3.5 the lattice planes cut an axis, for example the x-axis, into equal parts that have a length $a/1$, $a/2$, $a/3$, $a/4$ etc. The whole numbers 1, 2, 3, 4, ... are called indices. A set of lattice planes is determined by three indices h, k, and l, if the planes cut the x-axis in a/h, y in b/k, and z in c/l pieces. If a set of planes is parallel

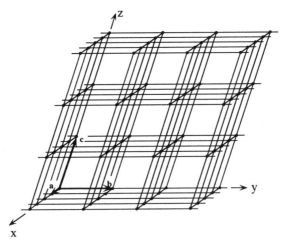

Figure 3.4. A crystal lattice is a three-dimensional stack of unit cells.

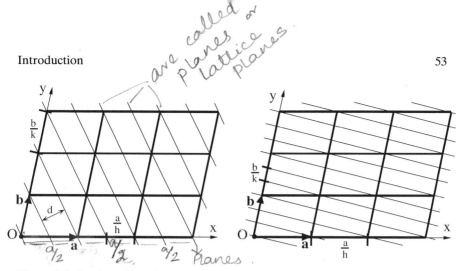

Figure 3.5. Lattice planes in a two-dimensional lattice. In the left figure, $h = 2$ and $k = 1$; in the right figure, $h = 1$ and $k = 3$.

to an axis that particular index is 0 (the plane intercepts the axis at infinity). Therefore, the unit cell is bounded by the planes (100), (010), and (001) (Figure 3.6). The flat faces of a crystal are always parallel to lattice planes (Figures 3.1 and 3.7). The parentheses in $(h\ k\ l)$ are used to distinguish a lattice plane from a line segment in the unit cell, which is given in brackets: for example, [100] is the line segment from the origin of the unit cell to the end of the a-axis and [111] is the body diagonal from the origin to the opposite corner.

From Figure 3.5 it is clear that the projection of a/h, b/k, and c/l on the line perpendicular to the corresponding lattice plane $(h\ k\ l)$ is equal to the lattice plane distance d. We have not yet discussed the choice of unit cell in the crystal. For example, in Figure 3.8 the choice could be either unit cell I, II, or III. Often the problem does not exist because of symmetry considerations in the crystal (Section 3.2). If the choice does

Figure 3.6. One unit cell bounded by the planes (100), (010), and (001). The directions along **a**, **b**, and **c** are indicated by [100], [010], and [001].

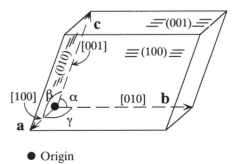

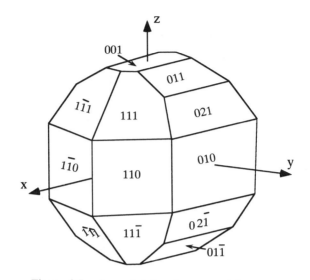

Figure 3.7. A crystal showing several faces.

exist then certain rules should be followed, given in *International Tables for Crystallography*, Vol. A, Chapter 9 (Hahn, 1983).

The main conditions are as follows:

1. The axis system should be right-handed.

2. The basis vectors should coincide as much as possible with directions of highest symmetry (Section 3.2).

3. The cell taken should be the smallest one that satisfies condition 2. This condition sometimes leads to the preference of a face-centered (A, B, C, or F) or a body-centered (I) cell over a primitive (P) smallest cell

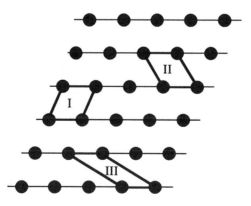

Figure 3.8. In this two-dimensional lattice the unit cell can be chosen in different ways: as I, as II, or as III.

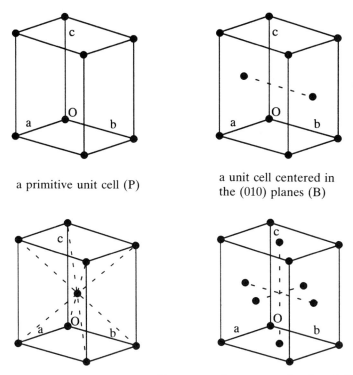

a primitive unit cell (P)

a unit cell centered in
the (010) planes (B)

a body-centered unit cell (I)   a face-centered unit cell (F)

Figure 3.9. Noncentered and centered unit cells.

(Figure 3.9). Primitive cells have only one lattice point per unit cell, whereas nonprimitive cells contain two or more lattice points per unit cell. They are designated A, B, or C if one of the faces of the cell is centered: it has extra lattice points on opposite faces of the unit cell, respectively, on the *bc* (A), *ac* (B), or *ab* (C) faces. If all faces are centered the designation is F (Figure 3.9).

4. Of all lattice vectors none is shorter than *a*.

5. Of those not directed along **a** none is shorter than *b*.

6. Of those not lying in the *a*, *b* plane none is shorter than *c*.

7. The three angles between the basis vectors **a**, **b** and **c** are either all acute ($<90°$) or all obtuse ($\geq90°$).

3.2. Symmetry

The search for a minimum free energy and as a consequence the regular packing of molecules in a crystal lattice often leads to a symmetric relationship between the molecules. As we have seen in the previous

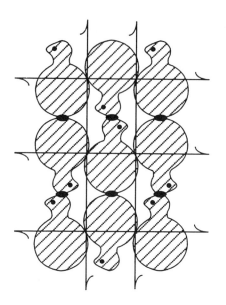

Figure 3.10. A two-dimensional lattice with 2-fold symmetry axes perpendicular to the plane of the figure and 2-fold screw axes in the plane.

section, a characteristic of a crystal is that it has unit translations in three dimensions, also called three-dimensional translational symmetry, corresponding to the repetition of the unit cells. Often, additional symmetry is encountered.

Examples are given in Figures 3.10 and 3.11, which show operations having 2- and 3-fold (screw) rotation axes as symmetry elements. Figures

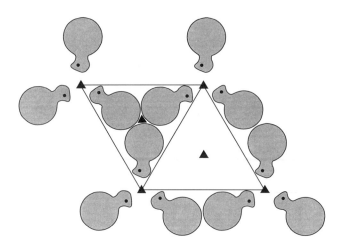

Figure 3.11. A two-dimensional lattice with 3-fold symmetry axes perpendicular to the plane of the figure.

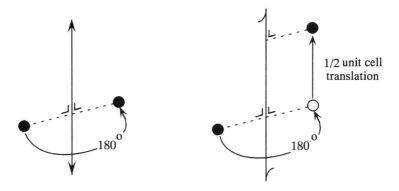

Figure 3.12. A 2-fold axis (left) and a 2-fold screw axis (right); the latter relates one molecule to another by a 180° rotation plus a translation over half of the unit cell.

3.12 and 3.13 give further examples of 2-fold and 3-fold symmetry operations. n-fold axes with $n = 5$ or $n > 6$ do not occur. The reason is that space cannot be filled completely with, for example, a 5-fold or a 7-fold axis. In addition to axes of symmetry, crystals can have mirror planes, inversion centers (centers of symmetry) (Figure 3.14), and rotation inversion axes, which combine an inversion and a rotation. Table 3.1 lists all possible symmetry operations together with their symbols and the translation operation.

Another way of looking at symmetry is the following. Application of the symmetry operators, such as rotations with or without translations,

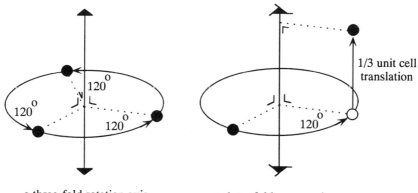

a three-fold rotation axis   a three-fold screw axis

Figure 3.13. A 3-fold axis (left) and a 3-fold screw axis (right); the latter relates one molecule to another by a 120° rotation and a translation over one-third of the unit cell.

Figure 3.14. The effect of a mirror and of an inversion center.

mirror plane

center of symmetry
or inversion center

leaves the entire crystal unchanged; it looks exactly as before. Therefore, the properties of the crystal, such as those of an electrical or mechanical nature, obey at least the same symmetry. There are 230 different ways to combine the allowed symmetry operations in a crystal, leading to 230 space groups. They can be found in the *International Tables of Crystallography*, Volume A (Hahn, 1983).

The restrictions given above for the symmetry axes of "ordinary" crystals are the consequence of their three-dimensional translation symmetry because this requires space to be filled entirely by identical repeating units. This symmetry is not present in so-called quasicrystals, which consist of two or more (but a finite number) different units. The position of the units in the lattice of a quasicrystal is determined according to a predictable sequence that never quite repeats: they are quasiperiodic and can have any rotational symmetry, including 5-fold and 7-fold, which are forbidden in "ordinary" crystals. So far no quasicrystalline protein crystals have been found and they will not be considered further.

3.3. Possible Symmetry for Protein Crystals

Not all 230 space groups are allowed for protein crystals. The reason is that in protein crystals the application of mirror planes and inversion

Table 3.1. Graphic Symbols for Symmetry Elements[a]

Symmetry axis or symmetry point	Graphic symbol	Screw vector of a right-handed screw rotation in units of the shortest lattice translation vector parallel to the axis	Printed symbol
Symmetry axes normal to the plane of projection (three dimensions) and symmetry points in the plane of the figure (two dimensions)			
Identity	None	None	1
Twofold rotation axis ⎫ Twofold rotation point (two dimensions) ⎭		None	2
Twofold screw axis: "2 sub 1"		$\frac{1}{2}$	2_1
Threefold rotation axis ⎫ Threefold rotation point (two dimensions) ⎭		None	3
Threefold screw axis: "3 sub 1"		$\frac{1}{3}$	3_1
Three fold screw axis: "3 sub 2"		$\frac{2}{3}$	3_2
Fourfold rotation axis ⎫ Fourfold rotation point (two dimensions) ⎭		None	4
Fourfold screw axis: "4 sub 1"		$\frac{1}{4}$	4_1
Fourfold screw axis: "4 sub 2"		$\frac{1}{2}$	4_2
Fourfold screw axis: "4 sub 3"		$\frac{3}{4}$	4_3
Sixfold rotation axis ⎫ Sixfold rotation point (two dimensions) ⎭		None	6

(continued)

Table 3.1. *Continued*

Symmetry axis or symmetry point	Graphic symbol	Screw vector of a right-handed screw rotation in units of the shortest lattice translation vector parallel to the axis	Printed symbol
Sixfold screw axis: "6 sub 1"		$\frac{1}{6}$	6_1
Sixfold screw axis: "6 sub 2"		$\frac{1}{3}$	6_2
Sixfold screw axis: "6 sub 3"		$\frac{1}{2}$	6_3
Sixfold screw axis: "6 sub 4"		$\frac{2}{3}$	6_4
Sixfold screw axis: "6 sub 5"		$\frac{5}{6}$	6_5
Center of symmetry, inversion center: "1 bar" } Reflection point, mirror point (one dimension)		None	$\bar{1}$
Twofold rotation axis with center of symmetry		None	$2/m$
Twofold screw axis with center of symmetry		$\frac{1}{2}$	$2_1/m$
Inversion axis: "3 bar"		None	$\bar{3}$
Inversion axis: "4 bar"		None	$\bar{4}$
Fourfold rotation axis with center of symmetry		None	$4/m$
"4 sub 2" screw axis with center of symmetry		$\frac{1}{2}$	$4_2/m$
Inversion axis: "6 bar"		None	$\bar{6}$

Description		Translation	Notation
Sixfold rotation axis with center of symmetry		None	$6/m$
"6 sub 3" screw axis with center of symmetry		$\frac{1}{2}$	$6_3/m$

Symmetry axes parallel to the plane of projection

Description		Translation	Notation
Twofold rotation axis		None	2
Twofold screw axis: "2 sub 1"		$\frac{1}{2}$	2_1
Fourfold rotation axis		None	4
Fourfold screw axis: "4 sub 1"		$\frac{1}{4}$	4_1
Fourfold screw axis: "4 sub 2"		$\frac{1}{2}$	4_2
Fourfold screw axis: "4 sub 3"		$\frac{3}{4}$	4_3
Inversion axis: "4 bar"		None	$\bar{4}$

Symmetry axes inclined to the plane of projection (in cubic space groups only)

Description		Translation	Notation
Twofold rotation axis		None	2
Twofold screw axis: "2 sub 1"		$\frac{1}{2}$	2_1
Threefold rotation axis		None	3
Threefold screw axis: "3 sub 1"		$\frac{1}{3}$	3_1
Threefold screw axis: "3 sub 2"		$\frac{2}{3}$	3_2
Inversion axis: "3 bar"		None	$\bar{3}$

[a] Reprinted from the *International Tables of Crystallography*, Volume A (Hahn, 1983), with permission of The International Union of Crystallography.

centers (centers of symmetry) would change the asymmetry of the amino acids: an L-amino acid would become a D-amino acid, but these are never found in proteins. This limitation restricts the number of space groups for proteins appreciably: only those without any symmetry (triclinic) or with exclusively rotation or screw axes are allowed. However, mirror lines and inversion centers do occur in projections of protein structures along an axis. For example, a projection along a 2-fold axis has an inversion center and mirror lines do occur in a projection of the structure on a plane parallel to a 2-fold axis.

3.4. Coordinate Triplets: General and Special Positions

The position of a point P in the unit cell is given by its position vector $\mathbf{r}$ (Figure 3.15). In terms of its fractional coordinates x, y, and z with respect to the crystal axes $\mathbf{a}$, $\mathbf{b}$, and $\mathbf{c}$, $\mathbf{r}$ is given by

$$\mathbf{r} = \mathbf{a}x + \mathbf{b}y + \mathbf{c}z \qquad (3.1)$$

The position of P can thus be described by its fractional coordinates, that is, by its coordinate triplet x, y, z. The coordinate triplets of the points P and P', related by the 2-fold axis along $\mathbf{c}$ in Figure 3.15, are x, y, z and $-x$, $-y$, z. If a molecule occupies position x, y, z then an identical molecule occupies position $-x$, $-y$, z. These molecules are said to occupy general positions.

If, however, the molecule itself has a 2-fold axis that coincides with the crystallographic 2-fold axis, half of the molecule is mapped onto the other half by the symmetry operation. This molecule occupies a "special" position. Each special position has a certain point symmetry. In the present example the point symmetry would be given by the symbol 2.

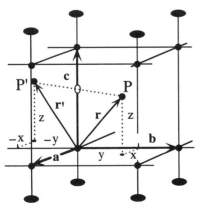

Figure 3.15. This crystal has a 2-fold axis along c. The point P with coordinate triplet x, y, z, is related by the symmetry operation to point P' with coordinate triplet $-x$, $-y$, z.

3.5. Asymmetric Unit

If the lattice has a level of symmetry higher than triclinic, then each particle in the cell will be repeated a number of times as a consequence of the symmetry operations. For example in space group $P2_12_12_1$ (space group number 19 in the *International Tables*) one can always expect at least four equal particles in the unit cell related by the symmetry operations (Figure 3.16). This unit cell has four asymmetric units. The number of molecules in a unit cell is not necessarily equal to the number of asymmetric units. There may be two or more independent molecules in each asymmetric unit. On the other hand if a molecule occupies a special position, e.g., if a symmetry axis passes through a molecule, relating one part of the molecule to one or more other parts in that molecule, the unit cell contains fewer molecules than anticipated from the number of asymmetric units.

It is important to note that molecules related by crystallographic symmetry are identical and have identical crystallographic environments. However, if two or more molecules occur in the asymmetric unit they do not have an identical environment and, moreover, they may differ in conformation.

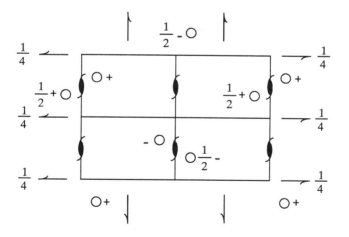

Figure 3.16. The projection of a $P2_12_12_1$ unit cell; it contains four asymmetric units. The circles represent identical particles related by the symmetry of the space group. The symbol + means that the particle is at a certain distance (for instance z) above the bottom plane of the unit cell and 1/2+ at distance z above the middle plane. The – sign indicates that the particle is at distance z below the bottom plane. With 1/2−, it is z below the middle plane. The screw axes 1/4 are located at height 1/4 (and 3/4) of the unit cell.

3.6. Point Groups

A characteristic feature of crystals is the presence of flat boundary faces with sharp edges between them. Internal symmetry in the crystal is reflected in the arrangement of these boundary faces. However, the translation components of the symmetry operations cannot be observed on the macroscopic level and for the outer shape of the crystals we are left with symmetry without translation, e.g., a 2-fold screw axis becomes a 2-fold axis. This limited group of symmetry elements forms a collection of point groups, because the symmetry elements always pass through one point. The absence of 5-fold, 7-fold etc. axes in crystals (because of the requirement of unit translation symmetry) limits the number of point groups to 32. The absence of a 5-fold crystallographic axis does not mean that a 5-fold axis never occurs. For instance, in virus particles no unit translation symmetry exists and 5-fold axes do occur.

3.7. Crystal Systems

With the choice of the unit cell according to the rules in Section 3.1, the 32 point groups can be assigned to seven and not more than seven crystallographic systems, shown in Table 3.2. This limitation to seven systems is due to the combination of symmetry elements. We shall show this with an example.

If the unit cell has one 2-fold axis, the system is clearly monoclinic. If a second 2-fold axis is added, the two axes must be perpendicular or make an angle of 30° or 60° with each other, because otherwise an unlimited number of 2-fold axes would be generated in the same plane (Figure 3.17). From this figure it is evident that two perpendicular 2-fold axes

Table 3.2. The Seven Crystal Systems

Crystal system	Conditions imposed on cell geometry	Minimum point group symmetry
Triclinic	None	1
Monoclinic	$\alpha = \gamma = 90°$ (b is the unique axis; for proteins this is a 2-fold axis or screw axis)	2
	or: $\alpha = \beta = 90°$ (c is unique axis; for proteins this is a 2-fold axis or screw axis)	
Orthorhombic	$\alpha = \beta = \gamma = 90°$	222
Tetragonal	$a = b$; $\alpha = \beta = \gamma = 90°$	4
Trigonal	$a = b$; $\alpha = \beta = 90°$; $\gamma = 120°$ (hexagonal axes)	3
	or: $a = b = c$; $\alpha = \beta = \gamma$ (rhombohedral axes)	
Hexagonal	$a = b$; $\alpha = \beta = 90°$; $\gamma = 120°$	6
Cubic	$a = b = c$; $\alpha = \beta = \gamma = 90°$	23

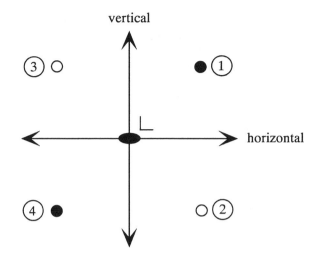

Figure 3.17. The two 2-fold axes in the plane of the page relate the four circles to each other. The open circles are above the plane, the filled circles below it. If a molecule is placed at position 1, the horizontal 2-fold axis generates a molecule in position 2 and the vertical axis in position 3. But then the horizontal axis also generates a molecule in position 4. It is evident that a third 2-fold axis is generated perpendicular to the first two.

generate a third one perpendicular to the plane of the first two axes. Therefore, crystal symmetry excludes the existence of a crystal system with one angle equal to 90° and two angles different from 90°.

The trigonal system can be treated either with hexagonal axes or with rhombohedral axes (Figure 3.18). In the hexagonal unit cell a and b are

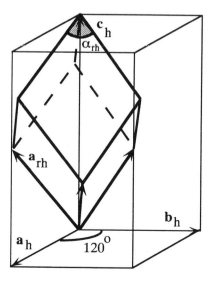

Figure 3.18. A rhombohedral unit cell and its corresponding hexagonal cell.

equal in length and have an angle of 120° to each other. c is perpendicular to the ab-plane and differs in length from a and b. The rhombohedral cell has three equal axes at angles not necessarily 90° with each other. It can be regarded as a cube either compressed or elongated along a body diagonal. The rhombohedral cell corresponds to a hexagonal cell centered at 2/3, 1/3, 1/3 and 1/3, 2/3, 2/3. The relation between the cell parameters a and c of the hexagonal cell and the parameters a' and α' of the rhombohedral cell are as follows:

$$a = a'\sqrt{2}\sqrt{1 - \cos\alpha'} = 2a'\sin\frac{\alpha'}{2}$$

$$c = a'\sqrt{3}\sqrt{1 + 2\cos\alpha'}$$

$$\frac{c}{a} = \sqrt{\frac{3}{2}}\sqrt{\frac{1 + 2\cos\alpha'}{1 - \cos\alpha'}} = \sqrt{\frac{9}{4\sin^2(\alpha'/2)} - 3}$$

$$a' = \frac{1}{3}\sqrt{3a^2 + c^2}$$

$$\sin\frac{\alpha'}{2} = \frac{3}{2\sqrt{3 + (c^2/a^2)}} \quad \text{or} \quad \cos\alpha' = \frac{(c^2/a^2) - 3/2}{(c^2/a^2) + 3}$$

It is important to realize that the conditions imposed on cell geometry are not sufficient to distinguish between the crystal systems. For instance, if a unit cell is found to have three angles of 90° it does not necessarily mean that the crystal belongs to the orthorhombic system. It could also be triclinic with three angles of 90° in the unit cell by coincidence. By being orthorhombic it is in addition required to have a minimum point group symmetry of 222 in the crystal, which expresses itself as mmm symmetry in the diffraction pattern ("mmm" means three perpendicular mirror planes).

3.8. Radiation Damage

In Chapter 2 it was mentioned that the high energy photons of X-rays have a harmful effect on living tissue. This also applies to crystals of biological macromolecules that undergo radiation damage if exposed to X-rays. Some systems are far more sensitive to it than others. Radiation damage is sometimes so serious that after only a few hours of exposure at room temperature the X-ray pattern dies away. The photons cause the formation of radicals, which leads to subsequent chemical reactions that gradually destroy the crystalline order.

   The radicals are produced in the biological macromolecules and the solvent. Some of these radicals, e.g. oxygen or hydroxy radicals, diffuse away and exercise their damaging effect at other sites in the crystal. Although the problem of radiation damage has been reduced with modern, sensitive X-ray detectors that allow relatively short exposure times, it is strongly reduced by cooling the crystals to cryogenic temperatures, that is the region between 100 and 120 °K (Section 1.7). At these temperatures radicals are still created by the X-ray photons but their diffusion through the crystal is eliminated. This allows for most biological macromolecules to collect a complete data set on one crystal, except for extremely strong undulator radiation.

3.9. Characterization of the Crystals

If crystals of sufficient size (≥ 0.2 mm) have been grown, they must first be characterized:

• What is their quality?
• What are the unit cell dimensions?
• To what space group do they belong?
• How many protein molecules are in the unit cell and in one asymmetric unit?

The quality of the crystals depends on the ordering of the molecules in the unit cell. Because of thermal vibrations and static disorder, the positions of the atoms are not strictly fixed. As a consequence the intensities of the X-ray reflections drop at higher diffraction angles. In Section 4.7.2 it will be shown that the diffracted X-ray beams can be considered as being reflected against lattice planes, and that the relation between the lattice plane distance d and the diffraction angle θ is given by $2d \sin \theta = \lambda$ (Bragg's law). Diffraction patterns with maximal observed resolution corresponding to a lattice spacing of 5 Å can be regarded as poor, of 2.0–2.5 Å as normal, and of 1.0–1.5 Å as very high quality.

Most crystals cannot be considered as ideal single crystals, because the regular repetition of the unit cells is interrupted by lattice defects. The diffraction pattern of such crystals can be regarded as the sum of the diffraction patterns originating from mosaic blocks with slightly different orientations. The mosaicity in good quality protein crystals is moderate, between 0.2 and 0.5 degrees.[1] The cell dimensions can easily be derived from the diffraction pattern collected on X-ray film, image plate, or measured with a diffractometer or area detector.

---

[1] The Program DENZO defines mosaicity as "The rocking angle in degrees, in both the vertical and the horizontal directions which would generate all the spots seen on a still diffraction photograph. It includes contributions due to X-ray bandwith, beam crossfire, etc." (Gewirth, 1997)

From the symmetry in the diffraction pattern and the systematic absence of specific reflections in that pattern, one can deduce the space group to which the crystal belongs. As discussed in Section 3.3, space groups with mirror planes or inversion centers do not apply to protein crystals. An inversion center does occur, however, *in the X-ray diffraction pattern* if anomalous scattering (Section 7.8) is absent.

An estimation of the number of molecules per unit cell (Z) can be made by a method proposed by Matthews. He found that for many protein crystals the ratio of the unit cell volume and the molecular weight is between 1.7 and 3.5 $Å^3$/Da with most values around 2.15 $Å^3$/Da. This number is called V_M (Matthews, 1968).

Example: Suppose a crystal belongs to space group C2 and has a unit cell volume of 319,000 $Å^3$. The molecular weight M_r of the protein is known to be 32,100. Then, for Z of 2, 4, or 8, V_M is, respectively, 5, 2.5, or 1.25 $Å^3$/Da. This crystal most likely has four molecules in the unit cell. Space group C2 (no. 5 in the *International Tables*) has four asymmetric units and, therefore, there is one protein molecule per asymmetric unit.

For a more accurate determination of the number of protein molecules per unit cell, the density of the crystal and its solvent content should be determined. The density can best be measured in a density gradient column, made up from organic solvents, or from a water solution of the polymer compound Ficoll in a concentration of 30–60%. Because of the high viscosity of the Ficoll solutions, these measurements must be done in a centrifuge (Westbrook, 1985).

The solvent content of the crystals could—in principle—be determined by weighing a number of crystals before and after drying. However, a problem arises because in the wet state, solvent around the crystals must be removed, but not internal solvent that fills the pores inside the crystals and that is part of the crystal structure. This separation is difficult to achieve. With the unit cell volume (V_{cell}), the density ρ, and the solvent content (weight fraction x) known, Z can be calculated:

$$\frac{Z \times M_r}{N} = (1 - x) \times \rho \times V_{cell} \qquad (3.2)$$

where N is Avogadro's number. From V_M, the volume fraction of the solvent in the crystal can be calculated in the following way:

$$
\begin{aligned}
V_{protein} &= \frac{\text{Volume of protein in the unit cell}}{V_{cell}} \\
&= \frac{(Z \times M_r \times \text{specific volume of the protein})/N}{V_M \times Z \times M_r} \\
&= \frac{\text{specific volume in cm}^3/\text{g}}{V_M \text{ in } Å^3/\text{Da} \times N \text{ mol}^{-1}}
\end{aligned}
$$

The specific volume of a protein molecule is always approximately $0.74\,\mathrm{cm}^3/\mathrm{g}$, and this gives

$$V_{\mathrm{protein}} = 1.23/V_{\mathrm{M}} \quad \text{and} \quad V_{\mathrm{solvent}} = 1 - 1.23/V_{\mathrm{M}}.$$

For the example given above, where $V_{\mathrm{M}} = 2.5$, $V_{\mathrm{solvent}} = 0.51$.

Summary

People have been fascinated by crystals since prehistoric times. However, before the introduction of X-ray diffraction in 1912 only their external properties could be studied, such as the angles between their faces or physical properties like hardness. X-ray diffraction not only showed the wave nature of X-rays, but also allowed investigation of the internal structure of crystals and of their constituent molecules or ions, in atomic detail. Therefore, for a protein crystallographer the beauty of protein crystals is not in their external faceted shape, but in the periodic arrangement of the molecules in the crystal. Without crystals there is no structure determination (at least not by X-ray crystallography), and the better the quality of the crystals, the more accurately the structure can be determined. Since crystals are the main tool for a crystallographer, familiarity with their properties, with lattice planes, symmetry, space groups and asymmetric units, is a requirement.

Chapter 4
The Theory of X-ray Diffraction by a Crystal

4.1. Introduction

The best way to learn protein X-ray diffraction is by practical work in the laboratory. However, it would be very unsatisfying to perform the experiments without understanding why they have to be done in such and such a way. Moreover, at several stages in the determination of protein structures it is necessary to decide what the next step should be. For instance, after growing suitable crystals and soaking these crystals in solutions of heavy atom reagents, applying the isomorphous replacement method, how do you obtain the positions of the heavy atoms in the unit cell and, if you do have them, how do you proceed? Questions such as these can be answered only if you have some knowledge of the theoretical background of protein X-ray crystallography. This is presented in this chapter. A slow path will be followed, and a student with a minimal background in mathematics, but the desire to understand protein X-ray crystallography should be able to work through the chapter. A working knowledge of differentiation and integration is required. If you further accept that an X-ray beam can be regarded as a wave that travels as a cosine function and if you know what a vector is, you have a good start. Derivations and explanations that are not absolutely necessary to follow the text are set off within rules; these can be skipped, if you want.

Chapter 1 introduced the diffraction of X-rays by a crystal. In analogy with the scattering of visible light by a two-dimensional grid, a crystal diffracts an X-ray beam in a great many directions. From the diffraction experiment with lysozyme it was concluded that diffraction by an X-ray beam can be understood as being derived from the intersection of an imaginary lattice, the reciprocal lattice, and a sphere, called the Ewald sphere. From the direction of the diffracted beams the dimensions of the unit cell can be derived. But we are, of course, more interested in the content of the unit cell, that is, in the structure of the protein molecules. The molecular structure and the arrangement of the molecules in the unit cell determine the intensities of the diffracted beams. Therefore, a relationship must be found between the intensities of the diffracted beams and the crystal structure. In fact, this relation is between the diffraction data and the electron density distribution in the crystal structure, because X-rays are scattered almost exclusively by the electrons in the atoms and not by the nuclei.

The scattering is an interaction between X-rays as electromagnetic waves and the electrons. If an electromagnetic wave is incident on a system of electrons, the electric and magnetic components of the wave exert a force on the electrons. This causes the electrons to oscillate with the same frequency as the incident wave. The oscillating electrons act as radiation scatterers and they emit radiation of the same frequency as the incident radiation. Energy from the incident wave is absorbed by the electrons and then emitted. Because of the attraction between the electrons and the atomic nucleus, an electrical restoring force exists for the electrons of an atom. However, in X-ray diffraction the electrons in an atom can be regarded, to a good approximation, as free electrons.

The wave scattered by the crystal may be described as a summation of the enormous number of waves, each scattered by one electron in the crystal. This may sound somewhat intimidating because a single unit cell in a protein crystal contains approximately 10,000 or more electrons, and there are a great many unit cells in a crystal. And all these waves must be added! It is clear that we need a convenient way to add waves. This method shall be presented first; familiarity with the technique will reveal how it simplifies the whole process. It is then fairly easy to derive an expression that relates each wave scattered by the crystal to the electron density distribution in the crystal and in its unit cells. The next step is to reverse this expression and derive the electron density distribution as a function of the scattering information.

4.2. Waves and Their Addition

An electromagnetic wave travels as a cosine function (Figure 4.1a). E is the electromagnetic field strength, λ is the wavelength of the radiation, and $\nu = c/\lambda$ is the frequency, with c the speed of light (and of any other electromagnetic radiation). A is the amplitude of the wave.

Let Figure 4.1a show the wave at time $t = 0$. The electric field strength at time $t = 0$ and position z is

$$E(t = 0; z) = A \cos 2\pi \frac{z}{\lambda}$$

During a time period t the wave travels over a distance $t \times c = t \times \lambda \times \nu$. Therefore, at time t the field strength at position z is equal to what it was at time $t = 0$ and position $z - t \times \lambda \times \nu$.

$$E(t;z) = A \cos 2\pi \frac{1}{\lambda}(z - t \times \lambda \times \nu)$$

$$= A \cos 2\pi\left(\frac{z}{\lambda} - \nu t\right) = A \cos 2\pi\nu\left(t - \frac{z}{c}\right)$$

At $z = 0$ the field strength is $E(t;z = 0) = A \cos 2\pi\nu t$, and substituting for convenience ω for $2\pi\nu$:

$$E(t;z = 0) = A \cos \omega t$$

Let us now consider a new wave with the same λ and the same amplitude A, but displaced over a distance Z with respect to the original wave (Figure 4.1b). Z corresponds with a phase shift $(Z/\lambda) \times 2\pi = \alpha$.

Original wave at $z = 0$ and time t: $E_{orig}(t;z = 0) = A \cos \omega t$

New wave at $z = 0$ and time t: $E_{new}(t;z = 0) = A \cos(\omega t + \alpha)$

Let us now consider only the new wave as representing any wave with a phase angle difference with respect to a reference wave:

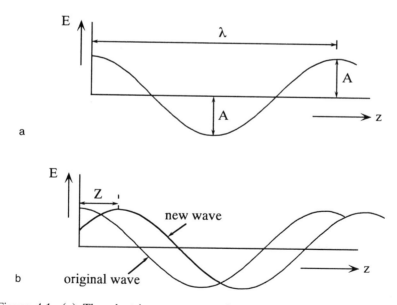

Figure 4.1. (a) The electric component of an electromagnetic wave. A is the amplitude and λ the wavelength. The accompanying magnetic component is perpendicular to the electric one, but we do not need to consider it here. (b) A new wave, displaced over a distance Z, is added.

Figure 4.2. The real component
$A \cos \alpha$ and the imaginary component $A \sin \alpha$ of vector **A** in an
Argand diagram.

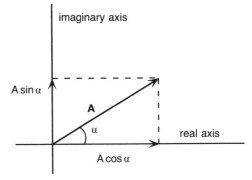

$$A \cos(\omega t + \alpha) = A \cos \alpha \cos \omega t - A \sin \alpha \sin \omega t$$
$$= A \cos \alpha \cos \omega t + A \sin \alpha \cos(\omega t + 90°)$$

Therefore, the wave $A \cos(\omega t + \alpha)$ can be regarded as being composed of
two waves: wave 1 of amplitude $A \cos \alpha$ and phase angle 0° and wave 2 of
amplitude $A \sin \alpha$ and phase angle 90°. Wave 1 is called the real part and
wave 2 the imaginary part of the total wave. To add several waves with
different phase angles, their real parts may be added together because
they all have phase angle 0. Similarly, their imaginary parts, with phase
angle 90°, may be added together. This can be represented conveniently
in an axial system called an Argand diagram, in which the real axis is
horizontal and the imaginary axis vertical (Figure 4.2). The wave itself is
represented by vector **A**, with projections $A \cos \alpha$ on the real axis and A
$\sin \alpha$ on the imaginary axis.[1] Addition of all the real parts and of all the
imaginary parts of several waves is the same as adding the several wave
vectors (like **A**) together. The plane with the real and the imaginary axis is
called the complex plane.

Conclusion: We have simplified the problem of adding waves with the
same frequency (or wavelength) by applying the following procedure:

- Represent each wave as a vector in a two-dimensional axial system.
  The length of each vector is equal to the amplitude of the wave. The
  vector makes an angle with the horizontal, or real axis equal to its
  phase with respect to a reference wave (angle α in Figure 4.2).
- The vector representing the total wave of a system is obtained by
  adding the vectors of the separate waves together.

Though the representation of waves as vectors in the Argand diagram is
extremely convenient, it would still require an enormous amount of work
to add thousands of these vectors together manually. This problem is
solved by writing the vectors in mathematical form. Consider the wave of

---

[1] A vector will be indicated by boldface type (**A**). The length of this vector is given by $|A|$.

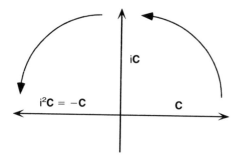

Figure 4.3. Multiplication of a vector **C** in the Argand diagram by i simply means rotating **C** 90° counterclockwise. Therefore, $i^2\mathbf{C} = -\mathbf{C}$.

Figure 4.2 with the two components, $A\cos\alpha$ (horizontal) and $A\sin\alpha$ (vertical). The entire wave is written in mathematical form as $A\cos\alpha + iA\sin\alpha$. i simply means that the $(A\sin\alpha)$ component points vertical, or i means that this component has been rotated 90° counterclockwise with respect to the positive direction of the horizontal axis. It is then clear that $i = \sqrt{-1}$ (Figure 4.3). A further simplification is to write $A\cos\alpha + iA\sin\alpha$ as $A\exp[i\alpha]$.

Properties of Exponential Terms

We shall not prove that $A\cos\alpha + iA\sin\alpha = A\exp[i\alpha]$. You must know, however, the properties of exponential terms:

$$\exp[a] \times \exp[b] = \exp[a + b] \quad ; \quad \frac{\exp[a]}{\exp[b]} = \exp[a - b]$$

$$\exp[k \cdot a] = \{\exp[a]\}^k \quad ; \quad \exp[0] = 1$$

$$\exp[a] \to +\infty \quad \text{for} \quad a \to +\infty$$

$$\exp[a] \to 0 \quad \text{for} \quad a \to -\infty$$

Now we are ready to work with waves and electrons. We shall start with a simple system of only two electrons.

4.3. A System of Two Electrons

The system in Figure 4.4 has only two electrons: e_1 and another electron e_2 at position **r** with respect to e_1. An X-ray beam, indicated by the wave vector $\mathbf{s}_0$ with length $1/\lambda$ hits the system and is diffracted in the direction of wave vector **s**, which also has a length of $1/\lambda$. The beam that passes along electron e_2 follows a longer path than the beam along electron e_1.

The path difference between the two beams $(p + q)$ depends on (1) the position of electron e_2 with respect to e_1 and (2) the direction of diffraction. Since $\mathbf{s}_0$ and **s** are wave vectors of magnitude $1/\lambda$ each, $p =$

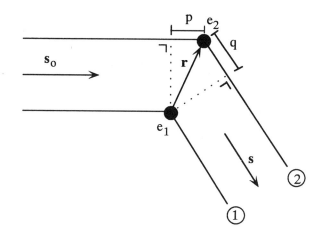

Figure 4.4. A system of two electrons: e_1 and e_2. The path difference between the scattered waves 1 and 2 is $p + q$.

$\lambda \cdot \mathbf{r} \cdot \mathbf{s_0}$ and $q = -\lambda \cdot \mathbf{r} \cdot \mathbf{s}$ (the minus sign is due to the fact that the projection of $\mathbf{r}$ on $\mathbf{s}$ has a direction opposite to $\mathbf{s}$).[2] The path difference is, therefore, $p + q = \lambda \cdot \mathbf{r} \cdot (\mathbf{s_0} - \mathbf{s})$.

The wave along electron e_2 is lagging behind in phase compared with the wave along e_1. With respect to wave 1, the phase of wave 2 is

$$-\frac{2\pi \mathbf{r} \cdot (\mathbf{s_0} - \mathbf{s}) \cdot \lambda}{\lambda} = 2\pi \mathbf{r} \cdot \mathbf{S}$$

where

$$\mathbf{S} = \mathbf{s} - \mathbf{s_0} \qquad (4.1)$$

It is interesting to note that the wave can be regarded as being reflected against a plane with θ as the reflecting angle and $|S| = 2(\sin \theta)/\lambda$ (Figure 4.5). The physical meaning of vector $\mathbf{S}$ is the following: Since $\mathbf{S} = \mathbf{s} - \mathbf{s_0}$,

Figure 4.5. The primary wave, represented by $\mathbf{s_0}$, can be regarded as being reflected against a plane. θ is the reflecting angle. Vector $\mathbf{S}$ is perpendicular to this plane.

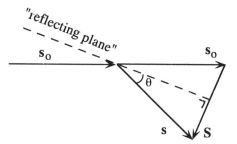

[2] $\mathbf{r \cdot s}$ is the scalar product of the vector $\mathbf{r}$ and $\mathbf{s}$.

with $|s| = |s_0| = 1/\lambda$, $\mathbf{S}$ is perpendicular to the imaginary "reflecting plane," which makes equal angles with the incident and reflected beam.

The Product of Two Vectors a and b

Let vectors $\mathbf{a}$ and $\mathbf{b}$, with lengths $|a|$ and $|b|$, be inclined at an angle θ.

Scalar product: Their scalar product is the number $\mathbf{a} \cdot \mathbf{b} = ab \cos\theta$ and $\mathbf{a} \cdot \mathbf{b} = \mathbf{b} \cdot \mathbf{a}$.

Vector product: Let $\mathbf{a}$ and $\mathbf{b}$ again be inclined at an angle θ with $0 \leq \theta \leq \pi$. Their vector product is a vector $\mathbf{c}$, which has a length $|c| = ab \sin\theta$ and points in a direction perpendicular to both $\mathbf{a}$ and $\mathbf{b}$, such that the vector system $\mathbf{a}$, $\mathbf{b}$, $\mathbf{c}$ is a right-handed triad; $\mathbf{c} = \mathbf{a} \times \mathbf{b} = -\mathbf{b} \times \mathbf{a}$.

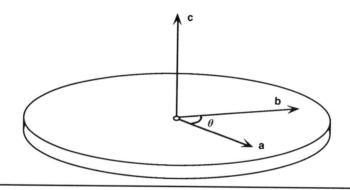

If we add the waves 1 and 2 in Figure 4.4, the Argand diagram shows two vectors **1** and **2** with equal length (amplitude) and a phase of $2\pi \mathbf{r} \cdot \mathbf{S}$ for wave **2** with respect to wave **1** (Figure 4.6). Vector $\mathbb{T}$ represents the sum of the two waves. In mathematical form: $\mathbb{T} = \mathbf{1} + \mathbf{2} = \quad 1 + 1$

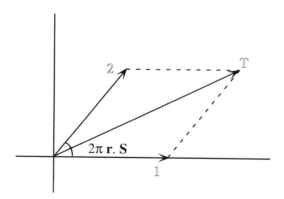

Figure 4.6. The summation of the two scattered waves in Figure 4.4 with the origin in electron e_1.

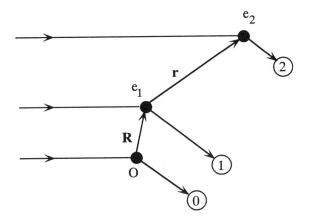

Figure 4.7. The origin, or reference point, for the scattered waves of the two-electron system is now located at O.

$\exp[2\pi i\mathbf{r} \cdot \mathbf{S}]$ if the length of the vectors equals 1. So far we had the origin of this two-electron system in e_1. Suppose we move the origin over $-\mathbf{R}$ from e_1 to point O (Figure 4.7). Then we obtain the following: With respect to a wave $\mathbb{0}$, wave $\mathbb{1}$ has a phase of $2\pi\mathbf{R} \cdot \mathbf{S}$, and wave $\mathbb{2}$ has a phase of $2\pi(\mathbf{r} + \mathbf{R}) \cdot \mathbf{S}$ (Figure 4.8).

$$\mathbb{T} = \mathbb{1} + \mathbb{2} = \exp[2\pi i\mathbf{R} \cdot \mathbf{S}] + \exp[2\pi i(\mathbf{r} + \mathbf{R}) \cdot \mathbf{S}]$$
$$= \exp[2\pi i\mathbf{R} \cdot \mathbf{S}]\{1 + \exp[2\pi i\mathbf{r} \cdot \mathbf{S}]\}$$

Conclusion: A shift of the origin by $-\mathbf{R}$ causes an increase of all phase angles by $2\pi\mathbf{R} \cdot \mathbf{S}$. The amplitude and intensity (which is proportional to the square of the amplitude) of wave $\mathbb{T}$ do not change.

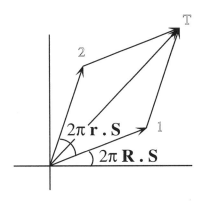

Figure 4.8. The summation of waves $\mathbb{1}$ and $\mathbb{2}$ with the origin of the two-electron system in position O.

From the two-electron system we shall now move to more complicated systems: first an atom, then a combination of atoms in a unit cell, and finally a crystal.

4.4. Scattering by an Atom

The electron cloud of an atom scatters an X-ray beam: the scattering is dependent on the number of electrons and their positions in the cloud. We wish to understand the scattering of an atom with the origin of the system in the nucleus (Figure 4.9) because the scattering by an atom located elsewhere will be the same, except for a phase shift, as was shown in the previous section. The electron density at position $\mathbf{r}$ is denoted by $\rho(\mathbf{r})$. The cloud is centrosymmetric around the origin, which means that $\rho(\mathbf{r}) = \rho(-\mathbf{r})$. From Figure 4.10 we can easily derive that the scattering by an atom is always real; the vector of the scattered wave is directed along the real axis in the Argand diagram.

The atomic scattering factor f is

$$f = \int_{\mathbf{r}} \rho(\mathbf{r}) \exp[2\pi i r \cdot \mathbf{S}] \, dr \qquad (4.2)$$

where the integration is over the entire space $\mathbf{r}$. From Figure 4.10:

$$f = \int_{\mathbf{r}} \rho(\mathbf{r})\{\exp[2\pi i r \cdot \mathbf{S}] + \exp[-2\pi i r \cdot \mathbf{S}]\} \, dr$$

$$= 2\int_{\mathbf{r}} \rho(\mathbf{r}) \cos[2\pi \mathbf{r} \cdot \mathbf{S}] \, dr$$

Now the integration is over half of the entire space. Generally in X-ray crystallography it is assumed that the electron cloud of an atom is spherically symmetric. Therefore, it does not make any difference if the orien-

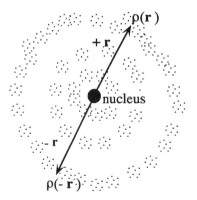

Figure 4.9. The electron cloud of an atom. $\rho(\mathbf{r})$ is the electron density. Because of the centrosymmetry $\rho(\mathbf{r}) = \rho(-\mathbf{r})$.

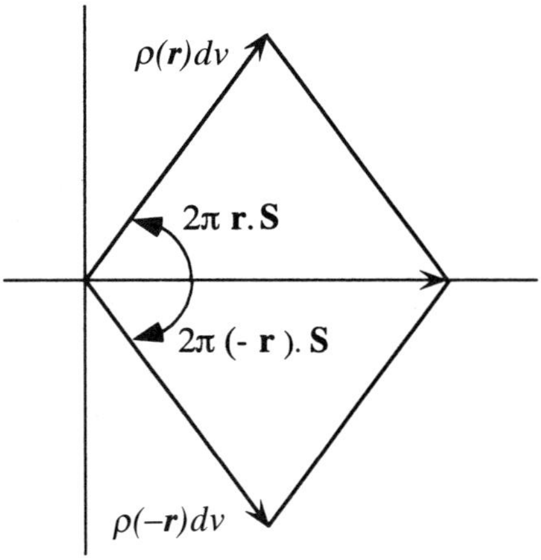

Figure 4.10. The scattering factor f of an atom is always real if we assume centrosymmetry of the electron cloud. The imaginary part of every scattering vector is compensated by the imaginary part of a vector with equal length but a phase angle of opposite sign.

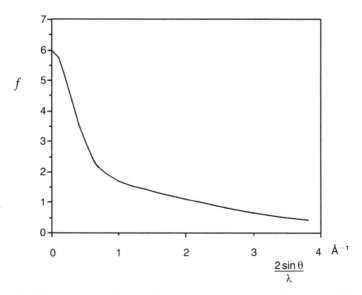

tation of the atom changes with respect to the direction of **S**. The atomic scattering factor is independent of the direction of **S**, but does depend on the length of **S**:

$$|S| = \frac{2 \sin \theta}{\lambda}$$

Values for the atomic scattering factor f can be looked up in tables, in which f is expressed as a function of $2(\sin \theta)/\lambda$ (see Figure 4.1).

Figure 4.11. The scattering factor f for a carbon atom as a function of $2(\sin \theta)/\lambda$. f is expressed as electron number and for the beam with $\theta = 0, f = 6$.

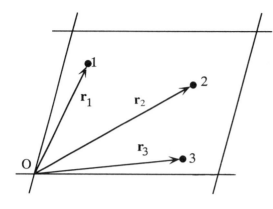

Figure 4.12. A unit cell with three atoms (1, 2, and 3) at positions $\mathbf{r}_1$, $\mathbf{r}_2$, and $\mathbf{r}_3$.

4.5. Scattering by a Unit Cell

Suppose a unit cell has n atoms at positions $\mathbf{r}_j$ ($j = 1, 2, 3, \ldots, n$) with respect to the origin of the unit cell (Figure 4.12). With their own nuclei as origins, the atoms diffract according to their atomic scattering factor. If the origin is now transferred to the origin of the unit cell, the phase angles change by $2\pi \mathbf{r}_j \cdot \mathbf{S}$. With respect to the new origin the scattering is given by

$$\mathbf{f}_j = f_j \exp[2\pi i \mathbf{r}_j \cdot \mathbf{S}]$$

where the $\mathbf{f}_j$s are the vectors in the Argand diagram. The total scattering from the unit cell is

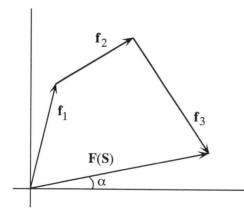

Figure 4.13. The structure factor $\mathbf{F(S)}$ is the sum of the scattering by the separate atoms in the unit cell.

$$F(S) = \sum_{j=1}^{n} f_j \exp[2\pi i r_j \cdot S] \qquad (4.3)$$

F(S) is called the *structure factor* because it depends on the arrangement (structure) of the atoms in the unit cell (Figure 4.13).

4.6. Scattering by a Crystal

Suppose the crystal has translation vectors **a**, **b**, and **c** and contains a large number of unit cells: n_1 in the **a** direction, n_2 in the **b** direction, and n_3 in the **c** direction (Figure 4.14). To obtain the scattering by the crystal we must add the scattering by all unit cells with respect to a single origin. We choose the origin O in Figure 4.14. For a unit cell with its own origin at position $t \cdot \mathbf{a} + u \cdot \mathbf{b} + v \cdot \mathbf{c}$, in which t, u, and v are whole numbers, the scattering is

$$F(S) \times \exp[2\pi i t \mathbf{a} \cdot S] \times \exp[2\pi i u \mathbf{b} \cdot S] \times \exp[2\pi i v \mathbf{c} \cdot S]$$

The total wave **K(S)** scattered by the crystal is obtained by a summation over all unit cells:

$$K(S) = F(S) \times \sum_{t=0}^{n_1} \exp[2\pi i t \mathbf{a} \cdot S] \times \sum_{u=0}^{n_2} \exp[2\pi i u \mathbf{b} \cdot S]$$

$$\times \sum_{v=0}^{n_3} \exp[2\pi i v \mathbf{c} \cdot S]$$

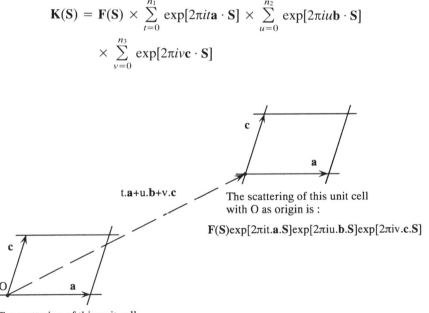

t.a+u.b+v.c

The scattering of this unit cell with O as origin is :

F(S)exp[2πit.a.S]exp[2πiu.b.S]exp[2πiv.c.S]

The scattering of this unit cell

with O as origin is **F(S)**

Figure 4.14. A crystal contains a large number of identical unit cells. Only two of them are drawn in this figure.

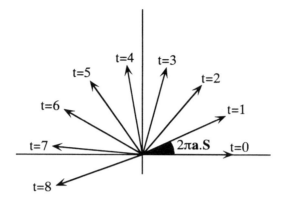

Figure 4.15. Each arrow represents the scattering by one unit cell in the crystal. Because of the huge number of unit cells and because their scattering vectors are pointing in different directions, the scattering by a crystal is in general zero. However in the special case that $\mathbf{a} \cdot \mathbf{S}$ is an integer h, all vectors point to the right and the scattering by the crystal can be of appreciable intensity.

Since n_1, n_2, and n_3 are very large, the summation $\Sigma_{t=0}^{n_1} \exp[2\pi it\mathbf{a} \cdot \mathbf{S}]$ and the other two over u and v are almost always equal to zero unless $\mathbf{a} \cdot \mathbf{S}$ is an integer h, $\mathbf{b} \cdot \mathbf{S}$ is an integer k, and $\mathbf{c} \cdot \mathbf{S}$ is an integer l. This is easy to understand if we regard $\exp[2\pi it\mathbf{a} \cdot \mathbf{S}]$ as a vector in the Argand diagram with a length of 1 and a phase angle $2\pi t\mathbf{a} \cdot \mathbf{S}$ (see Figure 4.15).

  Conclusion: A crystal does not scatter X-rays, unless

$$\mathbf{a} \cdot \mathbf{S} = h$$
$$\mathbf{b} \cdot \mathbf{S} = k \qquad\qquad (4.4)$$
$$\mathbf{c} \cdot \mathbf{S} = l$$

These are known as the Laue conditions. h, k, and l are whole numbers, either positive, negative or zero. The amplitude of the total scattered wave is proportional to the amplitude of the structure factor $\mathbf{F(S)}$ and the number of unit cells in the crystal.

4.7. Diffraction Conditions

4.7.1. Laue Conditions

For the formulation of the diffraction conditions we consider an infinitely large crystal. As we have just seen, the scattered wave has zero amplitude $[\mathbf{K(S)} = 0]$ unless the Laue conditions are fulfilled ($\mathbf{a} \cdot \mathbf{S} = h$, etc.), in which case all unit cells scatter in phase and $\mathbf{K(S)}$ is proportional to $\mathbf{F(S)}$.

4.7.2. Bragg's Law

In Section 4.3, we noted that vector **S** is perpendicular to a "reflecting" plane. With a chosen origin for the system, **r.S** is the same for all points in the reflecting plane. This is true because the projection of each **r** on **S** has the same length. Since **r.S** determines the phase angle the waves from all points in a reflecting plane reflect in phase. Choose the origin of the system in the origin O of the unit cell. The waves from a reflecting plane through the origin have phase angle 0 (**r.S** = 0). For a parallel plane with **r.S** = 1 they are shifted by 1 × 2π, etc. All parallel planes with **r.S** equal to an integer are reflecting in phase and form a series of Bragg planes. The plane with **r.S** = 1 cuts the **a**-axis at position $\mathbf{r} = \frac{\mathbf{a}}{\xi}$. Thus $\frac{\mathbf{a}}{\xi} \cdot \mathbf{S} = 1$. But from the Laue conditions we know that $\frac{\mathbf{a}}{h} \cdot \mathbf{S} = 1$. Therefore, $\xi = h$ and in the same way the reflecting plane cuts the **b**-axis at $\frac{\mathbf{b}}{k}$ and the **c**-axis at $\frac{\mathbf{c}}{\ell}$. The result is that the reflecting planes are the lattice planes. The projection of $\frac{\mathbf{a}}{h}$ on **S** has a length $\frac{1}{|S|}$ but this projection is also equal to the distance d between the lattice planes (Figure 4.16). From $\frac{1}{|S|} = d$ and $|S| = 2(\sin\theta)/\lambda$ (Section 4.3), the well-known Bragg law emerges:

$$\frac{2d\sin\theta}{\lambda} = 1 \qquad\qquad (4.5)$$

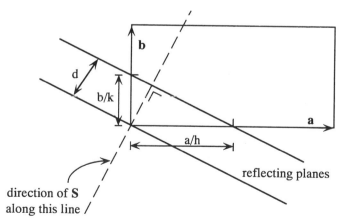

Figure 4.16. For simplicity a two-dimensional unit cell is drawn. The endpoints of the vectors **a**/h, **b**/k, and **c**/l form a lattice plane perpendicular to vector **S** (see the text). d is the distance between these lattice planes.

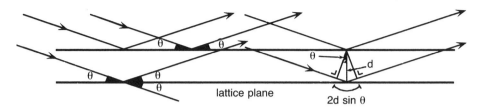

Figure 4.17. Two lattice planes are drawn separated by a distance d. The incident and the reflected beams make an angle θ with the lattice planes. Note that the beam is thus deflected through an angle of 2 θ relative to its incident direction.

The incident and reflected beam make an equal angle with the plane (Figure 4.17). In a series of parallel reflecting planes (Bragg planes), the phase difference between the radiation from successive planes is 2π. The diffraction of X-rays by lattice planes can easily form the impression that only atoms on lattice planes contribute to the reflection. This is completely wrong! All atoms in the unit cell contribute to each reflection, atoms on lattice planes and in between. The advantage of lattice plane reflection and Bragg's law is that it offers a visual picture of the scattering process.

4.8. Reciprocal Lattice and Ewald Construction

In Chapter 1 we noted the reciprocity between the crystal lattice and the diffraction pattern. The Ewald sphere was also introduced, as a convenient tool to construct the diffraction pattern. This will now be formulated in a more quantitative way. There is a crystal lattice and a reciprocal lattice. The crystal lattice is real, but the reciprocal lattice is an imaginary lattice.

Question: What is the advantage of the reciprocal lattice?
Answer: With the reciprocal lattice the directions of scattering can easily be constructed.

4.8.1. Construction of the Reciprocal Lattice

In Section 4.3 we noticed that vector **S** is related to the direction and angle of the reflected beam. Properties of **S**:

$$\mathbf{S} \perp \text{reflecting plane}$$

$$|S| = \frac{2\sin\theta}{\lambda} \text{ because } \mathbf{S}' = \mathbf{s} - \mathbf{s}_0 \quad \text{and} \quad |s_0| = |s| = \frac{1}{\lambda}$$

We also noted (Section 4.7) that the reflecting planes are in fact the lattice planes and that these lattice planes divide the **a**, **b**, and **c** axes into an integral number (h, k, and l) of equal pieces. Moreover, we found that $|S| = 1/d$, where d is the distance between the lattice planes in one set of planes.

We now pay attention to the special planes (100), (010), and (001). For (100), the indices are $h = 1$, $k = 0$, and $l = 0$. $\mathbf{S}(100)$ is perpendicular to this plane and has a length of $1/d(100)$; we call this vector $\mathbf{a}^*$. In the same way $\mathbf{S}(010) \perp$ plane (010) with length $1/d(010)$; we call this vector $\mathbf{b}^*$. And $\mathbf{S}(001) \perp$ plane (001) with length $1/d(001)$; we call this vector $\mathbf{c}^*$. Because $\mathbf{a}^* \perp$ plane (100), it is perpendicular to the b- and the c-axis, and, therefore, $\mathbf{a}^* \cdot \mathbf{b} = \mathbf{a}^* \cdot \mathbf{c} = 0$, but $\mathbf{a} \cdot \mathbf{a}^* = \mathbf{a} \cdot \mathbf{S}(100) = h = 1$. In the same way it can be shown that $\mathbf{b} \cdot \mathbf{b}^* = \mathbf{b} \cdot \mathbf{S}(010) = k = 1$ and $\mathbf{c} \cdot \mathbf{c}^* = \mathbf{c} \cdot \mathbf{S}(001) = l = 1$.

Why did we introduce the vectors $\mathbf{a}^*$, $\mathbf{b}^*$, and $\mathbf{c}^*$? The answer is that the endpoints of the vectors $\mathbf{S}(h\ k\ l)$ are located in the lattice points of a lattice constructed with the unit vectors $\mathbf{a}^*$, $\mathbf{b}^*$, and $\mathbf{c}^*$.

Proof: S can always be written as $\mathbf{S} = X \cdot \mathbf{a}^* + Y \cdot \mathbf{b}^* + Z \cdot \mathbf{c}^*$. Multiply by $\mathbf{a}$:

$$\mathbf{a.S} = X \times \mathbf{a.a}^* + Y \times \mathbf{a.b}^* + Z \times \mathbf{a.c}^*$$

$$= h \quad = X \times 1 \qquad = 0 \qquad = 0$$

It follows that $X = h$, and by the same token $Y = k$ and $Z = l$. Therefore, $\mathbf{S} = h \cdot \mathbf{a}^* + k \cdot \mathbf{b}^* + l \cdot \mathbf{c}^*$. The crystal lattice based on $\mathbf{a}$, $\mathbf{b}$, $\mathbf{c}$ is called the *direct* lattice and that based on $\mathbf{a}^*$, $\mathbf{b}^*$, $\mathbf{c}^*$ is called the *reciprocal* lattice. Each reflection ($h\ k\ l$) is denoted by a point ($h\ k\ l$) in the reciprocal lattice. The relation between direct and reciprocal unit cells is drawn schematically in Figure 4.18.

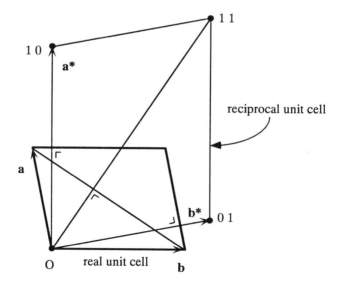

Figure 4.18. The relation between a real unit cell and the corresponding reciprocal unit cell. For simplicity a two-dimensional cell is chosen.

Table 4.1. The Relationship between the Axes and Angles in the Direct and the
Reciprocal Lattice in a Triclinic Space Group

$$a^* = \frac{bc \sin \alpha}{V} \qquad\qquad\qquad\qquad a = \frac{b^*c^* \sin \alpha^*}{V^*}$$

$$b^* = \frac{ac \sin \beta}{V} \qquad\qquad\qquad\qquad b = \frac{a^*c^* \sin \beta^*}{V^*}$$

$$c^* = \frac{ab \sin \gamma}{V} \qquad\qquad\qquad\qquad c = \frac{a^*b^* \sin \gamma^*}{V^*}$$

$$V = \frac{1}{V^*} = abc\sqrt{1 - \cos^2\alpha - \cos^2\beta - \cos^2\gamma + 2 \cos\alpha \cos\beta \cos\gamma}$$

$$V^* = \frac{1}{V} = a^*b^*c^*\sqrt{1 - \cos^2\alpha^* - \cos^2\beta^* - \cos^2\gamma^* + 2 \cos\alpha^*\cos\beta^*\cos\gamma^*}$$

$$\cos\alpha^* = \frac{\cos\beta\cos\gamma - \cos\alpha}{\sin\beta\sin\gamma} \qquad\qquad \cos\alpha = \frac{\cos\beta^*\cos\gamma^* - \cos\alpha^*}{\sin\beta^*\sin\gamma^*}$$

$$\cos\beta^* = \frac{\cos\alpha\cos\gamma - \cos\beta}{\sin\alpha\sin\gamma} \qquad\qquad \cos\beta = \frac{\cos\alpha^*\cos\gamma^* - \cos\beta^*}{\sin\alpha^*\sin\gamma^*}$$

$$\cos\gamma^* = \frac{\cos\alpha\cos\beta - \cos\gamma}{\sin\alpha\sin\beta} \qquad\qquad \cos\gamma = \frac{\cos\alpha^*\cos\beta^* - \cos\gamma^*}{\sin\alpha^*\sin\beta^*}$$

The relationship between the axes and the angles of the unit cell in
both lattices is given in Table 4.1 for a triclinic lattice. Note that the
volume V of the unit cell in the direct lattice is the reciprocal of the unit
cell volume V^* in the reciprocal lattice: $V = 1/V^*$.

If the magnitude of scattering $G(S)$, corresponding to each vector S, is
plotted at the tip of S in reciprocal space, a so-called *weighted reciprocal
lattice* is obtained. For crystals, $G(S)$ has nonzero values only at the lattice
points and then $G(S) = F(S)$. However, for nonperiodic objects $G(S)$ can
have a nonzero value anywhere in reciprocal space.

With the reciprocal lattice the diffraction directions can easily be constructed. The following procedure is applied:

Step 1: Direct the incoming (primary) X-ray beam (s_0) toward the origin
O of reciprocal space. Take the length of s_0 equal to $1/\lambda$. (Figure
4.19).

Step 2: Construct a sphere with the origin O of reciprocal space on its
surface, with center M on the line s_0, and radius $MO = 1/\lambda$. This
sphere is called the Ewald sphere.

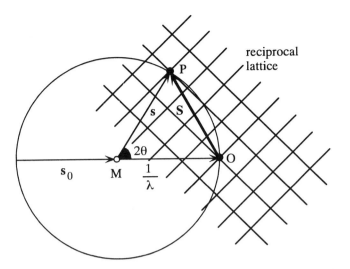

Figure 4.19. The Ewald sphere as a tool to construct the direction of the scattered beam. The sphere has radius $1/\lambda$. The origin of the reciprocal lattice is at O. s_0 indicates the direction of the incident beam; s indicates the direction of the scattered beam.

Step 3: If a wave vector **S** has its endpoint on the Ewald sphere, e.g., at point P, then MP is the scattered beam **s**. This is true because as shown in Figure 4.3, $s_0 + S = s$, or $S = s - s_0$.

4.8.2. Conclusions Concerning the Reciprocal Lattice

The reciprocal lattice is a convenient concept useful for constructing the directions of diffraction by a crystal. But remember that it exists only theoretically and not in reality.

Scattering occurs for all scattering vectors **S** having their endpoint P on the sphere and $G(S) \neq 0$. For crystals, $G(S) \neq 0$ implies that P must be a reciprocal lattice point. Crystal lattice planes for which the reciprocal lattice points $(h\ k\ l)$ do not lie on the sphere and thus are not in reflecting position can be brought to reflection by rotating the reciprocal lattice around O. From this construction with the imaginary reciprocal lattice we now return to the reality of the crystal lattice and note the following:

• The reciprocal lattice rotates exactly as the crystal does.
• The direction of the beam diffracted from the crystal is parallel to MP for the orientation of the crystal, which corresponds to the orientation of the reciprocal lattice.

From Figure 4.19 two properties of $S(h\ k\ l)$ can easily be derived:

1. The reciprocal space vector $\mathbf{S}(h\,k\,l) = \mathbf{OP}(h\,k\,l)$ is perpendicular to the reflecting plane $h\,k\,l$, which is in agreement with the definition of $\mathbf{S}$ in section 4.3.
2. $|S(h\,k\,l)| = 2(\sin\theta)/\lambda = 1/d$ and Bragg's law is fulfilled.

One more comment on lattice planes: If the beam $h\,k\,l$ corresponds to reflection against one face (let us say the front) of a lattice plane, then

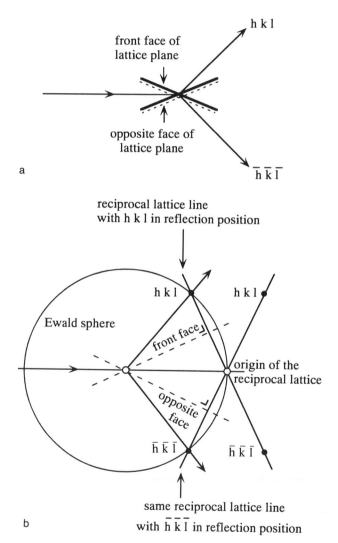

Figure 4.20. If the reflection $(h\,k\,l)$ corresponds to a reflection against the front face of a lattice plane, then $(\bar{h}\,\bar{k}\,\bar{l})$ corresponds to reflection against the opposite face of the plane. (a) Simple presentation of the reflections. (b) An explanation by means of the reciprocal lattice.

($\bar{h}\,\bar{k}\,\bar{l}$) [or ($-h$, $-k$, $-l$)] corresponds to the reflection against the opposite face (the back) of the plane (Figure 4.20).

4.9. The Temperature Factor

The size of the electron density cloud around an atomic nucleus is independent of the temperature, at least under normal conditions. This would suggest that X-ray scattering by a crystal would also be independent of the temperature. However, this is not true because the atoms vibrate around an equilibrium position. The X-rays do not meet identical atoms on exactly the same position in successive unit cells. This is similar to an X-ray beam meeting a smeared atom on a fixed position, the size of the atom being larger if the thermal vibration is stronger. This diminishes the scattered X-ray intensity, especially at high scattering angles. Therefore, the atomic scattering factor of the atoms must be multiplied by a temperature-dependent factor (Figure 4.21).

The vibration of an atom in a reflecting plane $h\,k\,l$ has no effect on the intensity of the reflection ($h\ k\ l$). Atoms in a plane diffract in phase and therefore a displacement in that plane has no effect on the scattered intensity. The component of the vibration perpendicular to the reflecting plane does have an effect. In the simple case in which the components of vibration are the same in all directions, the vibration is called *isotropic*.

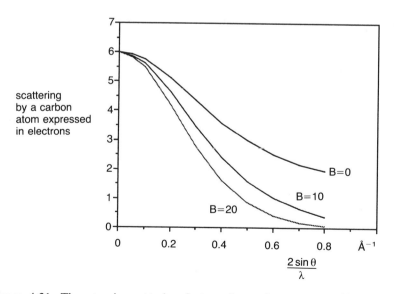

Figure 4.21. The atomic scattering factor of a carbon atom multiplied by the appropriate temperature factor.

Then the component perpendicular to the reflecting plane and thus along
S is equal for each $(h\,k\,l)$ and the correction factor for the atomic
scattering factor is

$$T(\text{iso}) = \exp\left[-B\frac{\sin^2\theta}{\lambda^2}\right] = \exp\left[-\frac{B}{4}\left(\frac{2\sin\theta}{\lambda}\right)^2\right]$$

$$= \exp\left[-\frac{B}{4}\left(\frac{1}{d}\right)^2\right] \tag{4.6}$$

Assuming isotropic and harmonic vibration it can be shown that the thermal parameter B is related to the mean square displacement $\overline{u^2}$ of the
atomic vibration:

$$B = 8\pi^2 \times \overline{u^2} \tag{4.7}$$

If the atomic vibration is split into three perpendicular components:
one perpendicular to the reflecting plane [vibr($\perp$)], and two in the plane,
vibr($\parallel_1$) and vibr($\parallel_2$), then vibr($\perp$) is the only one giving rise to the
temperature factor with parameter $B = 3 \times 8\pi^2 \times \overline{u^2}$ ($\perp$) (Garcia et al.,
1997). We do not observe the other two, but in isotropic vibration they
are just as large as the observed one and $\overline{u^2} = 3 \times \overline{u^2}$ ($\perp$).

For anisotropic vibration the temperature factor is much more compli-
cated. In this case $\overline{u^2}$ depends on the direction of **S**. It can be shown that
the temperature factor is given by

$$T(\text{aniso}; h\,k\,l) = \exp\left[-2\pi^2\left(\begin{array}{l} U_{11}h^2a^{*2} + U_{22}k^2b^{*2} + U_{33}l^2c^{*2} + \\ 2U_{12}hka^*b^* + 2U_{13}hla^*c^* + 2U_{23}klb^*c^* \end{array}\right)\right]$$

with U_{11} the $\overline{u^2}$ value along $\mathbf{a}^*$, U_{22} along $\mathbf{b}^*$, and U_{33} along $\mathbf{c}^*$. In general
$\overline{u^2}(\mathbf{e})$ along a unit vector $\mathbf{e}$ is given by

$$\overline{u^2}(\mathbf{e}) = U_{11} \cdot e_1^2 + U_{22} \cdot e_2^2 + U_{33} \cdot e_3^2 + 2U_{12} \cdot e_1e_2$$
$$+ 2U_{13} \cdot e_1e_3 + 2U_{23} \cdot e_2e_3$$

with e_1, e_2, and e_3 the components of $\mathbf{e}$ along unit axes $\mathbf{a}^*$, $\mathbf{b}^*$, and $\mathbf{c}^*$. The
points for which $\overline{u^2}(\mathbf{e})$ is constant form an ellipsoid: *the ellipsoid of
vibration*. For display purposes the constant can be chosen such that the
vibrating atom has a chance of, e.g., 50% of being within the ellipsoid
(Figure 4.22).

In protein structure determinations it is common to work with isotropic
temperature factors for the individual atoms. This is because of the restricted resolution and, consequently, the limited number of data. However, because of improvements in hardware an increasing number of
structures is determined at a sufficiently high resolution to use anisotropic
temperature factors. But in the normal situation one is restricted to isotro-

pic temperature factors. Then there are four unknown parameters per atom: x, y, z, and B. For a protein with 2000 atoms in the asymmetric unit, 8000 unknown parameters must be determined. To obtain a reliable structure the number of measured data (= reflection intensities) should well exceed the number of parameters. In general, because of the restricted number of data, this condition is fulfilled for the determination of isotropic, but not anisotropic, temperature factors. Average values for B in protein structures range from as low as a few Å^2 to 30 Å^2 in well-ordered structures. The highest values are found in more or less flexible surface loops.

The temperature factor is a consequence of the dynamic disorder in the crystal caused by the temperature-dependent vibration of the atoms in the structure. In addition to this *dynamic disorder*, protein crystals have *static disorder*: molecules, or parts of molecules, in different unit cells do not occupy exactly the same position and do not have exactly the same orientation. The effect of this static disorder on the X-ray diffraction pattern is the same as for the dynamic disorder and they cannot be distinguished, unless intensity data at different temperatures are collected. For $B = 30\ \text{Å}^2$ the root mean square displacement $\sqrt{\overline{u^2}}$ of the atoms from their equilibrium position is $\sqrt{30/8\pi^2} = 0.62$ Å. This gives an impression of the flexibility or disorder in protein structures, which are by no means completely rigid. Because of the disorder in the crystal the diffraction

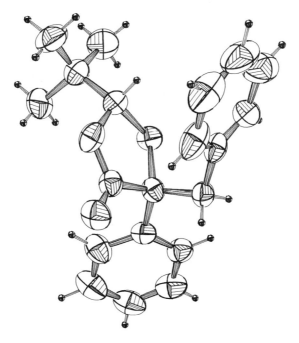

Figure 4.22. The plot of an organic molecule with 50% probability thermal ellipsoids. (Reproduced with permission from Strijtveen and Kellogg © 1987 Pergamon Press PLC).

pattern fades away at some diffraction angle θ_{max}. The corresponding lattice distance d_{min} is determined by Bragg's law:

$$d_{min} = \frac{\lambda}{2 \sin \theta_{max}}$$

d_{min} is taken as the resolution of the diffraction pattern and it is said that a structure has been determined to a resolution of, for instance, $2\,\text{Å}$. It is clear that the accuracy with which a structure can be determined depends strongly on the resolution of the diffraction pattern.

4.10. Calculation of the Electron Density ρ (x y z)

The intensity of the diffracted beam $(h\ k\ l)$ is proportional to the square of the amplitude of the structure factor $\mathbf{F}(h\ k\ l)$. The structure factor is a function of the electron density distribution in the unit cell:

$$\mathbf{F}(\mathbf{S}) = \sum_j f_j \exp[2\pi i r_j \cdot \mathbf{S}] \tag{4.3}$$

The summation is over all atoms j in the unit cell. Instead of summing over all separate atoms we can integrate over all electrons in the unit cell:

$$\mathbf{F}(\mathbf{S}) = \int_{cell} \rho(\mathbf{r}) \exp[2\pi i \mathbf{r} \cdot \mathbf{S}]\, dv \tag{4.8}$$

where $\rho(\mathbf{r})$ is the electron density at position $\mathbf{r}$ in the unit cell. If x, y, and z are fractional coordinates in the unit cell ($0 \leqslant x < 1$; the same for y and z) and V is the volume of the unit cell, we have:

$$dv = V \cdot dx\, dy\, dz$$

and

$$\mathbf{r} \cdot \mathbf{S} = (\mathbf{a} \cdot x + \mathbf{b} \cdot y + \mathbf{c} \cdot z) \cdot \mathbf{S} = \mathbf{a} \cdot \mathbf{S} \cdot x + \mathbf{b} \cdot \mathbf{S} \cdot y + \mathbf{c} \cdot \mathbf{S} \cdot z$$
$$= hx + ky + lz$$

Therefore, $\mathbf{F}(\mathbf{S})$ can also be written as $\mathbf{F}(h\ k\ l)$:

$$\mathbf{F}(h\ k\ l) = V \int_{x=0}^{1} \int_{y=0}^{1} \int_{z=0}^{1} \rho(x\ y\ z) \exp[2\pi i(hx + ky + lz)]\, dx\, dy\, dz \tag{4.9}$$

However, the goal of protein X-ray crystallography is not to calculate the diffraction pattern but to calculate the electron density ρ at every position x, y, z in the unit cell. How can this be done? The answer is by Fourier transformation.

$F(h\ k\ l)$ is the Fourier transform of $\rho(x\ y\ z)$ but the reverse is also true: $\rho(x\ y\ z)$ is the Fourier transform of $F(h\ k\ l)$ and therefore, $\rho(x\ y\ z)$ can be written as a function of all $F(h\ k\ l)$:

$$\rho(x\ y\ z) = \frac{1}{V}\sum_h\sum_k\sum_l F(h\ k\ l)\ \exp[-2\pi i(hx + ky + lz)] \quad (4.10)$$

The Laue conditions tell us that diffraction occurs only in discrete directions and therefore, in Eq. (4.10) the integration has been replaced by a summation. Because $\mathbf{F} = |F|\ \exp[i\alpha]$ we can also write:

$$\rho(x\ y\ z) = \frac{1}{V}\sum_h\sum_k\sum_l |F(h\ k\ l)|\exp[-2\pi i(hx + ky + lz) + i\alpha(h\ k\ l)] \quad (4.11)$$

It now seems easy to calculate the electron density $\rho(x\ y\ z)$ at every position $(x\ y\ z)$ in the unit cell. However, there is a problem. Although the $|F(h\ k\ l)|$s can be derived from the intensities $I(h\ k\ l)$, the phase angles $\alpha(h\ k\ l)$ cannot be derived straightforwardly from the diffraction pattern. Fortunately several methods have been developed to solve this problem; they will be discussed later.

_____

Fourier Transforms and the Delta Function

We want to know why Eq. (4.10) is true, or why $\rho(x\ y\ z)$ is the Fourier transform of $F(h\ k\ l)$. Take $F(h\ k\ l)$ in the form of Eq. (4.8):

$$F(\mathbf{S}) = \int_{\substack{\mathbf{r}\ over \\ the\ cell}} \rho(\mathbf{r})\ \exp[2\pi i\mathbf{r}\cdot\mathbf{S}]\,dv_r$$

Multiply by $\exp[-2\pi i\mathbf{r}'\cdot\mathbf{S}]$ and integrate over $\mathbf{S}$. dv_r is a small volume in real space ($\mathbf{r}$ space) and dv_S in reciprocal space ($\mathbf{S}$ space):

$$\int_{\mathbf{S}=-\infty}^{+\infty} F(\mathbf{S})\ \exp[-2\pi i\mathbf{r}'\cdot\mathbf{S}]\,dv_S$$

$$= \int_{\mathbf{S}=-\infty}^{+\infty} dv_s\int_{\mathbf{r}} \rho(\mathbf{r})\ \exp[2\pi i(\mathbf{r} - \mathbf{r}')\cdot\mathbf{S}]\,dv_r \quad (4.12)$$

Because $\rho(\mathbf{r})$ is independent of $\mathbf{S}$, Eq. (4.12) is equal to:

$$\int_{\mathbf{r}} dv_r\rho(\mathbf{r})\int_{\mathbf{S}=-\infty}^{+\infty} \exp[2\pi i(\mathbf{r} - \mathbf{r}')\cdot\mathbf{S}]\,dv_S \quad (4.13)$$

The integral $\int_{\mathbf{S}=-\infty}^{+\infty}\exp[2\pi i(\mathbf{r} - \mathbf{r}')\cdot\mathbf{S}]\,dv_S$ can be regarded as a summation of vectors in the Argand diagram. These vectors have a length equal to 1 but point in many different directions, because $\mathbf{S}$ has an infinite number of values in reciprocal space and hence the result of the integration is, in general, zero, except if $\mathbf{r} - \mathbf{r}' = 0$ or $\mathbf{r} = \mathbf{r}'$. In that case all vectors in the Argand diagram point to the right along the horizontal axis. Therefore, $\int_{S=-\infty}^{+\infty} \exp[2\pi i(\mathbf{r} - \mathbf{r}') \cdot \mathbf{S}] \, dv_S$ is a strange function: it is zero for $\mathbf{r} \neq \mathbf{r}'$ and infinite for $\mathbf{r} = \mathbf{r}'$. Such a function is called a delta function; here we have a three-dimensional delta function:

$$\int_{S=-\infty}^{+\infty} \exp[2\pi i(\mathbf{r} - \mathbf{r}') \cdot \mathbf{S}] \, dv_S = \delta(\mathbf{r} - \mathbf{r}')$$

A (one-dimensional) delta function has the property:

$$\int_{x=-\infty}^{+\infty} f(x)\delta(x - a) \, dx = f(a) \tag{4.14}$$

and in our three-dimensional case (4.13) becomes:

$$\int_{\mathbf{r}} \rho(\mathbf{r}) \cdot \delta(\mathbf{r} - \mathbf{r}') \, dv_r = \rho(\mathbf{r}')$$

This property will be proven further on. Equation (4.12) can now be written as:

$$\rho(\mathbf{r}') = \int_{S=-\infty}^{+\infty} F(\mathbf{S}) \exp[-2\pi i \mathbf{r}' \cdot \mathbf{S}] \, dv_S$$

or with $\mathbf{r}'$ replaced by $\mathbf{r}$:

$$\rho(\mathbf{r}) = \int_{S=-\infty}^{+\infty} F(\mathbf{S}) \exp[-2\pi i \mathbf{r} \cdot \mathbf{S}] \, dv_S \tag{4.15}$$

This proves the Fourier transformation.

   Because of the Laue conditions the integration in Eq. (4.15) can be replaced by summations over h, k, and l, giving a weight V^* to each term. V^* is the volume of the unit cell in reciprocal space and is equal to $1/V$, where V is the volume of the crystallographic unit cell (Section 4.8). With $\mathbf{r} \cdot \mathbf{S} = hx + ky + lz$, Eq. (4.15) is transformed into Eq. (4.10).

The One-Dimensional Delta Function
Consider the function $\delta_a(x)$, which has a constant value $1/2a$ between $-a$ and $+a$, and is zero outside this region.

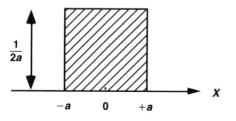

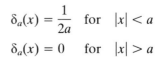

$$\delta_a(x) = \frac{1}{2a} \quad \text{for} \quad |x| < a$$

$$\delta_a(x) = 0 \quad \text{for} \quad |x| > a$$

The surface of the hatched area is $\int_{x=-\infty}^{+\infty} \delta_a(x)\,dx = 1$ and is independent of a.

Now let a tend to zero. Then $\lim_{a\to 0} \delta_a(x)$ is defined as $\delta(x)$ with $\delta(x) = 0$ for $x \neq 0$ and $\delta(x) = \infty$ for $x = 0$, but $\int_{x=-\infty}^{+\infty} \delta(x)\,dx = 1$. Although at $x = 0$ the function $\delta(x)$ is ∞, it is also infinitely thin and the surface area under the function is independent of a and remains 1. Of special importance for us is the following property of a delta function:

$$\int\limits_{x=-\infty}^{+\infty} f(x)\delta(x)\,dx = f(0) \tag{4.16}$$

where $f(x)$ is any normal function. This property can be proven by starting with $\delta_a(x)$:

$$\int\limits_{x=-\infty}^{+\infty} f(x)\delta_a(x)\,dx = \frac{1}{2a}\int\limits_{-a}^{+a} f(x)\,dx \tag{4.17}$$

If a tends to zero the left part of Eq. (4.17) becomes $\int_{x=-\infty}^{+\infty} f(x)\delta(x)\,dx$, and the right part:

$$\lim_{a\to 0}\frac{1}{2a}\int\limits_{-a}^{+a} f(x)\,dx = f(0) \tag{4.18}$$

That Eq. (4.18) is true can be understood in the following way: In a very narrow region around $x = 0$, we can assume for $f(x)$ the constant value $f(0)$. The surface area under the function between $-a$ and $+a$ $[\int_{-a}^{+a} f(x)\,dx]$ is then $f(0) \times 2a$ and, therefore, Eq. (4.18) is true. With a change of variables Eq. (4.16) can be written as:

$$\int\limits_{x=-\infty}^{+\infty} f(x)\delta(x - a)\,dx = f(a)$$

This proves Eq. (4.14).

We have already met the delta function in deriving the scattering by a crystal (Section 4.6) without mentioning it explicitly. There we had $\Sigma_{t=0}^{n} \exp[2\pi it\mathbf{a}\cdot\mathbf{S}]$ and two similar functions. In fact the presence of diffraction spots on an otherwise blank X-ray film or fluorescent plate can be regarded as the physical appearance of delta functions.

A special term in the Fourier summation of Eq. (4.11) is $F(000) \exp[0] = F(000)$; the phase angle $\alpha(000)$ is zero because there is no phase difference in the scattering by the electrons in the forward direction. The "reflection" (000) is not observable because it is in the line of the direct beam. Nevertheless $F(000)$ adds a constant term to the Fourier summation, which is equal to the total number of electrons in the cell. This requires that $F(000)$ and all other Fs are expressed in electrons. They are then on an *absolute scale.* Normally the measured intensities and the structure factor amplitudes derived from them are on an arbitrary scale. The resulting electron density map is then also on an arbitrary scale, but this does not prevent its interpretation. In Section 5.2 it is described how intensities can be put on the absolute scale.

The contribution to the electron density map by all other terms than $F(000)$ has an average value of zero over one full cycle. Therefore the average electron density in the unit cell is zero if $F(000)$ is not incorporated in the Fourier summation. But a constant term can be added to remove negative electron density and the absence of $F(000)$ does not inhibit the interpretation of an electron density map. The limited resolution of a protein X-ray diffraction pattern prevents the calculation of the electron density map at atomic resolution: although amino acid residues can be distinguished, atoms are not separated (Figure 4.23), except if the electron density map is calculated with very high resolution data (near 1 Å). Therefore, it seems surprising that the error in the atomic positions is not more than approximately 0.2 Å for a structure derived from a high resolution electron density map. The reason is that the structure of the building blocks—amino acids and small peptides—is known with high accuracy and one can safely assume that the atomic distances and bond angles are the same in the proteins as they are in the small compounds.

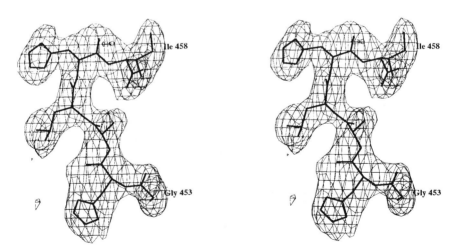

Figure 4.23. Stereo picture for the C-terminal residues 453–458 of the enzyme lipoamide dehydrogenase of *Pseudomonas putida* at 2.45 Å resolution. (Reproduced with permission from the thesis by Andrea Mattevi, University of Groningen, 1992.)

4.10.1. The Projection of Electron Density Along an Axis

The two-dimensional electron density projection along an axis can easily be calculated with a limited number of reflections. For protein structures this is usually not very useful because of the overlap of density and the inability to detect any interesting features in the projected density. Moreover, considering the speed and cost of modern computers and the speed of modern data collection methods, three-dimensional electron density information can easily be obtained. However, one should at least know the existence of the method. We shall discuss the projection of the electron density ρ along the c-axis onto the ab-plane. Expression (4.9) for the structure factor gives for the reflections (h k 0):

$$\mathbf{F}(h\ k0) = V \int_{xyz} \rho(x\ y\ z)\, dz\, \exp[2\pi i(hx + ky)]\, dx\, dy \qquad (4.19)$$

In Eq. (4.19) the integration over z is first performed. Let us assume that the unit cell has an angle of 90° between the c-axis and the ab-plane. The volume of the unit cell V is then equal to $A \times c$, in which A is the surface of the ab-plane. In Eq. (4.19), z and dz are fractional coordinates. Therefore, $V\, dz$ can be written as $A\, c\, dz$ and $\mathbf{F}(h\ k\ 0)$ as:

$$\mathbf{F}(h\ k\ 0) = A \int_{xyz} \rho(x\ y\ z)\, c\, dz\, \exp[2\pi i(hx + ky)]\, dx\, dy \qquad (4.20)$$

With $\int_z \rho(x\ y\ z)\, c\, dz = \rho(x\ y)$, the electron density in the projected structure in e/Å², we obtain:

$$\mathbf{F}(h\ k\ 0) = A \int_{xy} \rho(x\ y)\, \exp[2\pi i(hx + ky)]\, dx\, dy \qquad (4.21)$$

The projected electron density ρ is obtained by Fourier inversion:

$$\rho(x\ y) = \frac{1}{A} \sum_h \sum_k \mathbf{F}(h\ k\ 0)\, \exp[-2\pi i(hx + ky)] \qquad (4.22)$$

If the projection is along a 2-fold axis or screw axis along z, the expressions for $\mathbf{F}(h\ k\ 0)$ and $\rho(x\ y)$ are further simplified, because in that case $\rho(x\ y) = \rho(\bar{x}\ \bar{y})$:

$$\mathbf{F}(h\ k\ 0) = \frac{1}{2} A \int_{xy} \rho(x\ y)\{\exp[2\pi i(hx + ky)] + \exp[-2\pi i(hx + ky)]\}\, dx\, dy$$

$$= A \int_{xy} \rho(x\ y)\, \cos[2\pi(hx + ky)]\, dx\, dy \qquad (4.23)$$

and

$$\rho(x\ y) = \frac{1}{2}\frac{1}{A}\sum_h\sum_k F(h\ k\ 0)\{\exp[-2\pi i(hx + ky)] + \exp[2\pi i(hx + ky)]\}$$

$$= \frac{1}{A}\sum_h\sum_k F(h\ k\ 0)\cos[2\pi(hx + ky)] \qquad (4.24)$$

For such a centrosymmetric case $F(h\ k\ 0)$ always points along the real axis in the Argand diagram. It has a positive sign for a phase angle of $0°$ and a negative sign for a phase angle of $180°$.

4.11. Comparison of $F(h\ k\ l)$ and $F(\bar{h}\ \bar{k}\ \bar{l})$

$$F(h\ k\ l) = V \int_{\text{cell}} \rho(x\ y\ z)\ \exp[2\pi i(hx + ky + lz)]\,dx\,dy\,dz$$

[see Eq. (4.9)]. In the same way we can write for the reflection $(\bar{h}\ \bar{k}\ \bar{l})$:

$$F(\bar{h}\ \bar{k}\ \bar{l}) = V \int_{\text{cell}} \rho(x\ y\ z)\ \exp[2\pi i(-hx - ky - lz)]\,dx\,dy\,dz \quad (4.25)$$

$F(h\ k\ l)$ is obtained as the result of a vector summation in which the amplitudes of the constituent vectors are $\rho(x\ y\ z)\,dx\,dy\,dz$ and the phase angles $2\pi(hx + ky + lz)$. For $F(\bar{h}\ \bar{k}\ \bar{l})$ the amplitudes are the same but the phase angles have just the opposite value: $-2\pi(hx + ky + lz)$. The resulting vectors $F(h\ k\ l)$ and $F(\bar{h}\ \bar{k}\ \bar{l})$ also have, therefore, the same length, but opposite phase angles: α and $-\alpha$ in Figure 4.24. The consequence is that Eq. (4.11) reduces to

$$\rho(x\ y\ z) = \frac{2}{V}\sum_{hkl=0}^{+\infty}|F(h\ k\ l)|\cos[2\pi(hx + ky + lz) - \alpha(h\ k\ l)]$$

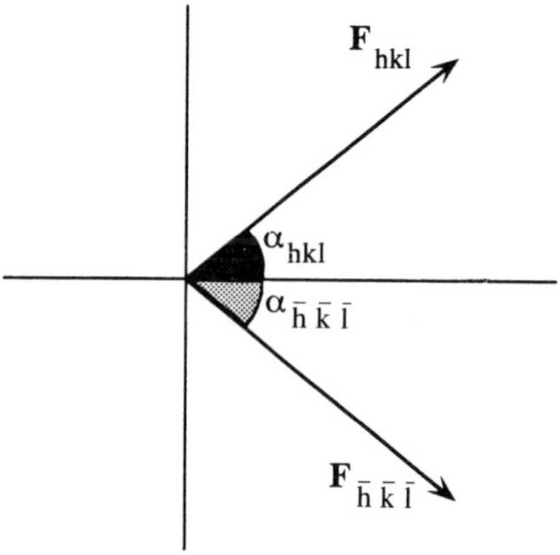

Figure 4.24. Argand diagram for the structure factors of the reflections $F(h\ k\ l)$ and $F(\bar{h}\ \bar{k}\ \bar{l})$.

This expression does not contain an imaginary term but is real, as expected for the electron density as a physical quantity.

Because the intensity of a diffracted beam is proportional to the square of its amplitude [$I(h\ k\ l)$ proportional to $|F(h\ k\ l)|^2$], the intensities $I(h\ k\ l)$ and $I(\bar{h}\ \bar{k}\ \bar{l})$ are also equal. The reflections $(h\ k\ l)$ and $(\bar{h}\ \bar{k}\ \bar{l})$ are called *Friedel* or *Bijvoet*[3] *pairs*. Their equal intensities give rise to a center of symmetry in the diffraction pattern, even if such a center is not present in the crystal structure. The $I(h\ k\ l) = I(\bar{h}\ \bar{k}\ \bar{l})$ equality is usually assumed to be true in crystal structure determinations. It depends, however, on the condition that anomalous scattering is absent (Section 7.8).

4.12. Symmetry in the Diffraction Pattern

In the previous section we noticed that the diffraction pattern from a crystal has a center of symmetry. In addition, more symmetry in the pattern can be present, depending on the symmetry elements in the crystal. After the application of a symmetry operation, the crystal structure looks exactly as before. For this reason, and because the diffraction pattern rotates with the crystal, the symmetry of the diffraction pattern must be at least the same as for the crystal. Reflections in the diffraction pattern related by symmetry will have the same intensity and only a portion of all data is unique in terms of intensity.

As we saw in Section 3.3, the only symmetry elements allowed in protein crystals are symmetry axes. In the next section we shall show what the effect on the diffraction pattern is of a 2-fold axis along the y-axis and a 2-fold screw axis along the y-axis.

4.12.1. A 2-Fold Axis Along y

If a 2-fold axis through the origin and along y is present, then the electron density $\rho(x\ y\ z) = \rho(\bar{x}\ y\ \bar{z})$ (Figure 4.25). Therefore,

$$\mathbf{F}(h\ k\ l) = V \int_{\substack{\text{asymm} \\ \text{unit}}} \rho(x\ y\ z)\{\exp[2\pi i(hx + ky + lz)]$$

$$+ \exp[2\pi i(-hx + ky - lz)]\}\,dx\,dy\,dz \qquad (4.26)$$

The integration in Eq. (4.26) is over one asymmetric unit (half of the cell), because the presence of the second term under the integral takes care of the other half of the cell.

---

[3] J.M. Bijvoet, 1892–1980, the famous Dutch crystallographer, was the first to determine the absolute configuration of an organic compound.

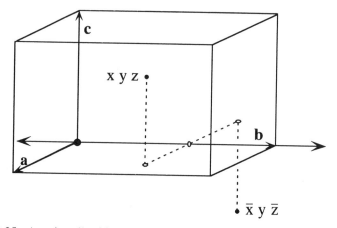

Figure 4.25. A unit cell with a 2-fold axis through the origin and along y. The electron density is equal at positions $x\,y\,z$ and $\bar{x}\,y\,\bar{z}$.

$$\mathbf{F}(\bar{h}\,k\,\bar{l}) = V \int_{\substack{\text{asymm}\\\text{unit}}} \rho(x\,y\,z)\{\exp[2\pi i(-hx + ky - lz)]$$

$$+ \exp[2\pi i(hx + ky + lz)]\}\,dx\,dy\,dz \qquad (4.27)$$

It follows that $\mathbf{F}(h\,k\,l) = \mathbf{F}(\bar{h}\,k\,\bar{l})$ and also $I(h\,k\,l) = I(\bar{h}\,k\,\bar{l})$, because the intensities I are proportional to $|F|^2$. Therefore, the diffraction pattern has the same 2-fold axis as the crystal.

4.12.2. A 2-Fold Screw Axis Along y

For a 2-fold screw axis along y (Figure 4.26):

$$\rho(x\,y\,z) = \rho\{\bar{x}(y + 1/2)\bar{z}\}$$

$$\text{term I } \downarrow$$

$$\mathbf{F}(h\,k\,l) = V \int_{\substack{\text{asymm}\\\text{unit}}} \rho(x\,y\,z)\{\exp[2\pi i(hx + ky + lz)]$$

$$+ \exp[2\pi i(-hx + k(y + 1/2) - lz)]\}\,dx\,dy\,dz \qquad (4.28)$$

$$\text{term II } \uparrow$$
$$\text{term III } \downarrow$$

$$\mathbf{F}(\bar{h}\,k\,\bar{l}) = V \int_{\substack{\text{asymm}\\\text{unit}}} \rho(x\,y\,z)\{\exp[2\pi i(-hx + ky - lz)]$$

$$+ \exp[2\pi i(hx + k(y + 1/2) + lz)]\}\,dx\,dy\,dz \qquad (4.29)$$

$$\text{term IV } \uparrow$$

In Eq. (4.28) term II is

$$\exp\{2\pi i[-hx + k(y + 1/2) - lz]\} = \exp[2\pi i(-hx + ky - lz + 1/2k)]$$

For k is even this is equal to term III in Eq. (4.29). The same is true for term IV in Eq. (4.29) and term I in Eq. (4.28). Therefore, when k is even: $\mathbf{F}(h\ k\ l) = \mathbf{F}(\bar{h}\ k\ \bar{l})$ and also $I(h\ k\ l) = I(\bar{h}\ k\ \bar{l})$. When k is odd, the terms I and IV have a difference of π in their phase angles: $2\pi(hx + ky + lz)$ and $2\pi(hx + ky + lz + 1/2k)$. The same is true for II and III. If we again regard these exponential terms as vectors in the Argand diagram with a length of 1 and appropriate phase angles, we easily see that when k is odd, $\mathbf{F}(h\ k\ l)$ and $\mathbf{F}(\bar{h}\ k\ \bar{l})$ have equal length but a phase difference of π (Figure 4.27a). But again $I(h\ k\ l) = I(\bar{h}\ k\ \bar{l})$ because $I(h\ k\ l) = |F(h\ k\ l)|^2$.

The result is that a 2-fold screw axis in the crystal is found as a 2-fold axis in the X-ray diffraction pattern. The phase angles of symmetry-related reflections are the same or differ by π, but this cannot be observed in the diffraction pattern. The other possible screw axes also express themselves in the X-ray diffraction pattern as normal nonscrew rotation axes. Although the screw character of a 2-fold screw axis is not detected in the symmetry of the X-ray diffraction pattern, it has an effect on the reflections along the corresponding 2-fold axis in the diffraction pattern. Take again a 2-fold screw axis along y:

$$\mathbf{F}(0\ k\ 0) = V \int_{\substack{\text{asymm} \\ \text{unit}}} \rho(x\ y\ z)\{\exp[2\pi iky] + \exp[2\pi ik(y + 1/2)]\}\, dx\, dy\, dz$$

$$(4.30)$$

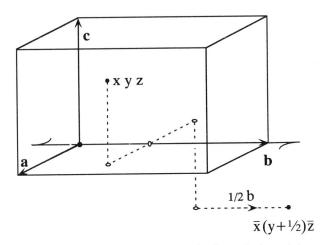

Figure 4.26. A unit cell with a 2-fold screw axis through the origin and along y. The electron density is equal at positions $x\ y\ z$ and $\bar{x}(y + 1/2)\bar{z}$.

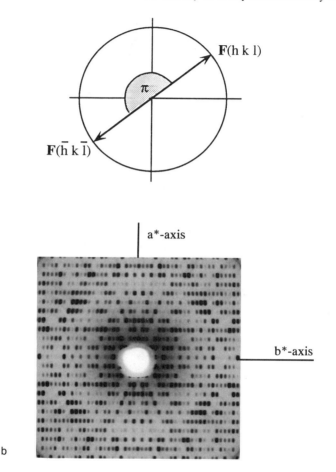

Figure 4.27. (a) The Argand diagram with the structure factors $F(h\ k\ l)$ and $F(\bar{h}k\ \bar{l})$ for a structure with a 2-fold screw axis along y with k odd. (b) A diffraction picture of a crystal of the enzyme papain. The crystal belongs to space group $P2_12_12_1$. Note the absence of reflections ($h00$) for h odd and of reflections ($0k0$) for k odd.

When k is even this is $2 \times V\int\rho(x\ y\ z)\ \exp[2\pi iky]\ dx\ dy\ dz$. However when k is odd the two terms in Eq. (4.30) cancel and $F(0\ k\ 0) = 0$ (Figure 4.27b).

The conditions for these reflections are called *serial reflection conditions*, because they apply to a set of reflections lying on a line through the origin in the reciprocal lattice. *Integral reflection conditions* involve all reflections; they are observed for centered lattices as discussed in the next section.

Figure 4.28. A unit cell centered in the *ab*-plane.

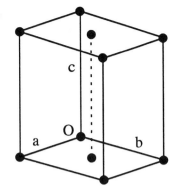

4.13. Integral Reflection Conditions for Centered Lattices

Consider for example, a lattice centered in the *ab*-plane, also called the *C*-plane (Figure 4.28). The following electron density relationship exists:

$$\rho(x\ y\ z) = \rho(x + 1/2,\ y + 1/2,\ z)$$

The structure factor **F** can now be written as

$$\mathbf{F}(h\ k\ l) = V \int_{\substack{0 \leqslant x < 1 \\ 0 \leqslant y < 1/2 \\ 0 \leqslant z < 1}} \rho(x\ y\ z)\{\exp[2\pi i(hx + ky + lz)]$$

$$+ \exp[2\pi i(h(x + 1/2) + k(y + 1/2) + lz)]\}\ dx\ dy\ dz \quad (4.31)$$

The second exponential term on the right in Eq. (4.31) is equal to

$$\exp\{2\pi i[hx + ky + lz + 1/2(h + k)]\}$$

If $h + k$ is odd the two contributions to **F** cancel each other because of opposite phase angles.

The conclusion is that in the diffraction pattern of a crystal that has its *C*-face centered, all reflections for which $h + k$ is odd have a zero value for their intensities. The *International Tables* give complete information about the extinction of reflections for all space groups.

4.14. The Intensity Diffracted by a Crystal

Starting from the wave $\mathbf{F}(\mathbf{S})$ diffracted by a unit cell, an expression can be derived for the integrated intensity $I(\text{int.},\ h\ k\ l)$ of the reflection $(h\ k\ l)$.

The crystal is not perfect; rather it is imperfect and of finite size. These imperfect crystals can be regarded as being composed of small mosaic

Figure 4.29. Most crystals are imperfect and can be regarded as being composed of small mosaic blocks.

blocks, which can be considered as optically independent fragments (Figure 4.29). For such an imperfect crystal the intensity profile of a reflection has a certain width because of the angular spread of the mosaic blocks; for a protein crystal generally 0.25–0.5°. Moreover each tiny mosaic block has a small intrinsic reflection width (less than 0.01°) because it does not strictly obey the reflection condition of an infinite number of unit cells. This implies that diffraction occurs not only for a sharp point $(h\ k\ l)$ in the reciprocal lattice, but also for a small region in reciprocal space around the sharp point.

With the assumptions that,

1. Apart from ordinary absorption, the intensity I_0 of the incident beam is the same throughout the crystal, and
2. The mosaic blocks are so small that a scattered wave is not scattered again (i.e., multiple scattering does not occur)

the expression for I (int., $h\ k\ l$), if the crystal is rotated with an angular velocity ω through the reflection position, is:

$$I(\text{int., } h\ k\ l) = \frac{\lambda^3}{\omega \cdot V^2} \times \left(\frac{e^2}{mc^2}\right)^2 \times V_{cr} \times I_0 \times L \times P \times T_r \times |F(h\ k\ l)|^2$$

$$(4.32)$$

The complete scattering Eq. (4.32) not only contains $|F(h\ k\ l)|^2$ but also has terms related to the scattering by the electrons themselves, to the wavelength λ, the volume V_{cr} of the crystal, and the volume V of the unit cell. We shall explain the various terms of Eq. (4.32) instead of deriving them from first principles. I_0 is the intensity of the incident beam and I is the integrated intensity of the reflected beam. e is the charge and m the mass of an electron and c the velocity of light. Thomson showed that if the electron is a completely free electron and scatters the X-ray photons elastically, that is with the same frequency as the incident beam, the following are true:

• The phase difference between the incident and the scattered beam is π. This is because the scattered radiation is proportional to the displacement of the electron which differs π in phase with its acceleration imposed by the electric vector.

• The amplitude of the electric component of the scattered wave at a distance r, large in comparison with the wavelength of the radiation, is

$$E_{el.} = E_o \frac{1}{r} \frac{e^2}{mc^2} \sin \alpha \qquad (4.33)$$

where E_o is the amplitude of the electric vector of the incident beam and α the angle between the oscillation direction of the electron and the scattering direction (Figure 4.31a). Note that the component of E_o perpendicular to the scattering direction is $E_o \sin \alpha$.

In terms of energy:

$$I_{el.} = I_o \frac{1}{r^2} \left(\frac{e^2}{mc^2} \right)^2 \sin^2 \alpha \qquad (4.34)$$

Per unit solid angle the scattered energy is:

$$I_{el.}(\Omega = 1) = I_{el.} \times r^2 \qquad (4.35)$$

It has been derived by Klein and Nishina (1929) [see also Heitler (1966)] that the scattering by an electron can be discussed in terms of the classical Thomson scattering if the quantum energy $h\nu \ll mc^2$. This is not true for very short X-ray wavelengths. For $\lambda = 0.0243\,\text{Å}$, $h\nu$ and mc^2 are exactly equal, but for $\lambda = 0.1\,\text{Å}$, $h\nu$ is 0.0243 times mc^2. Since in macromolecular crystallography wavelengths are usually in the range 0.8–2.5 Å, the classical approximation is allowed. It should be noted that:

• The intensity scattered by a free electron is independent of the wavelength.
• Thomson's equation can also be applied to other charged particles, e.g., a proton. Because the mass of a proton is 1800 × the electron mass, scattering by a proton and by atomic nuclei can be neglected.
• For an unpolarized beam $\sin^2\alpha$ is replaced by a suitable polarization factor. The polarization factor will be discussed in Section 4.14.1 and the transmission factor Tr in Section 4.14.2. The Lorentz factor L depends on the data acquisition system. If the electron is replaced by a unit cell of the crystal, the scattered amplitude is enhanced by a factor $|F(h\,k\,l)|$ and the intensity by $|F(h\,k\,l)|^2$.

The factor V_{cr}/V^2 in Eq. (4.32) requires some discussion. It depends on the existence of the independently scattering mosaic blocks. The larger the crystal, the more mosaic blocks and, therefore, the proportionality of the integrated intensity with V_{cr}. But why is it proportional to $1/V^2$? In reflection position all unit cells in a mosaic block scatter in phase. For a unit cell of volume V, the number of unit cells for a given volume of the mosaic blocks is proportional to $1/V$, as is the scattered amplitude. Therefore the intensity is proportional to $1/V^2$.

I (int., $h\,k\,l$) is proportional to λ^3. This can be understood as follows:

- Consider the diffracting region around the reciprocal lattice point P. For a fixed position of the crystal, scattering occurs for the intersection of the diffracting region with the Ewald sphere.
- The intensity scattered by an electron into all directions or, in other words, toward the complete surface $4\pi(1/\lambda^2)$ of the Ewald sphere (radius $1/\lambda$), has a fixed value and is independent of λ. Therefore, the scattering per unit surface, and also the scattering by the intersection considered, is proportional to λ^2.

This λ^2-dependence must be multiplied by a λ-dependent term, which is related to the time t it takes for the complete diffracting region to pass through the Ewald sphere. For the simple case in which the incident and the diffracted beams are perpendicular to the rotation axis, t is found as follows (Figure 4.30).

If the angular speed of rotation is ω, then for the reciprocal lattice point P at a distance $1/d$ from the origin O, the linear speed perpendicular to OP is $v = (1/d)\omega$. Its component $v_\perp$ along PM, and thus perpendicular to the surface of the sphere, should be regarded:

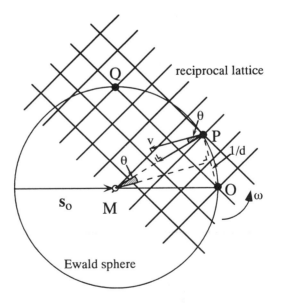

Figure 4.30. If the rotation of the reciprocal lattice is around an axis through the origin O and perpendicular to the plane of the drawing, point P will be in reflection condition much longer than point Q. Note that $OP = 1/d(P)$, $OQ = 1/d(Q)$, and $MP = s(P)$ and $MQ = s(Q)$.

$$v_\perp = \frac{1}{d}\omega\cos\theta$$

Because according to Bragg's law $d = \lambda/(2\sin\theta)$, we find:

$$v_\perp = \frac{2\sin\theta\cos\theta}{\lambda}\omega = \frac{\sin 2\theta}{\lambda}\omega$$

The time t required to pass the sphere is proportional to:

$$\frac{1}{v_\perp} = \frac{1}{\omega} \times \frac{\lambda}{\sin 2\theta}$$

Multiplying by the λ^2 dependence, found above, gives λ^3. The θ-dependent factor $1/\sin 2\theta$ represents the Lorentz factor L for the present case. For other acquisition techniques L may be a more complicated function of $(h\ k\ l)$. Note that for small θ values, $\lambda/\sin 2\theta$ is approximately equal to $\lambda/(2\sin\theta) = d$ (error less than 10% for $\theta < 25°$) and λ drops out. This implies that the intensity of reflections from protein crystals, with their small θ values, is not proportional to λ^3, but rather to λ^2.

4.14.1. The Polarization Factor

The polarization factor P in Eq. (4.32) originates from the fact that an electron does not scatter along its direction of vibration. In other directions electrons radiate with an intensity proportional to $(\sin\alpha)^2$ (Figure 4.31a). The effect is that if the incident X-ray beam is unpolarized and then reflected against a plane, the reflected beam is more or less polarized. This affects the intensity. This is illustrated in Figure 4.31b. The unpolarized beam is split into two other beams with equal intensity, one polarized in a direction parallel to the reflecting plane (beam ∥) and the other one in a direction perpendicular to the first one (beam ⊥). The polarization direction of incident beam ⊥ has, on a relative scale, a component $\sin 2\theta$ along the scattering direction (Figure 4.31c). The oscillation of electrons induced by this component does not contribute to the scattered intensity (α in Figure 4.31a is here 0). Only the other component ($\cos 2\theta$) does. Therefore, the scattered intensity is reduced by a factor $(1 - \sin^2 2\theta) = \cos^2 2\theta$. For incident beam ∥, which is polarized perpendicular to the scattering direction, no reduction of intensity occurs. Combining the intensities of the two reflected beams results in the θ-dependent polarization factor $(1 + \cos^2 2\theta)/2$. It should be emphasized that the polarization factor as given here is valid only if the crystal is radiated with an unpolarized beam. With a monochromator in the incident beam path this is not true and it is certainly not true for synchrotron radiation, which is strongly polarized. For a polarized beam the polarization factor is (Azaroff, 1955; Kahn et al., 1982),

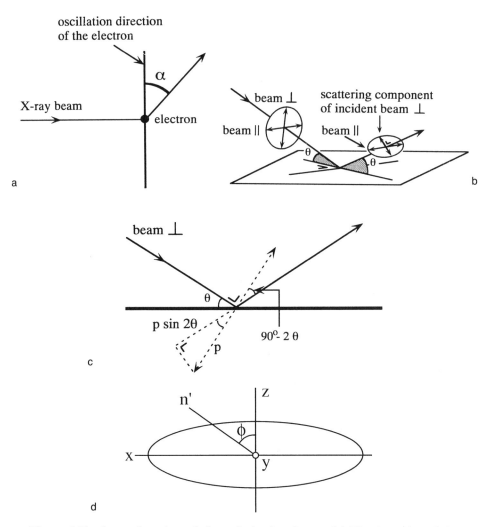

Figure 4.31. An explanation of the polarization factor. (a) The intensity of the radiation scattered by an oscillating electron is proportional to $(\sin \alpha)^2$. (b) An unpolarized incident beam can be split into two others with the same intensity: one polarized in a direction parallel to the reflecting plane (beam ∥) and the other one in a direction perpendicular to the first one (beam ⊥); both polarization directions lie in a plane perpendicular to the direction of the incident beam. (c) The dotted line (p) indicates the polarization direction of incident beam ⊥. Electrons induced by component $p \sin 2\theta$ do not contribute to the scattering. (d) Cross-section through a polarized beam with an intensity I_h in the horizontal direction and I_v in the vertical direction. The polarized beam is directed toward the reflecting plane along y with y perpendicular to the plane of the drawing. The ellips indicates the polarization. The normal n to the reflecting plane (not drawn) makes certain angles with x, y, and z. n' is the projection of n on the x–z—plane. ϕ is the angle between n' and the z-axis.

$$P = \frac{1}{2}\left(1 + \cos^2 2\theta - \tau \cos 2\phi \sin^2 2\theta\right)$$

For ϕ see Fig. 4.31 d; $\tau = \dfrac{I_h - I_v}{I_h + I_v}$ where I_h and I_v are the intensities of, respectively, the horizontal and vertical component of the polarized beam.

4.14.2. Absorption and Extinction

The transmission factor Tr is related to the absorption factor A: $Tr = 1 - A$. If an X-ray beam passes through matter, its intensity I diminishes as a consequence of absorption: $I = I_0 \exp[-\mu \cdot t]$. t(cm) is the path length in the matter and μ(cm^{-1}) is the total linear absorption coefficient. μ can be obtained as the sum of the atomic mass absorption coefficients μ_a because these are, to a good approximation, additive with respect to the elements composing the material (for a definition of μ_a see the legend to Figure 4.32):

$$\mu = \frac{1}{V}\sum_i n_i(\mu_a)_i$$

where n_i is the number of atoms of element i in volume V. The value of μ_a is more or less independent of the physical state of the material. In general, μ_a is larger for longer wave length and for atoms with a higher atomic number (Figure 4.32). A protein crystal of 0.5 mm has, for Cu

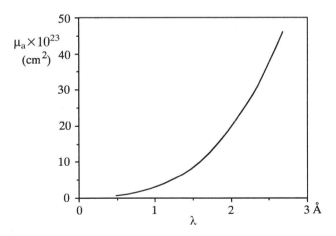

Figure 4.32. The atomic mass absorption coefficient μ_a of the element carbon as a function of the wavelength. The atomic mass absorption coefficient (cm^2) for element i is defined as $(\mu_a)_i = \mu_i/\rho_i \times A_i/N$, where μ_i is the linear absorption coefficient, ρ_i the density, and A_i the atomic weight; N is Avogadro's number.

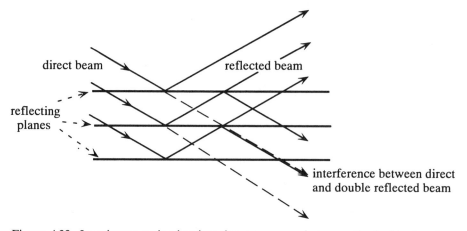

Figure 4.33. In primary extinction interference occurs between the incident and scattered waves. Because at each reflection there is a phase shift of $\pi/2$, the double reflected beam has a phase difference of π with the incident beam and reduces its intensity (Section 4.15).

radiation, a transmission of approximately 60%. For paper of 0.1 mm and for glass of 0.01 mm thickness, transmission is 93%. Air also absorbs X-ray radiation.

Absorption of X-rays is mainly caused by two effects:

1. Photoelectric absorption: The X-ray photon disappears completely. If the photon energy is sufficient to remove an electron from the atom, the absorption becomes particularly strong.

2. Scattering is the result of a collision between the X-ray photons and the electrons. One can distinguish two kinds of scattering: Compton scattering and Raleigh scattering. In Compton scattering, the photons loose part of their energy in the collision process (inelastic scattering) resulting in scattered photons with a lower energy and a longer wavelength. Compton scattering contributes to the background in an X-ray diffraction experiment. In Rayleigh scattering, the photons are elastically scattered, do not lose energy, and leave the material with an unchanged wavelength. In a crystal, they interfere with each other and give rise to Bragg reflections. Between the Bragg reflections there is no loss of energy because of elastic scattering and the incident beam is hardly reduced. The crystal is considered as an ideal mosaic if, in Bragg position, the reduction in intensity of the incident beam, as a result of elastic scattering, can still be neglected. For nonideal mosaic crystals (or nearly perfect crystals) the beam is reduced by *extinction*:

• The blocks are too large and multiple reflection occurs within a block. At each reflection process the phase angle shifts $\frac{\pi}{2}$ (Section 4.15). After two reflections the beam travels in the same direction as the incident beam but with a phase difference of π and this reduces the intensity. This is called "primary extinction" (Figure 4.33).

- The angular spread of the mosaic blocks is too small. Part of the incident beam is reflected by blocks close to the surface before it reaches lower lying blocks that are also in reflecting position. This reduces the intensity of the incident beam for the lower lying blocks and consequently also their diffracted intensity; it is called "secondary extinction." For the relatively weak reflections of protein crystals extinction does not play a significant role. Polikarpov and Sawyer (1995) noted that some protein crystals are rather perfect crystals, which is equivalent to large mosaic blocks and that primary extinction does play a role, especially for larger crystals and longer wavelength (1.5 instead of 1.0 Å). The correction can be as much as 13–15% for the strongest reflections.

4.15. Scattering by a Plane of Atoms[4]

A plane of atoms reflects an X-ray beam with a phase retardation of $\frac{\pi}{2}$ with respect to the scattering by a single atom. The difference is caused by the difference in path length, Source—atom—Detector, for the different atoms in the plane. Suppose the plane is infinitely large. The source is at S and the detector at D (Figure 4.34). The shortest connection between S and D via the plane is S–M–D. The plane containing S, M, and D is $\perp$ the reflecting plane and the lines S–M and M–D make equal angles with the reflecting plane. Moving outwards from atom M in the reflecting plane, for instance

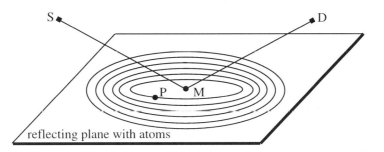

Figure 4.34. S is the X-ray source and D the detector. Scattering is by the atoms in a plane. The shortest distance between S and D via a point in the plane is through M. Path lengths via points in the plane outwards from M are longer, and when these beams reach the detector, they lag behind in phase with respect to the MD beam. The plane is divided in zones, such that from one zone to the next the path difference is $\lambda/2$.

---

[4]This text will also appear in the International Tables, Volume F, on Macromolecular Crystallography, to be published for the International Union of Crystallography.

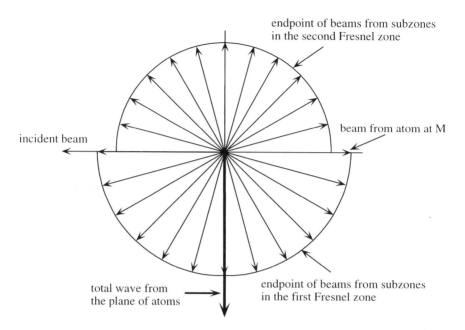

Figure 4.35. Schematic picture of the Argand diagram for the scattering by atoms in a plane. All electrons are considered as free electrons. The vector of the incident beam is pointing to the left. The electrons of atom M, regarded as free electrons, have a phase difference of π with respect to the incident beam. Subzones in the first Fresnel zone have the endpoint of their vectors on the lower half circle. For the next Fresnel zone they are on the upper half circle, which has a smaller radius because the amplitude decreases gradually for subsequent Fresnel zones (Kauzmann, 1957). The sum of all vectors is pointing down, meaning a phase lag of $\pi/2$ with respect to the beam scattered by the atom at M.

to P, the path length S–P–D is longer. At the edge of the first Fresnel zone the path is $\frac{\lambda}{2}$ longer (Figure 4.34). This edge is an ellipse with its center at M and its major axis on the line of intersection between plane S–M–D and the reflecting plane. Continuing outwards many more elliptic Fresnel zones are formed. Clearly the beams radiated by the many atoms in the plane interfere with each other. The situation is represented in the Argand diagram of Figure 4.35. Successive Fresnel zones can be subdivided into an equal number of subzones. If the distribution of electrons is sufficiently homogeneous, it can be assumed that the subzones in one Fresnel zone give the same amplitude at D. Their phases are spaced at regular intervals and their vectors in the Argand diagram lie in a half circle. In the lower part of Figure 4.35, this is illustrated for the first Fresnel zone. For the second Fresnel zone (upper part) the radius is slightly smaller because the intensity radiated by more distant zones decreases (Kauzmann, 1957). Therefore, the sum of vectors pointing upwards is shorter than for those pointing downwards and the resulting scattered wave lags $\frac{\pi}{2}$ in phase behind the scattering by the atom at M and has a phase difference of $\frac{\pi}{2}$ with the incident beam.

4.16. Choice of Wavelength, Size of Unit Cell, and Correction of the Diffracted Intensity

4.16.1. Choice of Wavelength

The wavelength λ has an appreciable influence on the intensity of the X-rays diffracted by a crystal [Eq. (4.32)]. Using a longer wavelength has the advantage of a stronger diffracted intensity, but the disadvantage of higher absorption. An optimal choice for protein crystallography is the Cu Kα wavelength of 1.5418 Å. However, if high intensity synchrotron radiation is available, a shorter wavelength, e.g., near 1 Å, has the advantage of lower absorption (Gonzalez, 1994; Helliwell et al., 1993). For an X-ray detector with a fluorescent screen, the optimum wavelength may also be below 1.5 Å, because more visible light is created per X-ray photon. However, part of the shorter wavelength X-ray beam may not be absorbed in the fluorescent layer. There is also more chance that the diffracted beams overlap. A further advantage of a short wavelength is that the "blind region" is smaller because of a larger Ewald sphere (section 2.7).

   Polikarpov et al. (1997) discussed the ultimate wavelength for protein crystallography from a theoretical point of view. Their conclusion is that it has no advantage to go below 0.9 Å. For very small crystals, where absorption is not a serious problem, a longer wavelength has the advantage of stronger reflection intensities. A special choice of wavelength is required if anomalous scattering is optimally used in protein phase angle determination (Chapter 9).

4.16.2. Effect of the Size of the Unit Cell on the Diffraction Intensity

If a crystal is larger, its diffraction (and also its absorption) is stronger. In protein crystallography a crystal size of 0.3–0.5 mm is regarded as optimal. Protein crystals are relatively weak scatterers for two reasons. First, because they consist only or mainly of light atoms: C, N, and O. The second and more important reason is the large size of their unit cells. This we can understand as follows. A crystal with a larger unit cell volume diffracts more weakly but has larger values of $|F(h\,k\,l)|^2$. We can combine these two effects by first calculating the mean square value

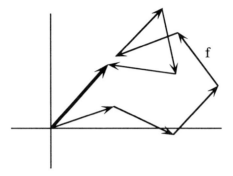

Figure 4.36. The displacement of a particle under the influence of Brownian motion. For n steps, where n is very large and each step has a length f, the final distance to the origin is $f\sqrt{n}$.

$\overline{|F(h\ k\ l)|^2}$. Suppose we have the simple situation of a crystal with one kind of atoms, each atom with a scattering factor f and n atoms in the unit cell. $\overline{|F(h\ k\ l)|^2}$ can be calculated by assigning random phases to the contributions f of the individual atoms in the Argand diagram. This problem is analogous to that of the displacement caused by Brownian motion (Figure 4.36): a number of n equal steps of length f gives a root mean square displacement from the origin of $f \times \sqrt{n}$. The root mean square value of $|F(h\ k\ l)|$ is then

$$\sqrt{\overline{|F(h\ k\ l)|^2}} = f \times \sqrt{n} \text{ and } \overline{|F(h\ k\ l)|^2} = f^2 \times n$$

Combining the effect of the unit cell volume V and $|F(h\ k\ l)|$ in the scattering Eq. (4.32) leads to

$$\overline{I(\text{int.}, h\ k\ l)} \text{ is proportional to } \frac{\overline{|F(h\ k\ l)|^2}}{V^2} = \frac{f^2}{V^2} \times n \quad (4.36)$$

If, through a reorganization of the molecular packing, a unit cell becomes two times as large and the number of molecules per unit cell doubles, V would become $2V$ and $n \rightarrow 2n$. Applying Eq. (4.36) results in an average intensity for the reflected beams that is half the original one. Note, however, that the total scattered intensity remains the same because the number of reflections doubles.

4.16.3. Correction of the Measured Intensity

In Eq. (4.32) $(\lambda^3/\omega V^2) \times (e^2/mc^2)^2 \times V_{cr} \times I_0$ is a constant for a given experiment. The intensity $|F(h\ k\ l)|^2$ is obtained on a relative scale by calculating I (int., $h\ k\ l)/(L \times P \times Tr)$. This is called "correction" of the measured intensity for L, P, and Tr with Tr $= 1 - $ A. In the following the resulting intensity will be called

$$I(h\ k\ l) = |F(h\ k\ l)|^2 \text{ on relative scale} \qquad (4.37)$$

The correction factors L and P are usually incorporated into the software package for the processing of the intensity data. Whether correction for absorption is required depends on the shape of the crystal, the wavelength, and the diffraction technique. For oscillation pictures the pathlengths of the primary and secondary beam in the crystal are not very different for all reflections on a particular image plate. As a result, the absorption is approximately the same for all those reflections. The absorption correction is then incorporated into the scaling factor for the exposures. It should be noted that inhomogeneously distributed mother liquor around the crystal mounted in an X-ray capillary can also have an appreciable absorption effect. If an absorption correction is applied, it is done in an empirical way in which absorption and extinction are considered simultaneously. For an area detector one can, for example, measure the intensity of symmetry-related reflections that follow different paths in the crystal. The solvent around the crystal not only has an effect on the absorption, but also causes a relatively strong but very diffuse ring in the diffraction pattern, to which the disordered solvent inside the crystal also contributes. The ring is in the region corresponding to 3–4 Å resolution, which is somewhat longer than has been found for pure water, which has its maximum near 3 Å (Narten and Levy, 1972).

As was pointed out by Blessing et al. (1996) the strong intensity at 3–4 Å resolution is also the result of ubiquitous atomic distances in proteins.

Summary

It requires some simple mathematics to understand the diffraction of X-rays by a crystal. In this chapter it is presented via the addition of waves, the scattering by a simple two-electron system, by an atom, by one unit cell, and by the arrangement of unit cells in a crystal. The crystal periodicity leads to the Laue diffraction conditions:

$$\mathbf{a} \cdot \mathbf{S} = h$$
$$\mathbf{b} \cdot \mathbf{S} = k$$
$$\mathbf{c} \cdot \mathbf{S} = l$$

and to Bragg's law:

$$2d \sin\theta = \lambda$$

They tell us that for suitable orientations a crystal diffracts an X-ray beam only in certain specific directions. With the reciprocal lattice formalism the required orientations and diffraction directions can be easily constructed.

A crystal structure is not static: the atoms vibrate around an equilibrium position, some more and others less. As a consequence, the intensity of the diffracted beams is weakened. This is expressed in the temperature factor.

The result of an X-ray structure determination is the electron density in the crystal and the fundamental equation for its calculation is

$$\rho(x\ y\ z) = \frac{1}{V}\sum_h\sum_k\sum_l |F(h\ k\ l)|\exp[-2\pi i(hx + ky + lz) + i\alpha(h\ k\ l)]$$

In this equation $\rho(x\ y\ z)$ is expressed as a Fourier transformation of the structure factors $\mathbf{F}(h\ k\ l)$. The amplitude of these structure factors is obtained from the intensity of the diffracted beam after application of certain correction factors: $I(h\ k\ l) = |F(h\ k\ l)|^2$. The phase angles $\alpha(h\ k\ l)$ cannot be derived in a straightforward manner, but can be found in an indirect way, which will be discussed in later chapters.

Chapter 5
Average Reflection Intensity and Distribution of Structure Factor Data

5.1. Introduction

A quick glance through this chapter indicates that it is short but that it is mainly of a mathematical nature. However, it is not as difficult as it seems.

In Section 4.16.2 we calculated the average reflection intensity for a structure consisting of identical atoms. In this chapter we shall extend this calculation to structures composed of different atoms and use the result to place the experimentally determined intensities on an absolute scale[1] and obtain a rough estimate of the temperature factor. In later chapters we shall need to know not only the average intensity, but also the probability distribution of the structure factors and their amplitudes. In the derivation of these distribution functions the Gaussian distribution function, also called the Gauss error function, plays an important role. This function is now presented.

The Gauss Error Function

Two Gauss error curves are drawn in Figure 5.1. They obey the equation

$$f(x) = \frac{1}{\sigma\sqrt{2\pi}} \exp\left[-\frac{(x-m)^2}{2\sigma^2}\right] \tag{5.1}$$

---

[1] Intensities are on absolute scale if the amplitudes of the structure factors $|F| = \sqrt{I}$ are expressed in electrons.

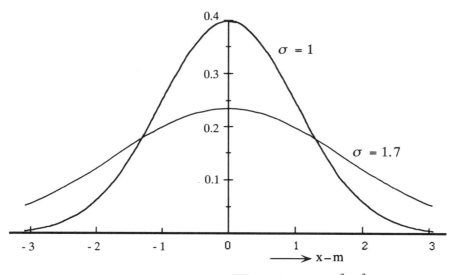

Figure 5.1. Gauss error functions: $(1/\sigma \sqrt{2\pi}) \exp[-(x - m)^2/2\sigma^2]$ plotted as a function of $x - m$ for $\sigma = 1$ and $\sigma = 1.7$.

We notice the following:

1. The probability of finding a value between x and $x + dx$ is equal to $f(x)\, dx$.

2. Because it is certain that x lies somewhere between $+\infty$ and $-\infty$, the probability of finding x between $+\infty$ and $-\infty$ is 1 and, therefore,

$$\int_{-\infty}^{+\infty} f(x)\, dx = 1$$

This is called normalization of the function $f(x)$.

3. For $x = m$ the value of the function is given by $f(m) = 1/\sigma \sqrt{2\pi}$.

4. The mean value of x is m: $\bar{x} = m$. This follows directly from the symmetry of the function $f(x)$ around m.

5. The spread of the curve is expressed in the variance σ^2 of x, which is defined as the average value of $(x - m)^2$:

$$\overline{(x - m)^2} = \sigma^2 = \int_{x=-\infty}^{+\infty} (x - m)^2 f(x)\, dx$$

σ is called the standard deviation. Figure 5.1 shows that the width of the curve increases with σ, as expected. The variance σ^2 in Eq. (5.1), is calculated for data x with a frequency distribution $f(x)$ and assuming random errors in the data.

6. In a large set of data with random errors, only a few measurements occur far away from the center, because the curve falls off on either side of the maximum. For a distance $\geqslant 3\sigma$ from m, the chance of finding a measurement is only 2.7‰. Therefore, it is assumed that a measurement greater than 3σ from m is significantly different from m.

7. If the frequency distribution is not known, we may still have a number of measurements. In such a case, the average $\bar{x}$ of the observations is determined, rather than the mean value m. The variance is then defined as:

$$\sigma^2 = \frac{\sum_i (x_i - \bar{x})^2}{N - 1}$$

where N is the number of measurements. The denominator is $N - 1$ and not N, because one degree of freedom is lost in calculating $\bar{x}$ from the measurements.

As an example in which the Gaussian distribution is used, we consider the measurement of an X-ray reflection intensity. Suppose that its actual value is I_{true}, but that values I_i are obtained by measuring the intensity a number of times or by considering the intensity of symmetry-related reflections. Because of experimental errors the values I_i show deviations $I_i - I_{true}$ from the actual value. If only random errors occur, the intensities observed for the particular reflection obey the Gauss error function, which for the present example is

$$f(I) = \frac{1}{\sigma[I]\sqrt{2\pi}} \exp\left[-\frac{(I - I_{true})^2}{2(\sigma[I])^2} \right]$$

in which $(\sigma[I])^2$ is the average of $(I - I_{true})^2$. Because in practice the number of measurements is too small to obtain the complete distribution function, we have to be satisfied with best estimates of I_{true} and $\sigma[I]$. For I_{true} the best estimate is the average value $\bar{I}$ of the measured intensities and for $\sigma[I]$ it is given by

$$\sigma_e[I] = \left[\frac{\sum_i (I_i - \bar{I})^2}{N - 1} \right]^{1/2}$$

In practice, the estimated standard deviation σ_e(ESD) is usually called σ. Outliers which differ more than $3\sigma[I]$ from the mean value are usually rejected.

For a small number of intensity measurements σ_e is not a good estimate for σ. However, even if there is only one measurement of the intensity, a standard deviation for the intensity I can still be given. This is based on counting statistics and the estimated standard deviation is given as $\sigma_e[I] = \sqrt{I}$; this is explained in Appendix 3.

The precision of the mean $\bar{I}$ is given by the standard error $\sigma[I]/\sqrt{N}$. The standard deviation is also frequently used in electron density plots. In such a plot, peak heights are often given as a multiple of the standard deviation. For instance, the electron density ρ of a significant peak should be at least $3 \times \sigma[\rho]$. Here, $\sigma[\rho]$ is obtained as:

$$\sigma[\rho] = \left[\frac{\sum_{i=1}^{N} \{\rho(x_i \, y_i \, z_i) - \bar{\rho}\}^2}{N} \right]^{1/2}$$

5.2. Average Intensity; Wilson Plots

Early in the process of determining a crystal structure it is possible to obtain a rough estimate of the value of the temperature factor and of the factor required for putting the intensities $I(\mathbf{S})$ on an absolute scale. To this end we calculate the average intensity for a series of reflections $\mathbf{S}$, starting from the expression for the structure factor [Eq. (4.3)]:

$$\mathbf{F(S)} = \sum_{i=1}^{n} f_i \exp[2\pi i \mathbf{r}_i \cdot \mathbf{S}]$$

where the scattering factor f_i of atom i includes the effect of thermal motion, and n is the number of atoms in the unit cell. On an absolute scale the intensity is given by

$$I(\text{abs}, \mathbf{S}) = \mathbf{F(S)} \cdot \mathbf{F^*(S)} = |F(\mathbf{S})|^2 = \sum_i \sum_j f_i f_j \exp[2\pi i (\mathbf{r}_i - \mathbf{r}_j) \cdot \mathbf{S}]^2$$

Suppose that we consider a series of reflections for which $\mathbf{S}$ varies so strongly that the values for the angles $[2\pi(\mathbf{r}_i - \mathbf{r}_j) \cdot \mathbf{S}]$ are distributed evenly over the range $0 - 2\pi$ for $i \neq j$. Then, the average value for all terms with $i \neq j$ will be zero. Only the terms with $i = j$ remain, and we obtain

$$\overline{|F(\mathbf{S})|^2} = \overline{I(\text{abs}, \mathbf{S})} = \sum_i f_i^2 \qquad (5.2)$$

This result is true for non-centric reflections (any phase angle) as well as for centric reflections (phase angles $0°$ or $180°$). We presented this result previously (Section 4.16.2) for a structure with identical atoms. There the equation was not rigorously derived, but was obtained by comparison with Brownian motion. To obtain a rough estimate of the temperature

---

[2] $\mathbf{F^*(S)} = \mathbf{F(-S)}$ is the conjugate complex of $\mathbf{F(S)}$

factor and of the scale factor, it is assumed that all atoms in the cell have the same isotropic thermal motion, or,

$$f_i^2 = \exp\left[-2B\frac{\sin^2\theta}{\lambda^2}\right] \times (f_i^0)^2$$

where f_i^0 is the scattering factor of atom i at rest. Comparison of the calculated values $\overline{I(\text{abs}, \mathbf{S})}$ with the experimental data $\overline{I(\mathbf{S})}$ requires a scale factor C:

$$\overline{I(\mathbf{S})} = C \times \overline{I(\text{abs}, \mathbf{S})} = C \exp\left[-2B\frac{\sin^2\theta}{\lambda^2}\right]\sum_i (f_i^0)^2$$

Both f_i^0 and the temperature factor depend on $(\sin\theta)/\lambda$. Therefore average intensities are calculated for reflections in shells of (almost) constant $(\sin\theta)/\lambda$. To determine B and C the equation is written in the form:

$$\ln\frac{\overline{I(\mathbf{S})}}{\sum_i (f_i^0)^2} = \ln C - 2B\frac{\sin^2\theta}{\lambda^2} \tag{5.3}$$

and $\ln \overline{I(\mathbf{S})}/\Sigma_i(f_i^0)^2$ is plotted against $(\sin^2\theta)/\lambda^2$. The result should be a straight line—the so-called Wilson plot (Wilson, 1942)—from which both the temperature factor and the absolute scale of the intensities can be derived (Figure 5.2). For proteins, the Wilson plot does not give a fully accurate result because the condition that the angles $[2\pi(\mathbf{r}_i - \mathbf{r}_j) \cdot \mathbf{S}]$ are distributed evenly over the $0 - 2\pi$ range is not fulfilled for shells of reflections with (almost) constant $(\sin\theta)/\lambda$ (see below).

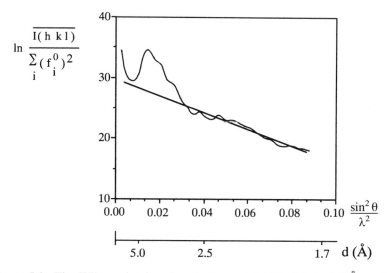

Figure 5.2. The Wilson plot for phospholipase A2 with data to 1.7 Å resolution. Only beyond 3 Å resolution is it possible to fit the curve to a straight line.

The Wilson Plot for a Protein Structure

For practical purposes, the condition that the angles $2\pi(\mathbf{r}_i - \mathbf{r}_j) \cdot \mathbf{S}$ are distributed evenly over the 0 to 2π range is replaced by the more relaxed condition that the range in the angular values should at least be 2π. In proteins, the shortest distance encountered between bonded atoms is approximately $1.5\,\text{Å}$. For this distance and given $\mathbf{S}$, the scalar product varies from $-3\pi|S|$ ($\mathbf{S}$ and $\mathbf{r}_i - \mathbf{r}_j$ antiparallel) to $+3\pi|S|$ ($\mathbf{S}$ and $\mathbf{r}_i - \mathbf{r}_j$ parallel). The condition that the range $6\pi|S|$ should be at least 2π requires that $|S| = 1/d$ is not smaller than $\frac{1}{3}\,\text{Å}^{-1}$ or $d < 3\,\text{Å}$ (Figure 5.3).

Therefore, we conclude that the Wilson plot gives reliable information only if reflections are used corresponding to Bragg spacings less than $3\,\text{Å}$. For longer distances Figure 5.2 shows deviations from the straight line. For instance, around $d = 4\,\text{Å}$ or $|S| = 0.25\,\text{Å}^{-1}$, there is a maximum that can be ascribed to the presence of many nonbonded distances around $4\,\text{Å}$ in the protein structure. Therefore, the angles $2\pi(\mathbf{r}_i - \mathbf{r}_j) \cdot \mathbf{S}$, with $i \neq j$, tend to cluster around 2π for $|S| = 0.25\,\text{Å}^{-1}$ and terms $\exp[2\pi i(\mathbf{r}_i - \mathbf{r}_j) \cdot \mathbf{S}]$, $i \neq j$, do not average to zero.

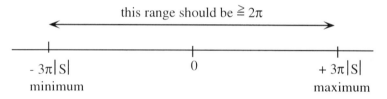

Figure 5.3. The range for $2\pi(\mathbf{r}_i - \mathbf{r}_j) \cdot \mathbf{S}$ in protein structures. The shortest value for $(\mathbf{r}_i - \mathbf{r}_j)$ is $1.5\,\text{Å}$. Therefore, the minimum value for $2\pi(\mathbf{r}_i - \mathbf{r}_j) \cdot \mathbf{S}$ is $-3\pi|S|$ and the maximum value $+3\pi|S|$. For longer distances, the border for the minimum moves to the left and for the maximum to the right; the problem is then less serious.

5.3. The Distribution of Structure Factors **F** and Structure Factor Amplitudes $|F|$

With the information that $\overline{|F(\mathbf{S})|^2} = \Sigma_i f_i^2$ [Eq. (5.2)] it is now easy to derive distribution functions for the structure factors and their amplitudes. However, we shall not need this information before Chapter 8 and, therefore, as an alternative, you can skip it now and read it later.

First we consider for general reflections the probability $p(\mathbf{F})d(\mathbf{F})$ of finding a structure factor between **F** and $\mathbf{F} + d\mathbf{F}$. In the two-dimensional

Argand diagram **F** is expressed as a vector with real component A and imaginary component B: $|F|^2 = A^2 + B^2$. Let A have the probability $p(A)d(A)$ of lying between A and $A + dA$ and similarly for B. Because the components A and B are independent of each other, the probability for **F** to point to the volume element $d\mathbf{F} = dA \times dB$ is given by

$$p(\mathbf{F})d(\mathbf{F}) = p(A)d(A) \times p(B)d(B)$$

A protein structure consists of a great many light atoms with an occasional heavy atom. For such a structure a Gaussian distribution can be assumed for the components A and B, if the reflections considered show a sufficiently large variation in their diffraction vectors **S**. Therefore, the distribution functions of A and B are determined by the average values $\bar{A}$ and $\bar{B}$, and by the variances $\sigma^2[A]$ and $\sigma^2[B]$. Because within the series of reflections all directions for the atomic scattering factors $\mathbf{f}_i$ are equally possible, we obtain:

$$\bar{A} = \bar{B} = 0$$

$$\sigma^2[A] = \overline{(A - \bar{A})^2} = \overline{A^2} \quad \text{and} \quad \sigma^2[B] = \overline{(B - \bar{B})^2} = \overline{B^2}$$

$$\overline{A^2} = \overline{B^2} = \frac{1}{2}\overline{|F|^2} = \frac{1}{2}\sum_i f_i^2 \quad \text{(Eq. 5.2)}$$

$$\sigma^2[A] = \sigma^2[B] = \frac{1}{2}\overline{[|F|]^2} = \frac{1}{2}\sum_i f_i^2$$

$$p(\mathbf{F})d(\mathbf{F}) = \frac{1}{2\pi \times \sigma(A) \times \sigma(B)} \exp\left[-\frac{A^2 + B^2}{2\sigma^2(A)}\right] d(A)\, d(B)$$

$$= \frac{1}{\pi \times \sum_{i=1}^{n} f_i^2} \exp\left[-\frac{|F|^2}{\sum_{i=1}^{n} f_i^2}\right] d(\mathbf{F}) \tag{5.4}$$

It is now easy to derive $p(|F|)d(|F|)$. This is the probability that the magnitude of a structure factor lies between $|F|$ and $|F| + d|F|$, or that the end of vector **F** in the Argand diagram is in an annulus between $|F|$ and $|F| + d|F|$:

$$p(|F|)\, d|F| = p(\mathbf{F}) \times 2\pi|F|\, d|F| = \frac{2|F|}{\sum_{i=1}^{n} f_i^2} \exp\left[-\frac{|F|^2}{\sum_{i=1}^{n} f_i^2}\right] d|F| \tag{5.5}$$

We shall also derive the expression for the probability distribution of centric reflections. Compared with the non-centric reflections we now have $B = 0$ and $A = F$. The average value of $F = 0$ and therefore,

$$\sigma^2[F] = \overline{F^2} = \sum_i f_i^2.$$

$$p(F)\,d(F) = \frac{1}{\sqrt{2\pi \sum_i f_i^2}} \exp\left[-\frac{F^2}{2\sum_i f_i^2}\right] d(F)$$

Note: Because of the decrease of f_i with $(\sin\theta)/\lambda$, the variance $\frac{1}{2}\sum_{i=1}^{n} f_i^2$ depends on $(\sin\theta)/\lambda$ and must be taken in shells of $(\sin\theta)/\lambda$. This problem does not exist if, instead of the structure factors $F(S)$, normalized structure factors $E(S)$ are used (Chapter 6):

$$E(S) = \frac{F(S)}{\left(\sum_i f_i^2\right)^{1/2}}$$

In the calculation of normalized structure factors, the atoms are regarded as points and their scattering is independent of $(\sin\theta)/\lambda$.

Summary

In this chapter we have derived an important equation for the average intensity:

$$\overline{I(\mathrm{abs}, S)} = \sum_i f_i^2$$

We shall need this result frequently. It has already been applied in this chapter in the derivation of distribution functions for the structure factors $F(S)$ and for their amplitudes $|F(S)|$. The Gauss error function played a central role in the derivation of these functions.

Chapter 6
Special Forms of the Structure Factor

6.1. Introduction

In this chapter some special forms of the structure factor will be presented. It is not essential reading to understand the following chapters, but it does provide an introduction to and definitions of *unitary structure factors* and *normalized structure factors*. The chapter is put in this position in the book because the material can be easily understood using the results presented in the section on the Wilson plot (Section 5.2). However, if this is your first introduction to protein X-ray crystallography, you can skip this chapter for the time being.

6.2. The Unitary Structure Factor

For statistical studies of structure factor amplitude distributions the normal form of the structure factor:

$$\mathbf{F}(\mathbf{S}) = \sum_j f_j \exp[2\pi i(\mathbf{r}_j \cdot \mathbf{S})]$$

is not quite suitable. $\mathbf{F}(\mathbf{S})$ decreases with $|S|$ because of the $|S|$ dependence of f_j and because of the temperature factor, and these effects are disturbing and must be eliminated. Therefore, the following modified structure factors have been introduced: the unitary structure factor $\mathbf{U}(\mathbf{S})$ and the normalized structure factor $\mathbf{E}(\mathbf{S})$.

The unitary structure factor is defined as

$$U(\mathbf{S}) = \frac{\mathbf{F}(\mathbf{S})_{pt}}{\sum_j Z_j} \tag{6.1}$$

where Z_j is the atomic number of atom j and $\Sigma_j Z_j = F(000)$. $\mathbf{F}(\mathbf{S})_{pt}$ is the structure factor on an absolute scale with the assumption that the individual scatterers are point atoms. Their scattering is independent of $|S|$ and is equal to the atomic number Z over the entire $|S|$ region. This also excludes any thermal motion. When there is only one type of atom in the unit cell but equal thermal motion for all atoms,

$$\mathbf{F}(\mathbf{S})_{pt} = \frac{Z \times \exp\left[B\left(\sin^2 \theta / \lambda^2\right)\right]}{f} \times \mathbf{F}_{obs} \tag{6.2}$$

where $\mathbf{F}_{obs}$ is the normal structure factor on absolute scale. For more than one type of atom but the same thermal parameter for all atoms, $\mathbf{F}(\mathbf{S})_{pt}$ is taken as

$$\mathbf{F}(\mathbf{S})_{pt} = \frac{\sum_j Z_j \times \exp\left[B\left(\sin^2 \theta / \lambda^2\right)\right]}{\sum_j f_j} \times \mathbf{F}_{obs} \tag{6.3}$$

combining (6.1) and (6.3):

$$U(\mathbf{S}) = \frac{\exp\left[B\left(\sin^2 \theta / \lambda^2\right)\right] \times \mathbf{F}_{obs}}{\sum_j f_j} \tag{6.4}$$

The exponential term eliminates the effect of the temperature factor and the division by $\Sigma_j f_j$ converts the atoms to point atoms.

    Clearly $|U(\mathbf{S})| \leq 1$. For proteins with a great many atoms scattered over the unit cell, the $U(\mathbf{S})$ values are rather small and if probability distributions of structure factors are discussed, it is more convenient to use normalized structure factors. They have the advantage of being independent of scaling factors between sets of reflections.

6.3. The Normalized Structure Factor

The normalized structure factor is

$$E(\mathbf{S}) = \frac{\mathbf{F}(\mathbf{S})}{\left(\sum_j f_j^2\right)^{1/2}} \tag{6.5}$$

Neither f_j nor $\mathbf{F}(\mathbf{S})$ includes the temperature factor.

$$\mathbf{F}(\mathbf{S}) = \mathbf{F}_{obs}(\mathbf{S}) \times \exp\left[B\frac{\sin^2 \theta}{\lambda^2}\right]$$

with $F_{obs}(S)$ including the temperature factor. Both $F(S)$ and $F_{obs}(S)$ are on an absolute scale here. According to Eq. (5.2)

$$\overline{|F(S)|^2} = \sum_j f_j^2$$

and, therefore,

$$\overline{|E(S)|^2} = \frac{\overline{|F(S)|^2}}{\left(\sum_j f_j^2\right)} = \frac{\overline{|F(S)|^2}}{\overline{|F(S)|^2}} = 1$$

Moreover, if $E(S)$ is written as

$$E(S) = \frac{F(S)}{(\overline{|F(S)|^2})^{1/2}}$$

we see that the scale factor is not important because the numerator and the denominator are on the same scale. $E(S)$ can be obtained from the experimental data in the following way:

$$E(S) = \frac{F(S)_{exp} \times \exp[B(\sin^2\theta/\lambda^2)]}{(\overline{|F(S)_{exp}|^2})^{1/2}} \qquad (6.6)$$

Sometimes a complication exists with the use of $E(S)$ values in probability distributions of X-ray intensities. This is caused by the fact that for some groups of reflections the E-values are higher than expected. An example will show this: Suppose the cell has an n-fold symmetry axis along z. For the $(00l)$ reflections only the z-coordinate counts and $2\pi(hx + ky + lz)$ $\rightarrow 2\pi lz$. The cell has N atoms but, because of the symmetry, there are $\frac{N}{n}$ groups, each of n symmetry related atoms. The contribution by one group of symmetry related atoms to $F(00l)$ is $n \times f_j \exp[2\pi i lz_j]$ instead of $f_1 \exp[2\pi i lz_1] + f_2 \exp[2\pi i lz_2] + \ldots f_n \exp[2\pi i lz_n]$. This contribution to the structure factor $F(00l)$ is equal to the contribution by a superatom with scattering factor $n f_j$. If we take the sum over all superatoms $\frac{N}{n}$ and calculate the average $\overline{|F(00l)|^2}$ over all reflections, we obtain: $\overline{|F(00l)|^2} = \sum_{j=1}^{N/n} (n f_j)^2 = n \sum_{j=1}^{N} f_j^2$ instead of $f_1^2 + f_2^2 + \ldots f_N^2 = \sum_{j=1}^{N} f_j^2$. $\overline{|E|^2}$, where the average is over all reflections, always remains 1. To allow statistical comparison of all reflections, the E-values for reflections belonging to the special groups with too high values of E are reduced by a factor $\sqrt{\varepsilon}$. ε is easily found by a procedure proposed by Stewart and Karle (1976). The more general form of $E(S)$ is

$$E(S) = \frac{F(S)_{exp} \times \exp[B(\sin^2\theta/\lambda^2)]}{(\varepsilon \times \overline{|F(S)_{exp}|^2})^{1/2}} \qquad (6.7)$$

Summary

In this chapter two alternative expressions for the structure factor have been introduced:

- The unitary structure factor:

$$U(S) = \frac{F(S)_{exp} \times \exp[B \sin^2\theta/\lambda^2]}{\sum_j f_j}$$

- The normalized structure factor:

$$E(S) = \frac{F(S)_{exp} \times \exp[B \sin^2\theta/\lambda^2]}{(\varepsilon \times \overline{|F(S)_{exp}|^2})^{1/2}}$$

They are more convenient in statistical studies of structure factor amplitude distributions than the common form $F(S)$ of the structure factor.

Chapter 7
The Solution of the Phase Problem by the Isomorphous Replacement Method

7.1. Introduction

As we have seen in Chapter 4 the electron density in a crystal can be obtained by calculating the Fourier summation:

$$\rho(x\ y\ z) = \frac{1}{V}\sum_{hkl}|F(h\ k\ l)|\ \exp[-2\pi i(hx + ky + lz) + i\alpha(h\ k\ l)] \qquad (7.1)$$

in which $|F(h\ k\ l)|$ is the structure factor amplitude of reflection $(h\ k\ l)$, including the temperature factor, and $\alpha(h\ k\ l)$ is the phase angle. x, y, and z are coordinates in the unit cell. From the diffraction pattern the values of $I(h\ k\ l)$ are obtained after applying the correction factors L, P, and A. Because $I(h\ k\ l) = |F(h\ k\ l)|^2$ the amplitudes $|F(h\ k\ l)|$ can be found. Unfortunately, no information is available on the phase angles. In principle, four techniques exist for solving the phase problem in protein X-ray crystallography:

1. The isomorphous replacement method, which requires the attachment of heavy atoms (atoms with high atomic number) to the protein molecules in the crystal.
2. The multiple wavelength anomalous diffraction method. It depends on the presence of sufficiently strong anomalously scattering atoms in the protein structure itself. Anomalous scattering occurs if the electrons in an atom cannot be regarded as free electrons.
3. The molecular replacement method for which the similarity of the unknown structure to an already known structure is a prerequisite.
4. Direct methods, the methods of the future, still in a stage of development toward practical application for proteins.

Molecular replacement, which will be discussed in Chapter 10, is the most rapid method for determining a protein structure. However, it requires the availability of a known model structure, e.g., of a homologous protein. If this does not exist, isomorphous replacement or multiple wavelength anomalous diffraction must be applied. They are the most general methods for determining protein phase angles and are used if nothing is, as yet, known about the three-dimensional structure of the protein.

The multiple wavelength anomalous diffraction method does not depend on the attachment of a heavy atom-containing reagent to the protein, but it does require the presence of an anomalously scattering atom and diffraction data collection at a number of X-ray wavelengths, in general, data collection with synchrotron radiation. This method for phase angle determination will be discussed in Section 9.5.

We shall first discuss the isomorphous replacement method. The initial step in this method requires the attachment of heavy atoms and subsequently the determination of the coordinates of these heavy atoms in the unit cell. A useful role in this process is played by the Patterson function. Therefore, we shall begin by discussing this function and its physical interpretation.

7.2. The Patterson Function

The Patterson function $P(\mathbf{u})$ or $P(u\ v\ w)$ is a Fourier summation with intensities as coefficients and without phase angles, or rather with all phase angles equal to zero.

$$P(u\ v\ w) = \frac{1}{V}\sum_{hkl}|F(h\ k\ l)|^2 \cos[2\pi(h\ u + k\ v + l\ w)] \qquad (7.2)$$

or shorter,

$$P(\mathbf{u}) = \frac{1}{V}\sum_{\mathbf{S}}|F(\mathbf{S})|^2 \cos[2\pi\mathbf{u} \cdot \mathbf{S}] \qquad (7.3)$$

u, v, and w are relative coordinates in the unit cell. To avoid confusion with the coordinates x, y, and z in the real cells, we use u, v, and w in the Patterson cell, which, however, has dimensions identical to the real cell. Note that the coefficients in the summations (7.2) and (7.3) are $|F(h\ k\ l)|^2$ and not $|F(h\ k\ l)|$ as in Eq. (7.1). Because all phase angles are zero in the Patterson function it can be calculated without any previous knowledge of the structure.

Further, it can be shown that the Patterson function $P(\mathbf{u})$ can alternatively be written as

$$P(\mathbf{u}) = \int_{\mathbf{r}_1} \rho(\mathbf{r}_1) \times \rho(\mathbf{r}_1 + \mathbf{u})\,dv \qquad (7.4)$$

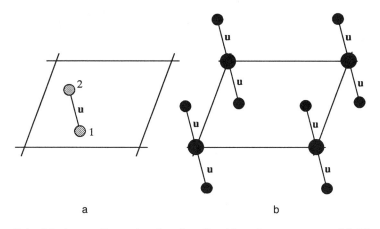

Figure 7.1. (a) A two-dimensional unit cell with only two atoms. (b) The corresponding Patterson cell.

The integration is for r_1 over all positions in the real unit cell. Assuming for the moment that this is true we can use this form of $P(u)$ to understand its physical interpretation: Move through the *real unit cell* with a vector u, multiply in every position of u the electron density ρ at the beginning of u (in position r_1) with the electron density at the end of vector u (in position $r_1 + u$) and take the integral of these values. Only if nonzero electron density is present at both the beginning and the end of u will the result of the multiplication be nonzero. What this leads to can best be understood from Figure 7.1, in which the real cell contains only two atoms.

$\rho(r_1) \times \rho(r_1 + u)$ has a significant value only if u starts in atom 1 and ends in atom 2, or the other way around. In the corresponding Patterson cell the vector u starts in the origin of the cell and points either in one direction (atom 1 to atom 2) or in the opposite direction (atom 2 to atom 1). A peak in a Patterson map at position u (or $u\ v\ w$), therefore, means that in the real cell atoms occur at a certain position x, y, and z, and at the position $x + u$, $y + v$, and $z + w$, or $x - u$, $y - v$, and $z - w$. So far the real atomic positions are not known, but the vectorial distance between them is clear from the Patterson map.

In simple structures with a limited number of atoms, the atomic positions can be derived fairly straightforwardly from the Patterson map. But this is impossible for complicated structures, like proteins. If a real unit cell contains N atoms, the corresponding Patterson map will show N^2 peaks, because one can draw N vectors from each atom. However, N vectors of the total number of N^2 vectors will have a length 0, because they go from an atom to the same atom. Therefore, the highest peak in a Patterson map is situated in the origin of the cell (Figure 7.2). The

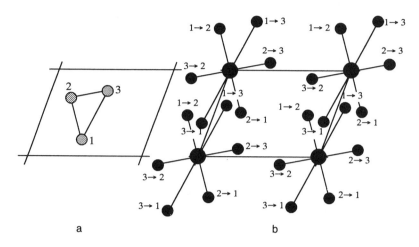

Figure 7.2. (a) A two-dimensional unit cell with three atoms. (b) The corresponding Patterson map. Note the large increase in the number of Patterson peaks compared with Figure 7.1. The total number of peaks is N^2, but the N self-peaks overlap at the origin and, therefore, $N(N - 1)$ nonorigin peaks are found in a Patterson map. Because of the centrosymmetry in the map the number of unique peaks is $[N(N - 1)]/2$; in this figure $1 \rightarrow 2$, $1 \rightarrow 3$, and $2 \rightarrow 3$ are unique peaks.

number of nonorigin peaks is $N^2 - N = N(N - 1)$. If the unit cell of a protein crystal contains 5000 nonhydrogen atoms, then the number of Patterson peaks would be 25×10^6. It is clear that such a Patterson map is uninterpretable. If, however, a limited number of heavy atoms in the large unit cell must be located, the Patterson function is extremely useful as you can see in Figure 7.4c.

We have presented two different expressions for the Patterson function: Eq. (7.2) or (7.3), which tells us how to calculate it and Eq. (7.4), which facilitates understanding of the physical meaning of the Patterson function.

We shall now prove that

$$P(\mathbf{u}) = \int_{\mathbf{r}_1} \rho(\mathbf{r}_1) \times \rho(\mathbf{r}_1 + \mathbf{u})dv = \frac{1}{V}\sum_{\mathbf{S}}|F(\mathbf{S})|^2 \cos[2\pi\mathbf{u} \cdot \mathbf{S}] \qquad (7.5)$$

and start with

$$\rho(\mathbf{r}_1) = \frac{1}{V}\sum_{\mathbf{S}}|F(\mathbf{S})| \exp[-2\pi i \mathbf{r}_1 \cdot \mathbf{S} + i\alpha(\mathbf{S})] \qquad (7.6)$$

$$\rho(\mathbf{r}_2) = \rho(\mathbf{r}_1 + \mathbf{u}) = \frac{1}{V}\sum_{\mathbf{S}'}|F(\mathbf{S}')| \exp[-2\pi i(\mathbf{r}_1 + \mathbf{u}) \cdot \mathbf{S}' + i\alpha(\mathbf{S}')] \qquad (7.7)$$

In Eq. (7.7) for $\rho(\mathbf{r}_2)$ we use $\mathbf{S}'$, just to distinguish it from $\mathbf{S}$ in the equation for $\rho(\mathbf{r}_1)$.

$$\rho(\mathbf{r}_1) \times \rho(\mathbf{r}_1 + \mathbf{u}) = \frac{1}{V^2} \sum_S \sum_{S'} |F(S)|\,|F(S')| \exp[-2\pi i \{\mathbf{r}_1 \cdot (S + S')$$
$$+ \mathbf{u} \cdot S'\} + i\alpha(S) + i\alpha(S')] \tag{7.8}$$

In Section 4.11 we have seen that $|F(h\,k\,l)| = |F(\bar{h}\,\bar{k}\,\bar{l})|$ or $|F(S)| = |F(-S)|$ and also that $\alpha(h\,k\,l) = -\alpha(\bar{h}\,\bar{k}\,\bar{l})$ or $\alpha(S) = -\alpha(-S)$. Therefore, in Eq. (7.8) the coefficients $|F(S)|\,|F(S')|$ are equal for S and $-S$, as well as for S' and $-S'$, whereas the exponential terms have just the opposite sign. This simplifies Eq. (7.8) to a summation of cos terms:

$$\rho(\mathbf{r}_1) \times \rho(\mathbf{r}_1 + \mathbf{u}) = \frac{1}{V^2} \sum_S \sum_{S'} |F(S)|\,|F(S')| \cos[2\pi \{\mathbf{r}_1 \cdot (S + S') + \mathbf{u} \cdot S'\}$$
$$- \alpha(S) - \alpha(S')] \tag{7.9}$$

The next step is an integration with $\mathbf{r}_1$ over the entire unit cell. In other words, $\mathbf{r}_1$ assumes different lengths and different directions:

$$P(\mathbf{u}) = \int_{\mathbf{r}_1} \rho(\mathbf{r}_1) \times \rho(\mathbf{r}_1 + \mathbf{u})\,dv$$

$$= \frac{1}{V^2} \sum_S \sum_{S'} |F(S)|\,|F(S')| \int_{\mathbf{r}_1} \cos[2\pi \{\mathbf{r}_1 \cdot (S + S') + \mathbf{u} \cdot S'\}$$
$$- \alpha(S) - \alpha(S')]\,dv \tag{7.10}$$

The integration with $\mathbf{r}_1$ over the entire unit cell means that the constant vector $\mathbf{u}$ must move through the unit cell and has its beginning in every position $\mathbf{r}_1$ of the unit cell. Since $\mathbf{r}_1$ can have different lengths and many different directions, the angles

$$2\pi \{\mathbf{r}_1 \cdot (S + S') + \mathbf{u} \cdot S'\} - \alpha(S) - \alpha(S')$$

can assume all values between 0 and 2π for $S + S' \neq 0$. Therefore, the integration

$$\int_{\mathbf{r}_1} \cos[2\pi \{\mathbf{r}_1 \cdot (S + S') + \mathbf{u} \cdot S'\} - \alpha(S) - \alpha(S')]\,dv$$

will in general lead to zero. However, in the special case when $S + S' = 0$ or $S' = -S$ and $\alpha(S') = -\alpha(-S') = -\alpha(S)$, a nonzero value will result:

$$P(\mathbf{u}) = \frac{1}{V^2} \sum_S |F(S)|^2 \cos[2\pi \mathbf{u} \cdot S] \int_{\mathbf{r}_1} dv = \frac{1}{V} \sum_S |F(S)|^2 \cos[2\pi \mathbf{u} \cdot S] \tag{7.11}$$

because $\int_{\mathbf{r}_1} dv = V$.

The Patterson function has the following properties:

1. The Patterson map has peaks at end points of vectors $\mathbf{u}$ equal to vectors *between* atoms in the real cell.

2. For every pair of atoms in the real cell, there exists a unique peak in the Patterson map.

3. A Patterson map is always centrosymmetric. Therefore, $P(\mathbf{u})$ (eq. 7.11) can also be written in exponential form:

$$P(\mathbf{u}) = \frac{1}{V}\sum_{S}|F(S)|^{2}\exp[2\pi i \, \mathbf{u}.S], \text{ or as}$$

$$P(\mathbf{u}) = \frac{1}{V}\sum_{S}|F(S)|^{2}\exp[-2\pi i \, \mathbf{u}.S]$$

This is allowed because $\exp[2\pi i \, \mathbf{u}.S] = \cos[2\pi \mathbf{u}.S] + i \sin[2\pi \mathbf{u}.S]$ and

$$\exp[-2\pi i \mathbf{u}.S] = \cos[2\pi \, \mathbf{u}.S] - i \sin[2\pi \mathbf{u}.S]$$

and the sin term disappears because of the centrosymmetry.

4. Screw axes in a real cell become normal axes in a Patterson cell. We shall prove this for a 2-fold screw axis along y.

In Section 4.12.2 we showed that for a 2-fold screw axis along y the diffraction pattern has a 2-fold axis along y:

$$I(h \; k \; l) = I(\bar{h} \; k \; \bar{l})$$

We must now prove that $P(u \; v \; w) = P(\bar{u} \; v \; \bar{w})$.

$$P(u \; v \; w) = \frac{1}{V}\sum_{hkl}|F(h \; k \; l)|^{2} \cos[2\pi(hu + kv + lw)]$$

This is exactly equal to

$$P(u \; v \; w) = \frac{1}{V}\sum_{hkl}|F(\bar{h} \; k \; \bar{l})|^{2} \cos[2\pi(\bar{h}u + kv + \bar{l}w)]$$

because the summation is still over all reflections $h\,k\,l$. We know already that $I(\bar{h} \; k \; \bar{l}) = I(h \; k \; l)$ or $|F(\bar{h} \; k \; \bar{l})|^{2} = |F(h \; k \; l)|^{2}$. Therefore, we can write $P(u \; v \; w)$ as

$$P(u \; v \; w) = \frac{1}{V}\sum_{hkl}|F(h \; k \; l)|^{2} \cos[2\pi(h\bar{u} + kv + l\bar{w})]$$

and this is precisely $P(\bar{u} \; v \; \bar{w})$. This proves that

$$P(u \; v \; w) = P(\bar{u} \; v \; \bar{w}).$$

5. Symmetry elements can cause a concentration of peaks in certain lines or planes: "Harker lines" or "Harker planes." Examples are given in Figures 7.3 and 7.4.

6. The Patterson function

$$P(\mathbf{u}) = \int_{\mathbf{r}_1} \rho(\mathbf{r}_1) \times \rho(\mathbf{r}_1 + \mathbf{u}) \, dv \qquad (7.12)$$

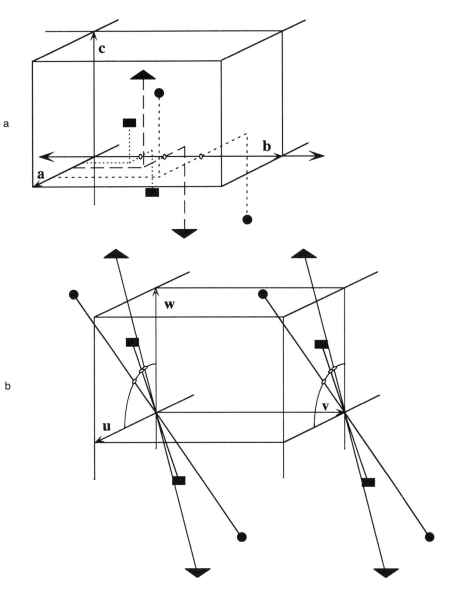

Figure 7.3. (a) A unit cell with a 2-fold axis along y; (b) the corresponding Patterson cell with Harker peaks in the $(u\ 0\ w)$ plane.

is the convolution of the structure and its inverse. The mathematical definition of the convolution $C(x)$ of two real, periodic functions $f(h)$ and $g(h)$ is

$$C(x) = \int_{\eta=0}^{1} f(\eta)g(x - \eta)\, d\eta \qquad (7.13)$$

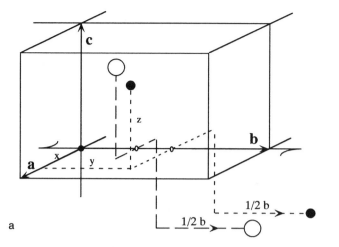

a

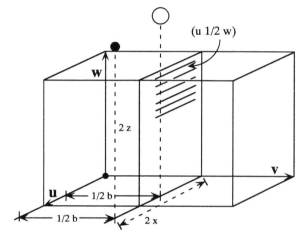

b

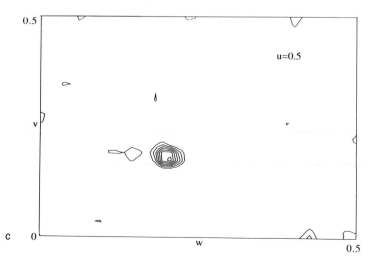

c

The Patterson function can be put into this form as follows: Replacing $r_1 + u$ by r'' in Eq. (7.12) gives

$$P(u) = \int_{r''} \rho(r'' - u)\rho(r'') \, dv \qquad (7.14)$$

With $\rho(r'' - u) = \rho_{inv}(u - r'')$ where $\rho_{inv}(u - r'')$ is the electron density distribution of the inverse structure, (7.14) transforms into[1]

$$P(u) = \int_{r''} \rho(r'')\rho_{inv}(u - r'') \, dv \qquad (7.15)$$

which is of the same form as (7.13) and this proves property 6.

A general property of the Fourier transform of a convolution is the following: If $F(h)$ is the transform of $f(\eta)$ in Eq. (7.13) and $G(h)$ is the transform of $g(\eta)$, then the product $F(h) \, G(h)$ is the Fourier transform of $C(x)$. Application to Eq. (7.15) gives the following: The product of the transform of $\rho(r'')$ [which is $F(S)$] and the transform of $\rho_{inv}(r'')$ [which is $F^*(S)$] is equal to the transform of $P(u)$ and, therefore, the transform of $P(u)$ is equal to $F(S)F^*(S) = |F(S)|^2$. $F^*(S)$ is the conjugate complex of $F(S)$; if $F = |F| \exp[i\alpha]$, then its conjugate complex is $F^* = |F| \exp[-i\alpha]$.

7. In locating Patterson peaks of heavy atoms in the isomorphous replacement method, it is useful to realize that the height of a peak is proportional to the product of the atomic numbers of the atoms that are responsible for the peak.

Convolution explained

The mathematical definition of a convolution is

$$C(x) = \int_{\eta=0}^{1} f(\eta)g(x - \eta)d\eta$$

f and g are both functions of x. The mathematical process can be illustrated with Figure 7.4.1. A fixed value of x is chosen and the variable is η which

---

[1] The electron density of an inverse structure at r is equal to the electron density in the original structure at $-r$: $\rho_{inv}(r) = \rho_{orig.}(-r)$ and $F_{inv}(S) = F^*_{orig.}(S)$ where $F^*_{orig.}(S)$ is the conjugate complex of $F_{orig.}(S)$.

◄ ─────────────

Figure 7.4. (a) A unit cell with a 2-fold screw axis along y; (b) in the $(u \; 1/2 \; w)$ plane of the corresponding Patterson cell a concentration of peaks is found; only two Harker peaks are indicated; (c) a realistic example showing the Harker section at $u = 0.5$ with $0 < v < 0.5$ and $0 < w < 0.5$. The high density peak indicates the end of the vector between the mercury atoms from a mercury-containing reagent attached to the protein hevamine. The crystals belong to space group $P2_12_12_1$. (Source: Anke Terwisscha van Scheltinga.)

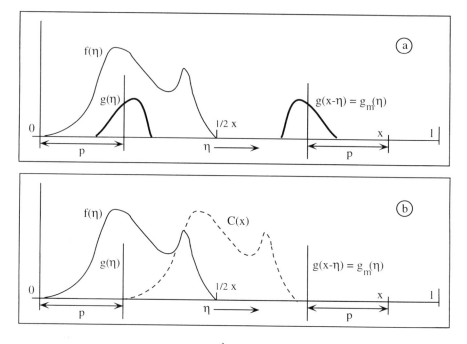

Figure 7.4.1 The convolution $C(x) = \int\limits_{\eta=0}^{1} f(\eta)g(x - \eta)\,d\eta$ of two functions f(η) and g(x-η). The moving variable for a fixed value of x is η. In (a), besides the functions f(η) and g(η), also g(x-η) is drawn for an arbitrarily chosen value of x. g(x-η) is obtained from g(η) by taking its mirror image with respect to the position 1/2 × x; it is called $g_m(\eta)$. In (a) C(x) is not drawn, but it is in (b). In that figure g(η) is a delta function. The convolution C(x) of this delta function with f(η) is the dashed line. C(x) is here a true representation of f(η) because g(η) is a delta function. The less sharp g(η) is, the more blurred C(x) will be.

moves from $\eta = 0$ to $\eta = 1$. In Figure 7.4.1.a arbitrary functions f(η) and g(η) are drawn. The convolution does not require g(η) but g(x-η). For the arbitrarily chosen value of x, the function g(x-η) is found on the right in Figure 7.4.1.a as the mirror image of g(η) with respect to the position 1/2 × x. This new function differs from g, and therefore it will be called $g_m(\eta)$. In the convolution process, f(η) and $g_m(\eta)$ must be multiplied for every value of η between 0 and 1. Summation of the products gives the value of C(x) for the chosen value of x. It is clear for the chosen value of x in Figure 7.4.1.a that the result will be zero because f(η) and $g_m(\eta)$ do not overlap. However, if x moves to a sufficiently small value, the two functions do overlap and the resulting C(x) will be different from zero.

A special situation exists if g(η) is a needle-sharp function, a delta function (Section 4.10) (Figure 7.4.1.b). If x moves to smaller values, the

$g_m(\eta)$ needle starts to scan $f(\eta)$ and the result will be an undistorted image of $f(\eta)$ but displaced over p to the right of the original $f(\eta)$. If the needle is blunt instead of sharp, the image of $f(\eta)$ will be blurred. This is comparable to the scanning in the atomic force microscope, in which an ideally-sharp probe gives an undistorted image of the specimen. With a blunt probe the image is the convolution of the probe shape and the object. Another example is the electron density of atoms with thermal motion. The density is the convolution of the fixed-atom structure and the function describing the thermal motion. Convolution is illustrated here with rather simple functions f and g, but the principle is the same for more complicated functions.

A convolution is often represented by the symbol "$*$," which is on the line: $C(x) = f(\eta) * g(\eta)$. The convolution symbol should not be mistaken for the symbol indicating the conjugate complex "$*$"; this is above the line.

7.3. The Isomorphous Replacement Method

Application of the isomorphous replacement method requires the X-ray diffraction pattern of the native protein crystal as well as that of the crystal of at least one heavy atom derivative. For perfect isomorphism the conformation of the protein, the position and orientation of its molecules as well as the unit cell parameters in the native and in the derivative crystals must be exactly the same. The intensity differences between the native and the other patterns are then exclusively due to the attached heavy atoms. For an example see Figure 7.5. From these differences the positions of the heavy atoms can be derived either manually or by an automatic Patterson search procedure, for instance SHELXS (Sheldrick et al., 1993); this is the starting point for the determination of the protein phase angles. Perfect isomorphism hardly ever occurs. Errors due to nonisomorphism are usually more serious than errors in the X-ray data. However, a modest change in the protein structure is not a great obstacle. Nonisomorphism often presents itself as a change in the cell dimensions. In a quick data collection one can determine whether the heavy atom has attached itself to the protein, by comparing the intensities of the reflections with those of the native crystal as well as whether this has seriously affected the cell dimensions and the quality of the diffraction pattern. A change in the cell dimensions of $d_{min}/4$, where d_{min} is the resolution limit, is tolerable. In principle nonisomorphism can occur without expressing itself in the cell dimensions, for instance, if a slight rotation of the protein molecules has occurred as a consequence of the heavy atom binding. This will later result in a poor refinement of the heavy atom parameters.

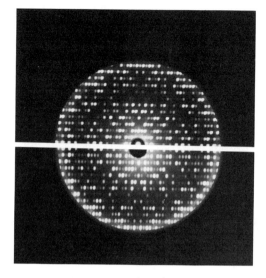

Figure 7.5. A comparison of the diffraction photographs of the same reciprocal lattice plane for a native papain crystal and a heavy atom derivative in which one mercury atom was attached to each protein molecule. Appreciable differences in intensity between corresponding diffraction spots can be seen.

The isomorphous replacement method requires the following steps:

1. Preparation of at least one, but preferably a few heavy atom containing derivatives of the protein in the crystalline state. A first check for isomorphism is measuring the cell dimensions.
2. X-ray intensity data must be collected for crystals of the native protein as well as for crystals of the derivatives.
3. Application of the Patterson function for the determination of the heavy atom coordinates.
4. Refinement of the heavy atom parameters and calculation of the protein phase angles.
5. Calculation of the electron density of the protein.

7.3.1. The Attachment of Heavy Atoms

The search for heavy atom derivatives is still basically an empirical method and very often dozens of reagents are tried before a few suitable ones are found. The preferred method is by soaking the protein crystal in a solution of the reagent. The composition of this solution is identical to the mother liquor, except for the added reagent, often with a slight increase of precipitant concentration. Cocrystallization is not commonly used, because of the risk that crystals will not grow or grow nonisomorphously. However, for covalently attached reagents, cocrystallization is sometimes an advantage because of the better control over the stoichiometry and the ability to prevent excess binding.

The soaking procedure depends on the existence of relatively wide pores in the crystal, wide enough to allow the reagent to diffuse into the crystal and to reach the reactive sites on the surface of all protein molecules in the crystal. An extremely high excess of reagent is commonly used, as the following example shows.

Let the protein have a molecular weight of 40,000 and the crystal a size of $0.5 \times 0.5 \times 0.5 \, mm^3$. This crystal contains approximately 2 nmol of protein. If the crystal is soaked in 1 ml solution with a reagent concentration of 10 mM, the amount of reagent in the solution is 10^4 nmol, an enormous excess in molarity of reagent with respect to protein. However, the position of the equilibrium is not determined by the total amount of the reagent and the protein, but by the concentrations of the reagent and the protein, and the value of the equilibrium constant K. For the reaction: reagent + protein $\rightarrow$ derivative,

$$K = \frac{[\text{derivative}]}{[\text{reagent}] \times [\text{protein}]}$$

The crystallographer is interested in the occupancy of the binding site:

$$\text{occupancy} = \frac{[\text{derivative}]]}{[\text{derivative}] + [\text{protein}]} = \frac{K \times [\text{reagent}]}{\{K \times [\text{reagent}]\} + 1}$$

The occupancy depends only on $K \times$ [reagent] and is close to 100% for $K \times$ [reagent] $> 10^2$. If the binding is not very strong and the occupancy of the protein binding site does not reach 100%, it is tempting to increase the reagent concentration. However, then the danger exists that the reagent will react with more sites and the chances of nonisomorphism or even crystal degradation are high.

The soaking time varies between hours and months. The minimum time required to reach the equilibrium of the reaction is determined by a number of factors. The diffusion of the reagent through the pores in the crystal is the first important step, and depends on the relative size of the pores and the reagent. Second, a slight conformational change in the protein may be required for a snug fitting of the reagent into its binding site. Finally, there is the chemical reaction itself. Sometimes it is an advantage to use a short soaking time, for example, if the protein molecule presents a great many binding sites to the reagent. If some of them are slow binding sites and others are fast, then, if the crystal is soaked for a short time, only the fast binding sites will react and the chances of maintaining the quality of the crystal are higher. In some cases it is better to soak the crystal for a long time, for example, if the reagent or the protein changes while standing and a suitable reagent or a reactive site on the protein develops in the course of weeks or months. For instance, Pt compounds can gradually change their ligands. If K_2PtCl_4 is kept in an ammonium sulfate solution, $[PtCl_4]^{2-}$ can exchange Cl^- for NH_3 and $[Pt(NH_3)Cl_3]^-$ or $[Pt(NH_3)_2Cl_2]$ or even $[Pt(NH_3)_4]^{2+}$, which has an opposite charge, can be formed with a concomitant change in reactivity. Protein modification with time may be caused by a chemical reaction in the protein, such as deamidation of asparagine or glutamine residues, or oxidation of sulfhydryl groups. This changes the overall charge of the protein and may influence its affinity for charged reagents. The solution can also change slowly, for example, if an ammonium sulfate solution loses ammonia.

7.3.2. Site of Attachment of Heavy Atoms

Although the search for a suitable heavy atom reagent is an empirical process, one should employ all available chemical and biochemical properties of the protein. If the protein contains a free sulfhydryl group, it is obvious that mercury-containing compounds should be tried. Even if sulfhydryl groups are absent, mercurials still have a chance. Histidine residues are frequently found as ligands to heavy atoms, but the pH should not be too low because it is the neutral histidine side chain that acts as the ligand. The sulfur atom in methionine is a preferred binding site for platinum compounds.

For proteins containing Ca^{2+} or Mg^{2+} ions, an attempt should always be made to replace the metal ion by a heavier one, notably rare earth ions. If Ca^{2+} is replaced by Sm^{3+}, for example, only the difference in electrons between the two elements is added: 41 and not 59 electrons. But anomalous scattering helps because it is strong for these heavy ions (see Section 7.8). The radius of the ions is also important, because the cavity containing the metal ion is least disturbed if the diameter of the introduced metal ions is close to the diameter of Ca^{2+} or Mg^{2+} ions. In Table 7.1 one can see that Ba^{2+} and Pb^{2+} are much larger than Ca^{2+} and are not good replacements for Ca^{2+}. However, this does not mean that

Table 7.1. Radii of Some Ions with 6-Coordination

	Ca^{2+}		Mg^{2+}		Ba^{2+}			Pb^{2+}
Electrons	18		10		54			80
Radius (pm)	114		86		149			133

	La^{3+}	Pr^{3+}	Sm^{3+}	Eu^{3+}	Gd^{3+}	Dy^{3+}	Er^{3+}	Tm^{3+}	Yb^{3+}
Electrons	54	56	59	60	61	63	65	66	67
Radius (pm)	117	113	110	109	108	105	103	102	101

they can never replace Ca^{2+}, because the flexibility of the binding site also plays a role. Ca^{2+} can best be replaced by one of the first rare earth ions because their radius is close to that of Ca^{2+}. The radius becomes smaller for the higher elements. Sometimes heavy atom derivatives of a biological substrate or cofactor are used, but this is not frequently done.

The pH of the solution should not be neglected. It has already been mentioned that histidine is a better ligand at higher pH values. The pH is also important in the binding of charged reagents such as HgI_4^{2-}, $Au(CN)^{2-}$, $PtCl_6^{2-}$, etc. At higher pH values, where the protein has a higher negative charge, these reagents react less readily with the protein. This is an advantage if the reaction is too strong at lower pH values when too many sites react and nonisomorphism occurs. On the other hand, if these negative ions do not bind at all or bind only poorly, the pH should be lowered a little. This is of course possible only if the protein crystal permits this change. Heavy atom salts which are easily hydrolyzed, such as UO_2^{2+} or Sm^{3+} salts, cannot be used at an alkaline pH.

If the medium is a water–organic solvent mixture, electrostatic forces are stronger because of the lower dielectric constant and the binding of ionic compounds will be stronger. However, the organic solvent might be a chelating agent for the heavy ion; this is true for 2-methyl-2,4-pentanediol (MPD), a popular organic solvent in protein crystallization experiments.

In the structure determination of protein/oligonucleotide complexes brominated or iodinated oligos are convenient heavy-atom compounds. Xenon as a heavy atom is a different story. The atoms of this inert gas can occupy holes in a protein molecule but the binding by purely Van der Waals interaction is rather weak, preventing a reasonable occupancy. To raise it close to 100%, pressurized xenon must be applied. The pressure is in the range of 0.5–5 MPa[2] and must be maintained during data collection. The requirement of pressurized equipment is an extra problem. Sauer et al. (1997) have circumvented this problem by performing data collection at cryotemperature, the most popular crystal mounting and data collection technique anyway (Section 1.7). The crystal in the cryoloop is inserted into a high-pressure cell and xenon pressure between 0.5 and 5 MPa applied for 1–30 minutes. After fast release of the xenon pressure the crystal is immediately shock-cooled and transferred to the X-ray diffraction instrument. At cryotemperature the vapour pressure of xenon is sufficiently low to prevent evaporation (according to Gmelin[3], 0.00026 atm = 0.26 hectoPa at 91 °K). Another advantage of the cryotechnique is that the strong X-ray absorption by the xenon gas in a pressurized cell is avoided.

---

[2] 1 MPa ≈ 10 atm.
[3] Handbook of Inorganic and Organometallic Chemistry GMELIN Reg. No. (GRN) 16318

7.3.3. Chemical Modification of the Native Protein

If straightforward soaking does not result in a useful complex, the situation is not completely hopeless. One may still try to modify the protein by covalently attaching a heavy atom containing reagent (see e.g. Spurlino et al., 1994) or a potential heavy atom binder, such as p-iodophenyliso-thiocyanate or p-iodophenylisocyanate. They react with the ε-NH$_2$ of lysine side chains to form a thiourea or urea derivative, at least at sufficiently high pH. Another chemical reaction is the iodination of tyrosine side chains (Sigler, 1970; Brzozowski et al., 1992; Derewenda, 1994; Spurlino et al., 1994). They can take up a maximum of three iodine atoms. Iodine has a reasonable number of electrons (53), but despite its successful use in a number of cases, it is not a very popular method. The reason is probably its tendency to react with other groups in the protein.

7.3.4. Genetic Modification of the Protein

Genetic engineering has opened up new areas for protein modification. For the preparation of heavy atom derivatives, it is sometimes useful to replace one of the amino acids by a cysteine (Tucker et al., 1989; Nagai et al., 1990). Of course this replacement helps only if the cysteine residue is not oxidized readily, which is a potential danger. Therefore, mutants should be treated with an antioxidant, such as dithiothreitol, before reaction with the mercury-containing compound (Nagai et al., 1991). The new cysteine residue should of course not disturb the protein structure, and should be accessible. This is difficult to predict beforehand. The best one can do is to replace a residue in a very polar region of the amino acid sequence, which hopefully is a loop at the surface of the molecule.

Another biological modification is to incorporate selenomethionine in place of methionine and solve the structure by the multiple wavelength anomalous diffraction (MAD) technique (see Section 9.5).

7.3.5. Problems Commonly Encountered in the Search for Heavy Atom Derivatives

7.3.5.1. Increased Radiation Damage

X-ray radiation damage is caused by radical formation and subsequent chemical reactions. This process can be slowed down by lowering the temperature of the crystal in the X-ray beam (Section 1.7). Even a few degrees lower helps (a temperature of 5°C instead of room temperature, for example).

7.3.5.2. The Insolubility of Phosphates

Phosphate buffers are often used for protein crystallization and soaking. However, some heavy metal phosphates are insoluble, including those of the rare earths and of uranyl ions. In such cases the phosphate buffer should be replaced by a suitable buffer, usually an organic one.

7.3.5.3. Ammonium Sulfate

Ammonium sulfate is a very popular precipitating agent. However, it can prevent the binding of heavy metals in two ways. First, it is in equilibrium with ammonia. At somewhat higher pH values the ammonia concentration in the solution is appreciable and this can act as a ligand for the heavy ion, which might prevent binding to the protein. The solution of this problem is to replace the ammonium sulfate by another salt, such as Li- or Cs-sulfate or K- or Na-phosphate, or by polyethylene glycol (PEG). The other problem with ammonium sulfate is its high ionic strength. This weakens electrostatic interactions and in this way can prevent the binding of a heavy ion. The solution is to change from ammonium sulfate to PEG.

7.4. Effect of Heavy Atoms on X-ray Intensities

Can the attachment of one or a few heavy atoms to a large protein molecule sufficiently change the intensities of the reflections? Suppose we have a protein with a molecular weight of 42,000. Each of its molecules contains about 3000 nonhydrogen atoms or $3000 \times 7 = 21,000$ electrons. In this ocean of 21,000 electrons a mercury atom adds only 80 electrons and yet its attachment changes the intensities of the X-rays diffracted by the crystal of the protein in a measurable way. This seems impossible, but it is nevertheless true, as we shall see.

Crick and Magdoff (1956) estimated the expected intensity changes resulting from heavy atom attachment and arrived at the following result: For centric reflections which have their structure factor along the real axis in the Argand diagram they found that the relative root mean square intensity change is

$$\frac{\sqrt{\overline{(\Delta I)^2}}}{\overline{I_P}} = 2 \times \sqrt{\frac{\overline{I_H}}{\overline{I_P}}} \tag{7.16}$$

and for acentric reflections

$$\frac{\sqrt{\overline{(\Delta I)^2}}}{\overline{I_P}} = \sqrt{2} \times \sqrt{\frac{\overline{I_H}}{\overline{I_P}}} \tag{7.17}$$

where $\overline{I_H}$ is the average intensity of the reflections if the unit cell would contain the heavy atoms only and $\overline{I_P}$ is the average intensity of the reflections for the native protein (see the derivation below).

Crick and Magdoff's Estimation of X-ray Reflection Intensity Changes if Heavy Atoms Are Attached to the Protein

First, centric reflections will be considered. If the origin of the system is placed in the center of symmetry, they have their structure factors **F** pointing along the horizontal axis in the Argand diagram, to the right for phase angle $\alpha = 0°$ and to the left for $\alpha = 180°$. Therefore, **F** can be expressed as a real number, equal to its amplitude, with positive sign for $\alpha = 0°$ and negative sign for $\alpha = 180°$, since for $\alpha = 0°$

$$\mathbf{F} = |F| \exp[i\alpha] = |F| (\cos \alpha + i \sin \alpha) = +F$$

and for $\alpha = 180°$

$$\mathbf{F} = |F| \exp[i\alpha] = |F| (\cos \alpha + i \sin \alpha) = -F$$

In the following discussion, P stands for the native protein, PH for the heavy atom derivative, and H for the heavy atoms. For centric reflections (Figure 7.6):

$$I_{PH} = (F_P + F_H)^2$$

Note that F_P and F_H can be either positive or negative.

$$I_P = F_P^2$$

$$\Delta I = 2F_P F_H + F_H^2$$

$$(\Delta I)^2 = 4F_P^2 F_H^2 + 4F_P F_H^3 + F_H^4$$

and the mean square change in intensity is

$$\overline{(\Delta I)^2} = 4\overline{F_P^2 F_H^2} + 4\overline{F_P F_H^3} + \overline{F_H^4}$$

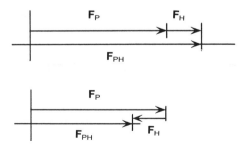

Figure 7.6. Structure factors in the isomorphous replacement method for centric reflections. $\mathbf{F_P}$ is for the protein, $\mathbf{F_{PH}}$ is for the derivative, and $\mathbf{F_H}$ is for the heavy atom contribution.

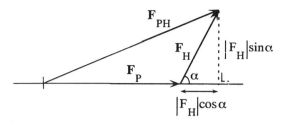

Figure 7.7. Structure factors in the isomorphous replacement method for non-centric reflections; the horizontal direction of $\mathbf{F_P}$ is arbitrary.

Since F_P and F_H are not correlated, $\overline{F_P F_H^3} = 0$ and $\overline{F_P^2 F_H^2} = \overline{F_P^2} \times \overline{F_H^2}$.

$$\frac{\sqrt{\overline{(\Delta I)^2}}}{I_P} = \frac{\sqrt{4\overline{F_P^2} \times \overline{F_H^2} + \overline{F_H^4}}}{\overline{F_P^2}}$$

$$= 2\sqrt{\frac{\overline{F_H^2}}{\overline{F_P^2}}}\sqrt{1 + \frac{\overline{F_H^4}}{4 \times \overline{F_P^2} \times \overline{F_H^2}}} \cong 2\sqrt{\frac{\overline{F_H^2}}{\overline{F_P^2}}} = 2\sqrt{\frac{I_H}{I_P}}$$

assuming that $\overline{F_H^4} \ll 4\overline{F_P^2} \times \overline{F_H^2}$. We shall now do the same for noncentric reflections (Figure 7.7):

$$\mathbf{F_{PH}} = \mathbf{F_P} + \mathbf{F_H}$$

$$I_{PH} = |\mathbf{F_P} + \mathbf{F_H}|^2 = |F_{PH}|^2 = |F_P|^2 + 2|F_P|\,|F_H|\cos\alpha + |F_H|^2$$

$$\Delta I = 2|F_P|\,|F_H|\cos\alpha + |F_H|^2$$

$$(\Delta I)^2 = 4|F_P|^2|F_H|^2\cos^2\alpha + 4|F_P|\,|F_H|^3\cos\alpha + |F_H|^4$$

$$\overline{(\Delta I)^2} = 2|F_P|^2|F_H|^2 + 4\overline{|F_P|\,|F_H|^3\cos\alpha} + \overline{|F_H|^4}$$

because $\overline{|F_P|}$ and $\overline{|F_H|}$ are not correlated and $\overline{\cos\alpha} = 0$:

$$\overline{(\Delta I)^2} = 2\overline{|F_P|^2} \times \overline{|F_H|^2} + 0 + \overline{|F_H|^4}$$

$$\frac{\sqrt{\overline{(\Delta I)^2}}}{I_P} = \sqrt{\frac{2\overline{|F_P|^2} \times \overline{|F_H|^2} + \overline{|F_H|^4}}{(\overline{|F_P|^2})^2}} = \sqrt{2} \times \sqrt{\frac{\overline{|F_H|^2}}{\overline{|F_P|^2}} + \frac{\overline{|F_H|^4}}{2(\overline{|F_P|^2})^2}}$$

$$= \sqrt{2} \times \sqrt{\frac{\overline{|F_H|^2}}{\overline{|F_P|^2}}} \times \sqrt{1 + \underbrace{\frac{\overline{|F_H|^4}}{2 \times \overline{|F_P|^2} \times \overline{|F_H|^2}}}_{\substack{\text{small with} \\ \text{respect to 1}}}}$$

$$\cong \sqrt{2} \times \sqrt{\frac{\overline{|F_H|^2}}{\overline{|F_P|^2}}} = \sqrt{2} \times \sqrt{\frac{I_H}{I_P}}$$

Table 7.2. Average Relative Change in Intensity for
the Acentric Reflections of a Protein Crystal if One
Mercury Atom is Attached per Protein Molecule[a]

Molecular Weight of the Protein	100% Occupancy	50% Occupancy
14,000	0.51	0.25
28,000	0.36	0.18
56,000	0.25	0.12
112,000	0.18	0.09
224,000	0.13	0.06
448,000	0.09	0.04

[a] The data are for $(\sin\theta)/\lambda = 0$. With increasing diffraction
angle the relative contribution of Hg is somewhat higher
because its atomic scattering falls off less rapidly than for the
light elements.

In Table 7.2 this result is used to calculate the size of the relative
change in intensity that can be expected for the acentric reflections of a
protein crystal, if one mercury atom is attached per protein molecule. $\bar{I}$
is obtained with the expression:

$$\bar{I} = \sum_i f_i^2$$

(see Section 5.2) f_i is 80 for a mercury atom and on the average 7 for a
typical protein atom. If we assume that the intensities can be determined
with an accuracy of 10%, then the practical limit for $\sqrt{\overline{(\Delta I)^2}}/\bar{I}_P$ is 0.10 ×
$\sqrt{2}$ = 0.14 (the factor $\sqrt{2}$ stems from the fact that the intensity difference
is the result of two measurements). From Table 7.2 we see that this
corresponds to a maximum molecular weight of 200,000 for full occupancy
by one mercury atom per protein molecule and 50,000 for half occupancy.
The changes are of course larger if more than one heavy atom is bound
per protein molecule but lower if the binding sites are not fully occupied.
The conclusion is that the isomorphous replacement method can be
applied successfully for the determination of a protein crystal structure,
even for large protein molecules. Two examples of extremely large
molecules for which the structure was determined by the isomorphous
replacement method are F_1-ATPase and β-galactosidase. Bovine heart
mitochondrial F_1-ATPase has a molecular weight of 371000. It crystallizes in space group $P2_12_12_1$ with cell dimensions of a = 286 Å, b = 108 Å
and c = 140 Å and one molecule of F_1-ATPase in the asymmetric unit
(Lutter et al., 1993). Its structure was solved with data from a single
heavy atom derivative (see Section 7.14) and was determined at a resolution of 2.8 Å, (Abrahams et al., 1994). The β-galactosidase molecule is a

tetramer with 222-point symmetry. The protein crystallizes in space group $P2_1$ with a = 107.9 Å, b = 207.5 Å, c = 509.9 Å and β = 94.7°. These crystals have four tetramers per asymmetric unit with a total molecular weight of $4 \times 465,412 = 1,861,648$!! The structure was solved with three heavy atom derivatives and refined to a resolution of 2.5 Å (Jacobson et al., 1994).

7.5. Determination of the Heavy Atom Parameters from Centrosymmetric Projections

In step 3 of the isomorphous replacement method the coordinates of the heavy atoms must be found. This is an easy procedure if the crystal has centrosymmetric projections, e.g., in space group $P2_12_12_1$, where the three projections along the axes of the unit cell are centrosymmetric because of the 2-fold screw axes. For centrosymmetric projections the vectorial summation (Figure 7.6)

$$\mathbf{F}_{PH} = \mathbf{F}_P + \mathbf{F}_H$$

is simplified to

$$|F_{PH}| = |F_P| \pm |F_H|$$

$$|F_H| = |F_{PH}| - |F_P|$$

or

$$|F_H| = |F_P| - |F_{PH}|$$

and

$$|F_H|^2 = (|F_{PH}| - |F_P|)^2$$

We have made the assumption that F_{PH} and F_P have the same sign, either both positive or both negative. With this assumption, the Patterson summation with the coefficients $(|F_{PH}| - |F_P|)^2$ will give a Patterson map of the heavy atom arrangement in the unit cell. For the majority of the reflections the assumption will be true, because in general F_H will be small compared with F_P and F_{PH}. If, however, F_P is small, F_{PH} could have the opposite sign and F_H would be $F_P + F_{PH}$. Fortunately this does not occur often enough to distort the Patterson map seriously.

To calculate $|F_H|$ the structure factor amplitudes $|F_{PH}|$ and $|F_P|$ should of course be put on the same scale. This can be done in an approximate way by applying the Wilson plot and putting the $|F_P|^2$ and $|F_{PH}|^2$ values on an absolute scale [Eq. (5.3)]. This gives the factors C_P and C_{PH} with B_P and B_{PH}. Alternatively, a relative Wilson plot is calculated:

$$\ln \frac{\overline{I_{PH}}}{\sum_i (f_i^0)^2 + (f_H)^2} - \ln \frac{\overline{I_P}}{\sum_i (f_i^0)^2} = \ln \frac{C_{PH}}{C_P} - 2(B_{PH} - B_P) \frac{\sin^2 \theta}{\lambda^2}$$

With the relative value C_{PH}/C_P and the difference between B_P and B_{PH}, $|F_{PH}|^2$ and $|F_P|^2$ can be put on the same scale. For the native protein $\Sigma_i(f_i^0)^2$ must be calculated for all protein atoms, but for the derivative an estimated heavy atom contribution $(f_H)^2$ should be added. In the subsequent process of refining the heavy atom parameters, the scale factor is refined together with the other parameters. Sometimes the (f_H) contribution to the structure factor F_{PH} is neglected and the $|F_P|$ and $|F_{PH}|$ values are put on the same scale by minimizing a least squares function E with respect to the relative scale factor (for the method of least squares, see Section 7.11):

$$E = \sum_h \frac{1}{\sigma_F^2}(k|F_{PH}| - |F_P|)^2 \tag{7.18}$$

where k is a scale factor and σ_F^2 is the variance to be chosen for either the $|F_{PH}|$ or the $|F_P|$ values and the summation is over all reflections h. The minimization of E with respect to k gives

$$k = \frac{\sum_h \frac{1}{\sigma_F^2}|F_{PH}| \times |F_P|}{\sum_h \frac{1}{\sigma_F^2}|F_{PH}|^2} \tag{7.19}$$

If the morphology of the crystal(s) used for collecting the native data set differs appreciably from the morphology of the derivative crystals, differences in absorption may affect the comparison of the two data sets. The isomorphous differences and $R_{deriv.}$ (see Appendix 2) appear larger than they really are. Therefore, it is always advisable to correct for absorption, such as by comparing and equalizing symmetry-related reflections within one data set or collect data at a shorter wavelength.

It is sometimes observed that after scaling, as just described, reflections that should have the same intensity tend to be stronger in one region of reciprocal space than in another. This can be due to problems of an instrumental nature or to poor absorption correction. In those cases local scaling must be applied, in which reciprocal space is divided into blocks of $h\,k\,l$ with an individual scaling factor for each block (Matthews and Czerwinski, 1975).

If the Patterson map of a centrosymmetric projection can be interpreted, it gives two coordinates of the heavy atoms and with a second projection the third coordinate as well. With the positions of the heavy atoms known, the structure factors F_H can be calculated, including their sign (for centrosymmetric projections) or their phase angle (for acentric reflections). In principle one heavy atom derivative is sufficient to determine the signs of centrosymmetric protein reflections. For example, if $|F_{PH}| > |F_P|$ and the sign of F_H is $+$, then F_P will have a $+$ sign. For a second heavy atom derivative it is not absolutely necessary to calculate a Patterson map, because the sign of the centric protein reflections in combination with the difference between $|F_P|$ and $|F_{PH}|$ of the second heavy atom derivative, gives the sign of F_H immediately. This allows the calculation of a Fourier summation, resulting in the projection of the heavy atom arrangement from which the coordinates can be obtained.

7.6. Parameters of Heavy Atoms Derived from Acentric Reflections

Not all space groups have centrosymmetric projections. The heavy atom positions can still be found in such cases. The two sets of known data are the amplitudes of the structure factors $\mathbf{F}_P$ and $\mathbf{F}_{PH}$. Their phase angles are not yet known. For each reflection there is a difference in length (Figure 7.8):

$$\Delta|F|_{iso} = |F_{PH}| - |F_P|$$

We will now see that the coordinates of the heavy atoms can generally be derived from a Patterson map calculated with $(\Delta|F|_{iso})^2$. The triangle ABC in Figure 7.8 expresses the vector sum: $\mathbf{F}_{PH} = \mathbf{F}_P + \mathbf{F}_H$. However, for the time being only the lengths of $\mathbf{F}_{PH}$ ($|F_{PH}|$) and that of $\mathbf{F}_P$ ($|F_P|$) are known, but not their directions. For $\mathbf{F}_H$ both the length and direction are unknown.

In Figure 7.8 CE = $|F_H| \cos(\alpha_{PH} - \alpha_H)$. In general $(\alpha_P - \alpha_{PH})$ is small, because for most reflections $|F_H| \ll |F_P|$ and $|F_{PH}|$. Therefore, CE $\cong \Delta|F|_{iso}$ and the result is that

$$\Delta|F|_{iso} = |F_H| \cos(\alpha_{PH} - \alpha_H) \tag{7.20}$$

The result is that a Patterson summation with $(\Delta|F|_{iso})^2$ as the coefficients will in fact be a Patterson summation with coefficients $|F_H|^2 \cos^2(\alpha_{PH} - \alpha_H)$. Since

$$\cos^2(\alpha_{PH} - \alpha_H) = \tfrac{1}{2} + \tfrac{1}{2}\cos 2(\alpha_{PH} - \alpha_H)$$

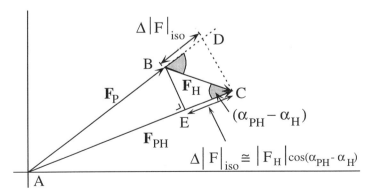

Figure 7.8. The structure factor triangle for isomorphous replacement: $\Delta|F|_{iso} = |F_{PH}| - |F_P|$.

we obtain

$$|F_H|^2 \cos^2(\alpha_{PH} - \alpha_H) = \tfrac{1}{2}|F_H|^2 + \tfrac{1}{2}|F_H|^2 \cos 2(\alpha_{PH} - \alpha_H)$$

Because the angles α_{PH} and α_H are not correlated, the second term on the right will contribute only noise to the Patterson map. However, the first term, $\tfrac{1}{2}|F_H|^2$, will give the Patterson function for the heavy atom structure on half the scale.

Because a Patterson map is centrosymmetric, the choice is between two sets of heavy atom positions, which are centrosymmetrically related. It is not yet known what the correct one is, but for the moment this does not matter and either of the two sets can be chosen. Later on, in determining the absolute configuration of the protein (Section 9.4), this problem will be discussed again.

Conclusion:

$$P(u\ v\ w) = \frac{1}{V}\sum_h (\Delta|F|_{iso})^2 \cos[2\pi(hu + kv + lw)] \qquad (7.21)$$

Calculation with the coefficients $(\Delta|F|_{iso})^2$ results (apart from the extra noise and the reduction in height) in a Patterson map of the heavy atom arrangement.

If the difference Patterson map cannot be interpreted in terms of the heavy atom positions, one can try to apply one of the direct methods for phase determination, as they are developed for the X-ray structure determination of small molecules, because the problem is very similar: for only a relatively small number of sites in the unit cell, the parameters must be found. SHELXS is such a program (Sheldrick et al., 1993).

We are now able to find the parameters of the attached heavy atoms in the crystal structure. We will discuss how a common origin for the coordinates of the heavy atoms from different derivatives can be found and how the heavy atom parameters can be refined. Subsequently the protein phase angles will be calculated. But before doing so we will discuss another extremely useful Fourier summation: "the difference Fourier." We shall also introduce so-called "anomalous scattering," because this can contribute to localization of the heavy atoms.

7.7. The Difference Fourier Summation

With this summation we can find the position of reagents attached to protein molecules in the crystal, either heavy atoms or other reagents, such as enzyme inhibitors. However, it is necessary that we know protein phase angles $\alpha_P(h\ k\ l)$!

$$\Delta\rho(x\ y\ z) = \frac{1}{V}\sum_{hkl} \Delta|F(h\ k\ l)|_{iso}\, \exp[-2\pi i(hx + ky + lz) + i\alpha_P(h\ k\ l)]$$

$$= \frac{1}{V}\sum_{hkl} \Delta|F(h\ k\ l)|_{iso}\, \cos[2\pi(hx + ky + lz) - \alpha_P(h\ k\ l)]$$

Written more compactly:

$$\Delta\rho(\mathbf{r}) = \frac{1}{V}\sum_{\mathbf{h}}\Delta|F(\mathbf{h})|_{\text{iso}}\exp[-2\pi i\mathbf{h}\cdot\mathbf{r} + i\alpha_P(\mathbf{h})] \qquad (7.22)$$

Thus a difference Fourier summation is calculated with the coefficients $\Delta|F|_{\text{iso}}$ and the phase angles α_P of the protein. Now we shall see what this leads to. Suppose the structure factors of the attached reagent are the still unknown vectors $\mathbf{F}_H$. In Section 7.6 we derived that

$$\Delta|F|_{\text{iso}} \approx |F_H|\cos(\alpha_{PH} - \alpha_H) \qquad (7.20)$$

Because $\exp[i\alpha] = \cos\alpha + i\sin\alpha$ and $\exp[-i\alpha] = \cos\alpha - i\sin\alpha$,

$$\Delta|F|_{\text{iso}} \approx \tfrac{1}{2}|F_H|\{\exp[i(\alpha_{PH} - \alpha_H)] + \exp[-i(\alpha_{PH} - \alpha_H)]\}$$

and

$$\Delta|F|_{\text{iso}}\exp[i\alpha_P] \approx \tfrac{1}{2}|F_H|\{\exp[i(\alpha_{PH} - \alpha_H)] \times \exp[i\alpha_P]$$
$$+ \exp[-i(\alpha_{PH} - \alpha_H)] \times \exp[i\alpha_P]\}$$

For $|F_H|\exp[i\alpha_H]$ we can write $\mathbf{F}_H$, and for $|F_H|\exp[-i\alpha_H]$ we write $\mathbf{F}_H^*$.

$$\Delta|F|_{\text{iso}}\exp[i\alpha_P] = \tfrac{1}{2}\mathbf{F}_H\exp[-i\alpha_{PH}] \times \exp[i\alpha_P]$$
$$+ \tfrac{1}{2}\mathbf{F}_H^*\exp[i\alpha_{PH}] \times \exp[i\alpha_P]$$

Since $\alpha_P \cong \alpha_{PH}$,

$$\Delta|F|_{\text{iso}}\exp[i\alpha_P] = \tfrac{1}{2}\mathbf{F}_H + \tfrac{1}{2}\mathbf{F}_H^*\exp[2i\alpha_P]$$

The term $\tfrac{1}{2}\mathbf{F}_H^*\exp[2i\alpha_P]$ will give noise in the Fourier map, because the vectors $\mathbf{F}_H^*$ and $\exp[2i\alpha_P]$ are not correlated in their direction in the Argand diagram. Therefore,

$$\Delta|F|_{\text{iso}}\exp[i\alpha_P] \approx \tfrac{1}{2}\mathbf{F}_H$$

If this result is combined with Eq. (7.22) then,

$$\Delta\rho(\mathbf{r}) = \frac{1}{V}\sum_{h}\frac{1}{2}\mathbf{F}_H\exp[-2\pi i\mathbf{h}\cdot\mathbf{r}]$$

$$= \frac{1}{2}\frac{1}{V}\sum_{h}|F_H(\mathbf{h})|\exp[-2\pi i\mathbf{h}\cdot\mathbf{r} + i\alpha_H] \qquad (7.23)$$

Conclusion: A difference Fourier map shows positive electron density at the site of atoms that were not present in the native structure and negative density at the positions of atoms present in the native, but not in the derivative structure. The height of the peaks is only half of what it would be in a normal Fourier map. With a difference Fourier map even small changes in the electron density can be observed, such as the attachment or removal of a water molecule. It is a powerful method, even if only preliminary values for the protein phase angles are known. For example, after the main heavy atom site has been found from a difference Patterson map additional weakly occupied sites can be detected with a difference Fourier map.

An improved difference Fourier map can be calculated if anomalous scattering data are incorporated in the calculation of the protein phase angles, which will be discussed in Chapter 9. Another useful Fourier summation has as coefficients $2|F_{PH}| - |F_P|$ and as phase angles the protein phase angles α_P:

$$\rho(\mathbf{r}) = \frac{1}{V}\sum_{h}(2|F_{PH}| - |F_P|)\exp[-2\pi i\mathbf{r}\cdot\mathbf{h} + i\alpha_P] \qquad (7.24)$$

The coefficients can be written as

$$2|F_{PH}| - |F_P| = 2(|F_{PH}| - |F_P|) + |F_P| = 2\Delta|F|_{iso} + |F_P|$$

Thus this electron density map will give the native protein structure and, apart from noise, the electron density of the attached atoms but now at full and not half height. Sometimes $\{3|F_{PH}| - 2|F_P|\}$ coefficients are preferred (Lamzin and Wilson, 1997).

Structural information in electron density maps is determined to a greater extent by the phase angles than by the Fourier coefficients. Therefore, the electron density maps calculated with native protein phase angles are biased toward the native protein. Read (1986) showed that the bias can be minimized if the electron density is calculated with protein phase angles α_P and as coefficients $2m|F_{PH}| - D|F_P|$, where m is the figure of merit (Section 7.12) and D is a multiplier equal to the Fourier transform of the probability distribution of $\overline{\Delta r}$, the mean error in the atomic positions (see Section 15.6).

The difference Fourier should not be mistaken for the residual Fourier, which can be calculated after the structure determination is almost complete. Its coefficients are $(|F_{PH}| - |\mathbf{F}_P + \mathbf{F}_H|)$. The values of $|F_{PH}|$ are the amplitudes measured for the derivative; $\mathbf{F}_P$ and $\mathbf{F}_H$ are the structure factors calculated for the present protein and heavy atom models.

$$\text{Res. Fourier} = \frac{1}{V}\sum_{h}(|F_{PH}| - |\mathbf{F}_P + \mathbf{F}_H|)\exp[-2\pi i\mathbf{h}\cdot\mathbf{r} + i\alpha_{PH}] \quad (7.25)$$

The phase angles α_{PH} are calculated for the present model of the derivative. $|F_{PH}|\exp[i\alpha_{PH}]$ represents the actual structure as far as its amplitude is concerned, but its phase corresponds with the present model of the structure. If the difference between the actual structure and the model is not too large, the residual Fourier can be shown to give this difference on half the scale. The map can be used in the search for undetected heavy atom sites.

Ideally, any Fourier summation, whether it is a Patterson map, an electron density map, or a difference Fourier, should be calculated with an infinite number of terms. In practice, this is never true and only a limited number of terms is used (truncation). This series termination causes ripples around maxima in the map.

7.8. Anomalous Scattering

If the absorption by an element, such as copper is plotted as a function of the X-ray wavelength λ, a typical curve is obtained (Figure 7.9). The sharp change in the curve is called an absorption edge. It is caused by photon absorption: an electron is ejected from an atom by the photon energy of the X-ray beam. For copper the K-absorption edge is at $\lambda = 1.380\,\text{Å}$. At this wavelength an electron is ejected from the K-shell to a state in the continuous energy region. Copper emits at its characteristic wavelength of $K_\alpha = 1.5418\,\text{Å}$, somewhat above the K-absorption edge, because now the electron falls back from the L-shell into the K-shell and this is a smaller energy difference. Remember that $E = hc/\lambda$. (See Figure 2.3 for the energy levels).

So far we have always regarded the electrons in an atom as free electrons. However, this is no longer true if the X-ray wavelength approaches

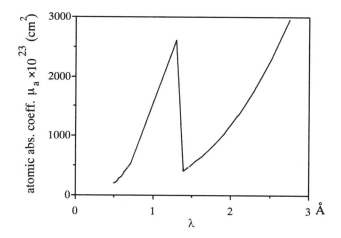

Figure 7.9. The atomic absorption coefficient for copper. The K-absorption edge is at $1.380\,\text{Å}$. The atomic absorption coefficient or "atomic cross-section" for absorption, μ_a (cm^2), is defined by $\mu_a = (\mu/\rho) \times (A/N)$, where μ/ρ is the mass absorption coefficient (cm^2/g), ρ is the density of the absorber, A is its atomic weight, and N is Avogadro's number. μ (cm^{-1}) is the total linear absorption coefficient defined by $I = I_0 \exp[-\mu t]$ with t the thickness of the material in cm, I_0 the intensity of the incident and I the intensity of the transmitted beam.

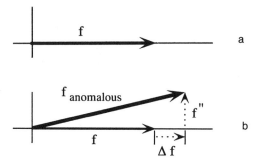

Figure 7.10. The atomic scattering factor for a completely free electron (a) and for a bound electron (b). The anomalous contribution consists of two parts: a real part Δf and an imaginary part if''.

an absorption edge wavelength. Classical dispersion theory has derived the scattering power of an atom by supposing that it contains dipole oscilators. In units of the scattering of a free electron, the scattering of an oscillator with eigen frequency ν_n and moderate damping factor κ_n was found to be a complex quantity:

$$f_n = \frac{\nu^2}{\nu^2 - \nu_n^2 - i\kappa_n\nu} \qquad (7.26)$$

where ν is the frequency of the incident radiation (James, 1965; International Tables for Crystallography, Vol.C, 1995). For $\nu \gg \nu_n$, Eq. 7.26 approaches $f_n = 1$, the scattering by a free electron, and for $\nu \ll \nu_n$ to $f_n = 0$, the lack of scattering from a fixed electron. Only for $\nu \cong \nu_n$, the imaginary part has an appreciable value.

Fortunately, quantum mechanics arrives at the same result by adding a rational meaning to the damping factors and interpreting ν_n as absorption frequencies of the atom (Hönl, 1933). For heavy atoms the most important transitions are to a continuum of energy states with $\nu_n \geq \nu_K$ or $\nu_n \geq \nu_L$, etc., the frequencies of the K, L, etc. absorption edges. Whereas free electrons have a phase difference of 180° with respect to the incident beam, the diffracted beam of the inner shell electrons of the heavy atoms does not differ 180° in phase from the incident beam. The situation is drawn in Figure 7.10. The atomic scattering factor for a completely free electron is drawn in Figure 7.10a and for a bound electron in Figure 7.10b; its scattering is anomalous and can be written as

$$f_{anom.} = f + \Delta f + if'' = f' + if''[4]$$

---

[4] The nomenclature of f' can be confusing. What here is called Δf is in other publications often indicated by f'.

Table 7.3. The Anomalous Scattering in Electrons of
Hg for Copper Radiation

	$(\sin\theta)/\lambda$		
	0 $(\infty)^a$	0.4 (Å^{-1}) $(1.25\,\text{Å})^a$	0.6 (Å^{-1}) $(0.83\,\text{Å})^a$
f	80	53	42
Δf	-5	-5	-5
f''	8	8	8

^{a} Corresponding lattice spacing.

Δf is the change in the electron scattering factor along the horizontal axis in the Argand diagram and if'' along the vertical axis (remember that i indicates a counterclockwise rotation of 90° (Section 4.2). For higher scattering angles the anomalous scattering becomes relatively more important. The reason is that it is nearly constant as a function of $(\sin\theta)/\lambda$, because it stems mainly from the inner electrons (Table 7.3). The effect of anomalous scattering is that in the Argand diagram the atomic scattering vector is rotated counterclockwise. As a consequence the structure factors $\mathbf{F}_{\text{PH}}(h\,k\,l)$ and $\mathbf{F}_{\text{PH}}(\bar{h}\,\bar{k}\,\bar{l})$ for the heavy atom derivative of the protein are no longer equal in length and have a different phase angle (Figure 7.11). This difference can be used separately or in combination with the isomorphous replacement method in the search for the heavy atom positions. We define:

$$\Delta|F|_{\text{ano}} = \{|F_{\text{PH}}(+)| - |F_{\text{PH}}(-)|\}\frac{f'}{2f''} \qquad (7.27)$$

$|F_{\text{PH}}(+)|$ represents the amplitude of the structure factor for a reflection $(h\,k\,l)$, and $|F_{\text{PH}}(-)|$ is the amplitude for the reflection $(\bar{h}\,\bar{k}\,\bar{l}) \equiv (-h, -k, -l)$. $\Delta|F|_{\text{ano}}$ is the difference between the amplitudes of the structure factor for the reflections $h\,k\,l$ and $\bar{h}\,\bar{k}\,\bar{l}$ (Bijvoet or Friedel pairs), scaled up with the factor $f'/2f''$.

From the anomalous Patterson map, calculated with $(\Delta|F|_{\text{ano}})^2$, the location of anomalous scatterers can be derived. These anomalous scatterers can be either extra heavy atoms attached to the protein or heavy atoms such as copper, iron, or even sulfur that are already present in the native protein structure. The anomalous scattering by sulfur atoms is relatively weak and it can be used only with rather small protein molecules. If the sulfur in the protein can be replaced by selenium the situation is more favorable (see Table 7.4) and even more favorable if a wavelength close to the K-absorption edge of Se (0.98 Å) is chosen. This replacement of sulfur by selenium can be done biologically by incorporation of selenomethionine instead of methionine.

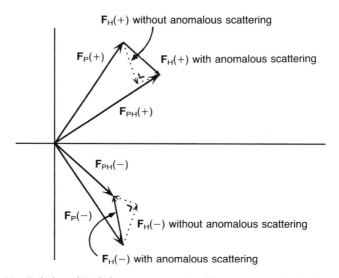

Figure 7.11. $\mathbf{F}_P(+)$ and $\mathbf{F}_P(-)$ are symmetric with respect to the horizontal axis in the Argand diagram, assuming the absence of anomalous scatterers in the protein. Without anomalous scattering, $\mathbf{F}_H(+)$ and $\mathbf{F}_H(-)$ are also symmetric. The imaginary part of the anomalous scattering contribution has been exaggerated. In this example $|F_{PH}(+)| > |F_{PH}(-)|$.

In Sections 9.2 and 9.3 we shall see how information from anomalous scattering can help in the determination of the protein phase angles if only one wavelength is used. In Section 9.5 protein phase angles are derived from multiple wavelength anomalous dispersion. In the next sections it will be shown that

1. $(\Delta|F|_{ano})^2$ as the coefficients in a Patterson summation will result in a Patterson map of the anomalous scatterers (the heavy atoms).

Table 7.4. The Anomalous Scattering in Electrons for CuK$_\alpha$ Radiation by Mercury, Sulfur, and Selenium for $(\sin \theta)/\lambda = 0$[a]

	f	Δf	f''
Hg	80	-5.0	7.7
S	16	0.3	0.6
Se	34	-0.9	1.1

[a] It is assumed that the anomalous scattering is isotropic. This is not true if the wavelength is close to an absorption edge corresponding to an electron ejection to an upper state with nonspherical symmetry; for a detailed discussion see Templeton and Templeton (1991).

2. $(\Delta|F|_{iso})^2 + (\Delta|F|_{ano})^2$ as coefficients will give the same Patterson map, but with less noise in the background than for separate Patterson maps.
3. Various methods exist to find the position of the heavy atoms in the different derivatives with respect to each other.

Moreover, as will be shown in Section 9.4, the absolute configuration of the protein can be determined with $\Delta|F|_{ano}$.

If the protein crystal contains anomalous scatterers it is essential to differentiate between the reflections $(+h, +k, +l)$ and $(-h, -k, -l)$. This can be done by applying the following rules:

1. Choose the unit cell angles α, β, and γ between the positive axes (**a**, **b**, and **c**) $\geqslant 90°$.
2. Use a right-handed coordinate system (see, for example, Figure 3.3).

As a result, the set of axes is always chosen in the same way with respect to the content of the unit cell for all crystals of the protein or protein derivative.

7.9. The Anomalous Patterson Summation

The heavy atom contribution to the structure factor consists of a normal part, $\mathbf{F_H}$, and an anomalous part, $\mathbf{F''_H}$. In Figure 7.12 this is drawn for a reflection $(h\,k\,l)$ and for $(\bar{h}\,\bar{k}\,\bar{l})$. However, for convenience, the structure factors for $(\bar{h}\,\bar{k}\,\bar{l})$ have been reflected with respect to the horizontal axis. It can be derived that a Patterson summation with the coefficients $(\Delta|F|_{ano})^2$

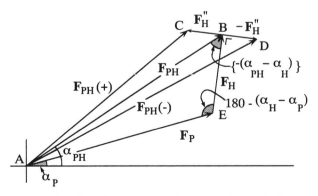

Figure 7.12. In this drawing the structure factors $\mathbf{F_P}(-)$, $\mathbf{F_{PH}}(-)$, $\mathbf{F_H}(-)$, and $\mathbf{F''_H}(-)$ have been reflected with respect to the horizontal axis and combined with the structure factors for the reflection $(h\,k\,l)$. Note that α_H is the phase angle for the nonanomalous part of $\mathbf{F_H}$.

can be approximated by a summation with the coefficients $|F_H|^2 \sin^2 (\alpha_{PH} - \alpha_H) = \frac{1}{2}|F_H|^2 - \frac{1}{2}|F_H|^2 \cos 2(\alpha_{PH} - \alpha_H)$. This will give a Patterson map of the anomalous scatterers (the heavy atoms) (see below).

The Patterson Summation with the Coefficients $(\Delta|F|_{ano})^2$

To simplify the derivation, the structure factors $\mathbf{F}_P(-)$, $\mathbf{F}_{PH}(-)$, $\mathbf{F}_H(-)$, and $\mathbf{F}_H''(-)$ have been reflected with respect to the horizontal axis in the Argand diagam and combined with the structure factors for the reflection $h\,k\,l$. The structure factors $\mathbf{F}_{PH}(+)$ and $\mathbf{F}_{PH}(-)$ have the imaginary part of the anomalous scattering included and $\mathbf{F}_{PH}$ has it excluded. The application of the cos rule in triangle ABC gives

$$|F_{PH}(+)|^2 = |F_{PH}|^2 + |F_H''|^2 - 2|F_H''| \times |F_{PH}| \cos\{90° + (\alpha_{PH} - \alpha_H)\}$$
$$= |F_{PH}|^2 + |F_H''|^2 + 2|F_H''| \times |F_{PH}| \sin(\alpha_{PH} - \alpha_H)$$

and in triangle ABD:

$$|F_{PH}(-)|^2 = |F_{PH}|^2 + |F_H''|^2 - 2|F_H''| \times |F_{PH}| \cos\{90° - (\alpha_{PH} - \alpha_H)\}$$
$$= |F_{PH}|^2 + |F_H''|^2 - 2|F_H''| \times |F_{PH}| \sin(\alpha_{PH} - \alpha_H)$$
$$|F_{PH}(+)|^2 - |F_{PH}(-)|^2 = 4|F_H''| \times |F_{PH}| \sin(\alpha_{PH} - \alpha_H)$$

but

$$|F_{PH}(+)|^2 - |F_{PH}(-)|^2 = (|F_{PH}(+)| + |F_{PH}(-)|) \times (|F_{PH}(+)| - |F_{PH}(-)|)$$
$$\cong 2|F_{PH}| \times (|F_{PH}(+)| - |F_{PH}(-)|)$$

and it follows that

$$|F_{PH}(+)| - |F_{PH}(-)| = 2|F_H''| \sin(\alpha_{PH} - \alpha_H) \qquad (7.28)$$

The scattering factor of a heavy atom j is $f_j = f_j' + if_j''$. Assuming that the proportion f_j''/f_j' is the same for all heavy atoms in the structure:

$$|\mathbf{F}_H''| = \left| \sum_j f_j'' \exp[2\pi i\mathbf{h} \cdot \mathbf{r}_j] \right| = \left| \sum_j \frac{f_j''}{f_j'} \times f_j' \exp[2\pi i\mathbf{h} \cdot \mathbf{r}_j] \right|$$
$$= \frac{f_j''}{f_j'} \times \left| \sum_j f_j' \exp[2\pi i\mathbf{h} \cdot \mathbf{r}_j] \right| = \frac{f''}{f'} \times |F_H|$$

$$|F_{PH}(+)| - |F_{PH}(-)| = \frac{2f''}{f'}|F_H| \sin(\alpha_{PH} - \alpha_H) \qquad (7.29)$$

or

$$\Delta|F|_{ano} = |F_H| \sin(\alpha_{PH} - \alpha_H) \qquad (7.30)$$
$$(\Delta|F|_{ano})^2 = |F_H|^2 \sin^2(\alpha_{PH} - \alpha_H)$$
$$= \frac{1}{2}|F_H|^2 - \frac{1}{2}|F_H|^2 \cos 2(\alpha_{PH} - \alpha_H)$$

As in the derivation of the $(\Delta|F|_{iso})^2$ Patterson, we obtain two terms. The second term leads to noise in the Patterson map and the first one leads to the Patterson peaks of the heavy atoms.

It is interesting to compare the coefficients for the isomorphous and the anomalous Patterson:

$$(\Delta|F|_{iso})^2 = |F_H|^2 \cos^2(\alpha_{PH} - \alpha_H)$$
$$\frac{(\Delta|F|_{ano})^2 = |F_H|^2 \sin^2(\alpha_{PH} - \alpha_H)}{(\Delta|F|_{iso})^2 + (\Delta|F|_{ano})^2 = |F_H|^2}$$

It is clear that a Patterson summation calculated with the coefficients

$$(\Delta|F|_{iso})^2 + (\Delta|F|_{ano})^2$$

will give, within the framework of the approximations made, an exact Patterson map of the heavy atoms, or at least a map with lower noise than the isomorphous or anomalous Patterson summation themselves.

7.10. One Common Origin for All Derivatives

If a number of heavy atom derivatives have been used for the phase determination, the position of the heavy atoms in all derivatives should be determined with respect to the same origin. In some space groups it is not a serious problem to find a common origin because crystal symmetry limits the choice of origin. An example is given in Figure 7.13. The origin can be chosen in the positions I, II, III, or IV at $z = 1/4$ or $3/4$. The choice has an effect on the phase angles. For instance, for the $(h\,k\,0)$ reflections:

Origin in I:

$$\mathbf{F}_{hk0} = \sum_j f_j \exp[2\pi i(hx_j + ky_j)]$$

Origin in II:

$$\mathbf{F}_{hk0} = \sum_j f_j \exp\left[2\pi i\left(hx_j + k\left\{y_j - \frac{1}{2}\right\}\right)\right]$$
$$= \sum_j f_j \exp[2\pi i(hx_j + ky_j) - i\pi k]$$

The phase angles stay the same if k is even but change by π when k is odd. By comparing the sets of protein phase angles obtained from dif-

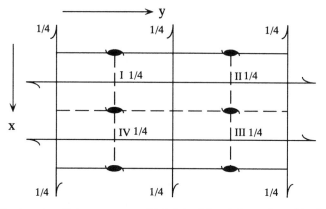

Figure 7.13. A unit cell (space group $P2_12_12_1$) with origin at the midpoint of three nonintersecting pairs of parallel 2-fold screw axes: at position I, II, III, or IV at $z = 1/4$ or 3/4.

ferent heavy atom derivatives, one can easily determine a common origin. In another method, the protein phase angles are derived for a centrosymmetric projection from one derivative and then a difference Fourier is calculated for a second derivative. These preliminary heavy atom positions are now also fixed with respect to the same origin. For space groups with lower symmetry the origin problem is different. For instance, in monoclinic space groups with the 2-fold axis along b, a common origin can easily be found in the x, z plane, but not along y. It is somewhere between $y = 0$ and 1. A very straightforward method of finding a common origin is to calculate a difference Fourier summation with $\{|F_{PH_2}| - |F_P|\}$ $\exp[\alpha_P]$, where the phase angles α_P are derived from derivative 1. $|F_{PH_2}|$ are the structure factor amplitudes for derivative 2. The "common origin" problem could also be solved, at least in principle, if a derivative could be prepared in which heavy atoms of type 1 and type 2 are jointly attached to the protein. A difference Patterson summation would then correspond to a structure containing both the heavy atoms 1 and 2. The coefficients for calculating this Patterson summation are

$$\mathbf{F}_{H_1+H_2}\mathbf{F}^*_{H_1+H_2} = |\mathbf{F}_{H_1+H_2}|^2 = (\mathbf{F}_{H_1} + \mathbf{F}_{H_2})(\mathbf{F}^*_{H_1} + \mathbf{F}^*_{H_2})$$

$$= \mathbf{F}_{H_1}\mathbf{F}^*_{H_1} + \mathbf{F}_{H_2}\mathbf{F}^*_{H_2} + \mathbf{F}_{H_1}\mathbf{F}^*_{H_2} + \mathbf{F}^*_{H_1}\mathbf{F}_{H_2} \quad (7.31)$$

$$-\,\text{I}\,\text{------}\,\text{II}\,\text{------}\,\text{IIIa}\,\text{------}\,\text{IIIb}-$$

The coefficients I give rise to the self-Patterson peaks of the heavy atoms 1, the coefficients II to the self-Patterson peaks of the atoms 2, and the coefficients III to cross-Patterson peaks between 1 and 2. IIIa leads to cross-peaks on positions $\mathbf{r}_{H_1} - \mathbf{r}_{H_2}$ and IIIb to the centrosymmetrically related cross-peaks $\mathbf{r}_{H_2} - \mathbf{r}_{H_1}$. For the derivation see below.

The Peaks in a Patterson Map if Two Kinds of Heavy Atoms Have Been Attached to the Protein Simultaneously

To obtain the value of the Patterson at position $\mathbf{u}$ we scan in principle through the electron density in the real unit cell with a vector $\mathbf{u}$ and calculate

$$P(\mathbf{u}) = \int_{\mathbf{r}} \rho(\mathbf{r}) \times \rho(\mathbf{r} + \mathbf{u}) \, dv$$

$P(\mathbf{u})$ has a high value (a peak), if at both $\mathbf{r}$ and $\mathbf{r} + \mathbf{u}$ the electron density is high. For the cross-peaks, $P(\mathbf{u})$ is high if the vector $\mathbf{u}$ begins at a high density in the first heavy atom structure (ρ_1) and ends at a high density in the second heavy atom structure (ρ_2).

$$P(\mathbf{u}) = \int_{\mathbf{r}_1} \rho_1(\mathbf{r}_1) \times \rho_2(\mathbf{r}_2) \, dv$$

with

$$\mathbf{r}_2 = \mathbf{r}_1 + \mathbf{u} \quad \text{or} \quad \mathbf{u} = \mathbf{r}_2 - \mathbf{r}_1$$

$$\rho_1(\mathbf{r}_1) = \frac{1}{V} \sum_{\mathbf{S}} \mathbf{F}_1(\mathbf{S}) \exp[-2\pi i \mathbf{r} \cdot \mathbf{S}]$$

$$\rho_2(\mathbf{r}_1 + \mathbf{u}) = \frac{1}{V} \sum_{\mathbf{S}'} \mathbf{F}_2(\mathbf{S}') \exp[-2\pi i (\mathbf{r}_1 + \mathbf{u})\mathbf{S}']$$

Replace $\mathbf{S}'$ by $-\mathbf{S}'$ and $\mathbf{F}_2(-\mathbf{S}')$ by $\mathbf{F}_2^*(\mathbf{S}')$:

$$\rho_2(\mathbf{r}_1 + \mathbf{u}) = \frac{1}{V} \sum_{\mathbf{S}'} \mathbf{F}_2^*(\mathbf{S}') \exp[2\pi i (\mathbf{r}_1 + \mathbf{u})\mathbf{S}']$$

$$\rho_1(\mathbf{r}_1) \times \rho_2(\mathbf{r}_1 + \mathbf{u}) = \frac{1}{V^2} \sum_{\mathbf{S}} \sum_{\mathbf{S}'} \mathbf{F}_1(\mathbf{S})\mathbf{F}_2^*(\mathbf{S}') \exp[2\pi i \mathbf{r}_1(\mathbf{S}' - \mathbf{S}) + 2\pi i \mathbf{u} \cdot \mathbf{S}']$$

$$\int_{\mathbf{r}_1} \rho_1(\mathbf{r}_1) \times \rho_2(\mathbf{r}_1 + \mathbf{u}) \, dv = P(\mathbf{u})$$

$$= \frac{1}{V^2} \sum_{\mathbf{S}} \sum_{\mathbf{S}'} \mathbf{F}_1(\mathbf{S})\mathbf{F}_2^*(\mathbf{S}') \exp[2\pi i \mathbf{u} \cdot \mathbf{S}'] \int_{\mathbf{r}_1} \exp[2\pi i \mathbf{r}_1(\mathbf{S}' - \mathbf{S})] \, dv$$

The integral on the right is 0, unless

$$\mathbf{S}' - \mathbf{S} = 0 \quad \text{or} \quad \mathbf{S}' = \mathbf{S}$$

Therefore,

$$P(\mathbf{u}) = \frac{1}{V^2} \sum_{\mathbf{S}} \mathbf{F}_1(\mathbf{S})\mathbf{F}_2^*(\mathbf{S}) \exp[2\pi i \mathbf{u} \cdot \mathbf{S}] \int_{\mathbf{r}_1} 1 \, dv$$

$$= \frac{1}{V} \sum_{\mathbf{S}} \mathbf{F}_1(\mathbf{S})\mathbf{F}_2^*(\mathbf{S}) \exp[2\pi i \mathbf{u} \cdot \mathbf{S}]$$

$$= \frac{1}{V} \sum_{\mathbf{S}} \mathbf{F}_1^*(\mathbf{S})\mathbf{F}_2(\mathbf{S}) \exp[-2\pi i \mathbf{u} \cdot \mathbf{S}]$$

$\mathbf{S}$ has been replaced by $-\mathbf{S}$ because a Fourier summation is calculated with $\exp[-2\pi i \mathbf{u} \cdot \mathbf{S}]$. The coefficients in this Fourier summation are

$$\mathbf{F}_1^*(\mathbf{S})\mathbf{F}_2(\mathbf{S}) = |F_1(\mathbf{S})| \times |F_2(\mathbf{S})| \exp[i\{\alpha_2(\mathbf{S}) - \alpha_1(\mathbf{S})\}]$$

(see Figure 7.14). This result tells us that the Fourier summation will give a noncentrosymmetric structure of $P(\mathbf{u})$ with peaks at the positions $\mathbf{u} = \mathbf{r}_2 - \mathbf{r}_1$. If we use the coefficients

$$\mathbf{F}_1(\mathbf{S})\mathbf{F}_2^*(\mathbf{S}) = |F_1(\mathbf{S})| \times |F_2(\mathbf{S})| \exp[-i\{\alpha_2(\mathbf{S}) - \alpha_1(\mathbf{S})\}]$$

then $P(\mathbf{u})$ has no peaks at the positions $\mathbf{r}_2 - \mathbf{r}_1$ but instead at the centrosymmetrically related positions $\mathbf{r}_1 - \mathbf{r}_2$.

Summarizing the four coefficients of the Patterson summation for the structure that would contain the heavy atoms of type 1 and type 2:

I. $\mathbf{F}_{H_1}\mathbf{F}_{H_1}^*$: phase angle 0; Patterson of atoms $1 \rightarrow 1$.
II. $\mathbf{F}_{H_2}\mathbf{F}_{H_2}^*$: phase angle 0; Patterson of atoms $2 \rightarrow 2$.
IIIa. $\mathbf{F}_{H_1}\mathbf{F}_{H_2}^*$: phase angle $\alpha_1 - \alpha_2$; Fourier with peaks at the end of the vectors $\mathbf{u}$ that begin at a density on position $\mathbf{r}_2$ in structure 2 and have their end on position $\mathbf{r}_1$ in structure 1: $\mathbf{u} = \mathbf{r}_1 - \mathbf{r}_2$.

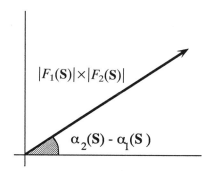

Figure 7.14. A cross-Patterson function between two heavy atom derivatives has Fourier terms with coefficients $|F_1(\mathbf{S})| \times |F_2(\mathbf{S})| \exp[-i\{\alpha_2(\mathbf{S}) - \alpha_1(\mathbf{S})\}]$.

IIIb. $F^*_{H_1}F_{H_2}$: phase angle $\alpha_2 - \alpha_1$; Fourier with peaks at the end of the vectors $\mathbf{u} = \mathbf{r}_2 - \mathbf{r}_1$.

7.11. Refinement of the Heavy Atom Parameters Using Preliminary Protein Phase Angles

After the heavy atom structure has been derived from a Patterson map, the heavy atom parameters can be improved (refined) by modifying them in such a way that $|F_{PH}(\text{obs})|$ and $|F_{PH}(\text{calc})|$ approach each other as close as possible. The refinement is carried out by means of the method of least squares or maximum likelihood. The least squares method will be discussed briefly in this Section. For a more extensive discussion see Section 13.2.1 and for the maximum likelihood method see Section 13.2.2.

The Method of Least Squares

In crystallography the measured data set has for each reflection $(h\,k\,l)$ an intensity $I(h\,k\,l)$ from which the amplitude of the structure factor $|F_{obs}(h\,k\,l)|$ can be derived. From the preliminary model, values for the structure factors $\mathbf{F}_{calc}(h\,k\,l)$ can be calculated and in the refinement procedure the values of $|F_{calc}(h\,k\,l)|$ should be brought as close as possible to $|F_{obs}(h\,k\,l)|$ for all reflections $(h\,k\,l)$. $|F_{calc}|$ can be varied by changing the parameters of the model. For some reflections $|F_{calc}|$ will be larger than $|F_{obs}|$ and for others it is just the other way around. We assume that the $|F_{obs}|$ values are free from systematic errors and distributed as in a Gaussian error curve (see Section 5.1) around their real values $|F_{real}|$, which means that the probability P of finding a value $|F_{obs}(h\,k\,l)|$ for the reflection $(h\,k\,l)$ between $|F_{obs}(h\,k\,l)|$ and $|F_{obs}(h\,k\,l)| + d|F_{obs}(h\,k\,l)|$ is

$$P(h\,k\,l) = \frac{1}{\sigma\sqrt{2\pi}}\exp\left[-\frac{\{|F_{obs}(h\,k\,l)| - |F_{real}(h\,k\,l)|\}^2}{2\sigma^2}\right]d|F_{obs}(h\,k\,l)|$$

σ^2 is the variance caused by arbitrary errors in the measurements:

$$\sigma^2 = \int_{-\infty}^{+\infty}\{|F_{obs}(h\,k\,l)| - \overline{|F_{obs}(h\,k\,l)|}\}^2 P(h\,k\,l)\,d|F_{obs}(h\,k\,l)|$$

Normalization requires

$$\int_{-\infty}^{+\infty} P(h\,k\,l)\,d\{|F_{obs}(h\,k\,l)|\} = 1$$

With the assumption that the errors in the $|F_{obs}(h\,k\,l)|$ values for different reflections are independent of each other, the total probability P for finding a certain set of $|F_{obs}(h\,k\,l)|$ is

$$P = \prod_{hkl} P(h\,k\,l)$$

$$= \prod_{hkl} \frac{1}{\sigma(h\,k\,l)\sqrt{2\pi}} \exp\left[-\frac{\{|F_{obs}(h\,k\,l)| - |F_{real}(h\,k\,l)|\}^2}{2\sigma^2(h\,k\,l)}\right] d|F_{obs}(h\,k\,l)|$$

$$= \exp\left[-\sum_{hkl} \frac{\{|F_{obs}(h\,k\,l)| - |F_{real}(h\,k\,l)|\}^2}{2\sigma^2(h\,k\,l)}\right] \times$$

$$\prod_{hkl} \frac{1}{\sigma(h\,k\,l)\sqrt{2\pi}} d|F_{obs}(h\,k\,l)|$$

The problem is that the real values of the $F(h\,k\,l)$s are unknown. However, it is assumed that these real values can be approximated by the calculated values. The goal is to bring the set of $|F_{calc}|$s as close as possible to the $|F_{obs}|$s. In the method of least squares this is defined as occurring at the maximum value of P. In other words, the optimal set of $|F_{calc}|$s is the one that has the highest probability P. A maximum for P is obtained for a minimum of

$$\sum_{hkl} \frac{\{|F_{obs}(h\,k\,l)| - |F_{calc}(h\,k\,l)|\}^2}{2\sigma^2(h\,k\,l)}$$

This is the principle of least squares.

The least squares minimum is found by varying the $|F_{calc}(h\,k\,l)|$s. This is done by differentiating with respect to the parameters of the atoms and setting the derivatives equal to zero.

The parameters of the heavy atoms, as derived from the difference Pattersons, can be improved (refined) with the "lack of closure" method (Dickerson et al., 1968). This requires preliminary values for the protein phase angles α_P. With the value of α_P known, the vector triangle $\mathbf{F}_P + \mathbf{F}_H = \mathbf{F}_{PH}$ (Figure 7.15a) can be drawn for each reflection. The length and direction of $\mathbf{F}_P$ and $\mathbf{F}_H$ are known and $\mathbf{F}_{PH}$ is then pointed to the end of vector $\mathbf{F}_H$. In practice the observed amplitude $|F_{PH}|$ will be too short or too long to exactly reach the endpoint of $\mathbf{F}_H$ (Figure 7.15b). The difference is called the "lack of closure error" ε. It is due to measurement errors and non-isomorphism. The goal of the refinement is to make the errors ε as small as possible. Or, in other words, the criterium of the refinement is to bring $|F_{PH}|_{calc}$ as close as possible to $|F_{PH}|_{obs}$. For each reflection, $|F_{PH}|_{calc}$ is calculated with the cosine rule:

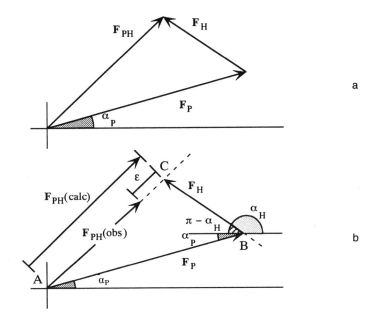

Figure 7.15. (a) The ideal isomorphous situation in which the vector triangle $\mathbf{F_P}$ + $\mathbf{F_H}$ = $\mathbf{F_{PH}}$ closes exactly. Normally, this is not true and the observed and calculated values of $|F_{PH}|$ differ by the lack of closure error ε (b).

$$|F_{PH}|_{calc} = \{|F_P|^2 + |F_H|^2 + 2|F_P| \times |F_H| \cos(\alpha_H - \alpha_P)\}^{1/2}$$

For the heavy atom derivative j, ε_j is defined as

$$\varepsilon_j = \{k_j(|F_{PH}|_{obs})_j - (|F_{PH}|_{calc})_j\} \qquad (7.32)$$

where k_j is a scaling factor. Original refinement procedures were based on least squares, minimizing the function

$$E_j = \sum_{hkl} m_{hkl}\varepsilon_j(h\ k\ l)^2 \qquad (7.33)$$

The variables are the parameters of the heavy atoms in derivative j: atomic coordinates, occupancy, temperature factor. Their value determines the length and direction of $\mathbf{F_H}$ and therefore its endpoint in the Argand diagram. Moreover the scale factor k_j can be refined. m_{hkl} is a weighting factor that indicates the quality of the phase angle; m is the figure of merit, defined in the next section. After one or a few refinement cycles new values for the protein phase angles α_P are calculated as will be described in the next section. It is advised to calculate the protein phase angles without using the isomorphous derivative j under consideration.

Otherwise the result depends to some extent on the input data. This is especially important if one derivative dominates the phase angle determination (Blow and Matthews, 1973).

Least squares refinement of diffraction data is not always the best procedure for finding optimal agreement between model data and experimental data. This will be pointed out in Section 13.2.2. In principle, a better technique employs a maximum likelihood algorithm in which data are replaced by their probabilities. For instance, in the program MLPHARE for the refinement of heavy-atom positions, the single value of the protein phase angle is replaced by an integrated value of all phase angles, weighted by their likelihood (Otwinowski, 1991). This will be further discussed in Section 13.2.3 after an introduction to the principle of maximum likelihood (Section 13.2.2).

7.12. Protein Phase Angles

After the refinement of heavy atom positions, protein phase angles can be determined. The principle, due to Harker (1956), is as follows. Draw a circle with radius $|F_P|$. From the center of the circle vector $-\mathbf{F}_H$ is drawn and next a second circle with radius $|F_{PH}|$ and with its center at the end of vector $-\mathbf{F}_H$ (Figure 7.16). The intersections of the two circles correspond to two equally probable protein phase angles, because for both points the triangle $\mathbf{F}_{PH} = \mathbf{F}_P + \mathbf{F}_H$ closes exactly. With a second heavy atom derivative one can, in principle, distinguish between the two alternatives. However, because of errors an exact intersection of the three circles with radii $|F_P|$, $|F_{PH_1}|$, and $|F_{PH_2}|$ will usually not be obtained, and some uncertainty as to the correct phase angle α_P remains. The errors are

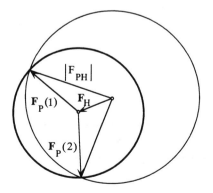

Figure 7.16. Harker construction for protein phase determination. In the isomorphous replacement method each heavy atom derivative gives two possibilities for the protein phase angle α_P, corresponding to the two vectors $\mathbf{F}_P(1)$ and $\mathbf{F}_P(2)$.

introduced in X-ray intensity data collection or by poor isomorphism. In practice more than two derivatives are used, if they are available, and, therefore, the method is called multiple isomorphous replacement (MIR).

For mathematical reasons the best procedure to follow is the following: The vector triangle, $\mathbf{F_{PH}} = \mathbf{F_P} + \mathbf{F_H}$, closes exactly only at the two intersection points of the circles with radii $|F_P|$ and $|F_{PH}|$ (see Figure 7.16). These intersection points correspond to two values of α_P. For all other values of α_P a closure error remains:

$$\varepsilon = A\,C - |F_{PH}|_{obs}$$

(see Figure 7.15b). It is assumed that all errors concern the length of $\mathbf{F_{PH}}$ and that both $\mathbf{F_H}$ and $\mathbf{F_P}$ are error-free. For every value of the protein phase angle α_P, the length of $A\,C$ in Figure 7.15b can be calculated with the cosine rule in triangle $A\,B\,C$, because $|F_P|$ and $\mathbf{F_H}$ are known. $\varepsilon(\alpha)$ is the difference between this calculated $A\,C$ and the observed value for $|F_{PH}|$ for a protein phase angle α. If $\varepsilon(\alpha)$ is smaller, chances for having the correct protein phase angle α are higher.

A Gaussian probability distribution is assumed for ε, and for each reflection in the diffraction pattern of one derivative:

$$P(\alpha) = P(\varepsilon) = N \exp\left[-\frac{\varepsilon^2(\alpha)}{2E^2} \right]$$

N is a normalization factor. It is related to the fact that the phase angle α is somewhere between 0 and 2π:

$$\int_{\alpha=0}^{2\pi} P(\alpha)\, d\alpha = 1$$

E^2 is the "mean square" value of ε. If E is small, the probability curves will have sharp peaks and the protein phase angle is well-determined, but for large E-values they are very poorly determined. For each reflection, $P(\alpha)$ is obtained as a function of α by calculating it for every few degrees around the phase circle. This function is symmetrical around a point D in Figure 7.17. The two equally high peaks to the left and right of D correspond to the intersection points of the circles with radii $|F_P|$ and $|F_{PH}|$. These curves can be calculated for each reflection and each of the n heavy atom derivatives. The total probability for each reflection is obtained by multiplying the separate probabilities (see below):

$$P(\alpha) = \prod_{j=1}^{n} P_j(\alpha) = N' \exp\left[-\sum_j \frac{\varepsilon_j^2(\alpha)}{2E_j^2} \right] \tag{7.34}$$

The total $P(\alpha)$ curve will normally not show the symmetry of Figure 7.17, but will look like Figure 7.18a or b.

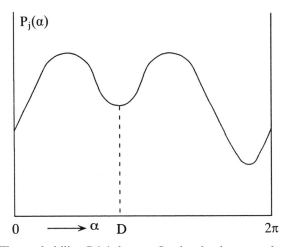

Figure 7.17. The probability $P_j(\alpha)$ for a reflection having α as the correct phase angle derived from derivative j is shown as a function of all angles between 0 and 2π.

The fundamental rule for the combination of probabilities is as follows:

If events are independent of each other, the probability of their occurring at the same time is the *product* of the probabilities of their separate occurrence.

Probabilities should be *added* if the events exclude each other. For instance, the chance that a particular plane of a die is on top is 1/6, but the chance that a 1 or a 5 is on top is 1/6 + 1/6 = 1/3.

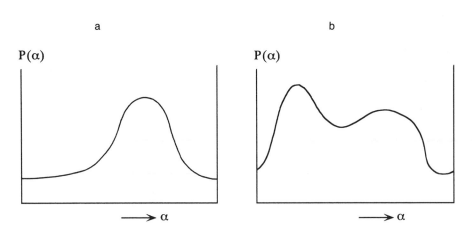

Figure 7.18. (a, b) Two examples of the total probability $P(\alpha)$ for the phase angle α of a reflection as derived from more than one derivative.

If an electron density map for a protein is calculated with a certain set of phase angles $\alpha(h\ k\ l)$, the probability that the particular map is correct is equal to $\Pi_{hkl}P_{hkl}(\alpha_{hkl})$. But how can we find the best Fourier map? At first thought, one would choose for each reflection the value $\alpha(h\ k\ l)$ for which $P_{hkl}(\alpha_{hkl})$ has a maximum and, therefore, their product $\Pi_{hkl}P_{hkl}(\alpha_{hkl})$ also has a maximum value. Accepting the maximum value of $\Pi_{hkl}P_{hkl}(\alpha_{hkl})$ would result in the most probable electron density map. However, this most probable map is not necessarily the best electron density map. This is defined as the map with the minimum mean square error in the electron density due to errors in the phase angles. The reason is that the function

$$P_{hkl}(\alpha) = \prod_{j=1}^{n} P_j(\alpha_{hkl})$$

does not always have a single maximum, but sometimes two, or is asymmetric around its maximum (Figure 7.18). In calculating the most probable electron density map, this is completely neglected. The best solution for this problem is as follows. For the structure factor $\mathbf{F}_{hkl} = |F_{hkl}| \exp[i\alpha_{hkl}]$ the amplitude is known, whereas from the experimental data of the derivatives the probability distribution $P_{hkl}(\alpha)$ for the phase angle α can be derived. The probability for the structure factor $\mathbf{F}_{hkl}$ to be $|F_{hkl}| \exp[i\alpha]$ is thus $P_{hkl}(\alpha)$. The best estimate, $\mathbf{F}_{hkl}(\text{best})$, for the actual structure factor on the basis of the present experimental data is given by the least-squares criterion:

$$Q = \int_{\alpha} \{P_{hkl}(\alpha)|F_{hkl}| \exp[i\alpha] - \mathbf{F}_{hkl}(\text{best})\}^2 \, d\alpha$$

should be a minimum. With

$$\frac{dQ}{d\{\mathbf{F}_{hkl}(\text{best})\}} = 0$$

it is found that

$$\mathbf{F}_{hkl}(\text{best}) = \int_{\alpha} \{P_{hkl}(\alpha)|F_{hkl}| \exp[i\alpha]\} \, d\alpha = |F_{hkl}|\mathbf{m} \qquad (7.35)$$

with

$$\mathbf{m} = \int_{\alpha} \{P_{hkl}(\alpha) \exp[i\alpha]\} \, d\alpha; \quad 0 \leqslant |m| \leqslant 1$$

As a result, the best value of $\mathbf{F}_{hkl}$ is obtained by taking the weighted average over the range of possible $\mathbf{F}_{hkl}$s. $\mathbf{F}_{hkl}(\text{best})$ points to the center of gravity ("centroid") of the probability distribution of $\mathbf{F}$ (Figure 7.19). In practice the integration is replaced by a summation in steps of, say, 5° and

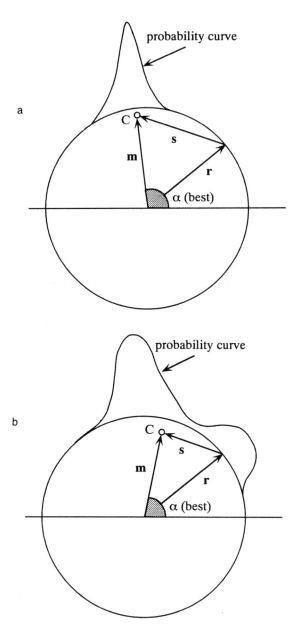

Figure 7.19. Three probability curves for the protein phase angle. The baseline for the curves is the circle with radius $|r| = 1$. C is the centroid of the probability distribution and **m** the vector that connects the center of the circle with C. In other words, the more spread-out the probability curve is, the poorer is the determination of the phase angle, and the shorter is **m**. (a) The sharp peak of the probability curve positions point C close to the circle; (b) it is somewhat further away; (c) it is close to the center of the circle.

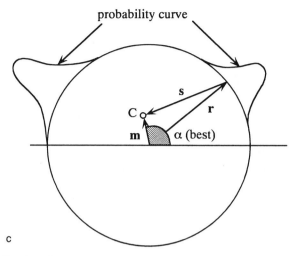

probability curve

s

C

r

m   α (best)

c

Figure 7.19. *Continued*

$$F_{\text{best}} = \frac{\Sigma P(\alpha)\mathbf{F}(\alpha)}{\Sigma P(\alpha)} \qquad (7.36)$$

With $\mathbf{m} = m \exp[i\alpha(\text{best})]$ Eq. (7.35) becomes

$$\mathbf{F}_{hkl}(\text{best}) = |F_{hkl}|m \exp[i\alpha(\text{best})] \qquad (7.37)$$

and

$$m = \frac{|F_{hkl}(\text{best})|}{|F_{hkl}|} \qquad (7.38)$$

m is called the "figure of merit."

If the $\mathbf{F}_{hkl}(\text{best})$ values are used in the calculation of the electron density distribution, a map is obtained that minimizes the mean square error in the electron density due to errors in the phase angles. The interpretation of the isomorphous replacement map in terms of a polypeptide chain produces a first model of the protein molecular structure that is certainly not the final model. That final model is obtained after refinement whose aim is adjusting the model to find a closer agreement between the structure factors calculated on the basis of the model and the observed structure factors (Chapter 13). In the procedure described above, which is due to Blow and Crick (1959), errors in the intensity measurements and in the heavy atom model are lumped together and treated as Gaussian errors in the measured $|F_{\text{PH}}|$.

This is of course an approximation and more detailed treatments of the propagation of the various errors into the protein phase angles have been given (Read, 1990a). The difference between these other procedures,

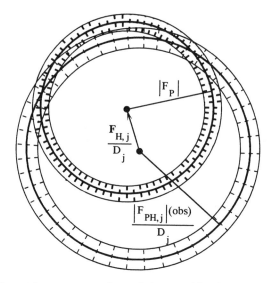

Figure 7.20. Schematic representation of the combination of information from native and derivative data. The shaded areas give an impression of the Gaussian distribution for errors in $|F_P|$ and $|F_{PH}|$. Where the shaded areas cross, a high joint probability exists for the true structure factor conditional on the observed structure factor amplitude for the native protein (dark dashes) and conditional on the observed structure factor amplitude $|F_{PH,j}|$ and the calculated structure factor $\mathbf{F}_{H,j}$ for the heavy atom derivative j (light dashes). D_j is a multiplication factor that is 1 for perfect isomorphism, but is usually less than 1; D_j will be discussed in Section 15.6.

which include errors, and the Blow and Crick procedure with sharp circles is schematically represented in Figure 7.20. Bella and Rossmann (1998) wrote a program WHALESQ that treats all $|F_P|$ and $|F_{PH}|$ observations in an equivalent manner. Moreover, the parameter refinement is not restricted to isomorphously attached heavy atoms but it can combine MIR and MAD information (Section 9.5).

7.13. The Remaining Error in the Best Fourier Map

The isomorphous replacement method provides us with a preliminary model of the protein structure; a final model is obtained only after refinement of the structure. Therefore, from a practical point of view we are not particularly interested in the errors that are present in a best Fourier map as calculated in the previous section. From an instructional point of view, however, it is worthwhile to consider these errors.

In Figure 7.19 phase circles are drawn with radius $|r| = 1$. The curve that describes the probability that the phase angle is correct is drawn on the circle as a baseline. The centroid of the probability distribution is not on the circle, but at C somewhat closer to the center of the circle. If the probability distribution is sharp, as in Figure 7.19a, the centroid is nearly on the circle and the phase angle is well defined. If the distribution is as shown in Figure 7.19c, the centroid is near the center of the circle and the phase angle is extremely poorly defined. m is the vector from the center of the circle to the centroid C; α(best) is the best phase angle. If α is an arbitrary phase angle

$$\mathbf{F}(\text{best}) = \frac{\Sigma P(\alpha)\mathbf{F}(\alpha)}{\Sigma P(\alpha)}$$

dividing by $|F|$ gives

$$m \cdot \exp[i\alpha(\text{best})] = \mathbf{m} = \frac{\Sigma P(\alpha)\mathbf{r}(\alpha)}{\Sigma P(\alpha)}$$

and because $\mathbf{r}(\alpha) = 1 \cdot \exp[i\alpha]$,

$$\mathbf{m} = \frac{\Sigma P(\alpha)\,\exp[i\alpha]}{\Sigma P(\alpha)} \tag{7.39}$$

The real part of vector $\mathbf{m}$ is its projection on the horizontal axis in the Argand diagram:

$$m\cos\alpha(\text{best}) = \frac{\Sigma P(\alpha)\,\cos(\alpha)}{\Sigma P(\alpha)}$$

and its imaginary part is

$$m\sin\alpha(\text{best}) = \frac{\Sigma P(\alpha)\,\sin(\alpha)}{\Sigma P(\alpha)}$$

If for every reflection the origin is moved from $\alpha = 0$ to $\alpha = \alpha$(best), then $\cos\alpha(\text{best}) \to 1$ and $\sin\alpha(\text{best}) \to 0$

$$m = \frac{\sum P\{\alpha - \alpha(\text{best})\}\cos\{\alpha - \alpha(\text{best})\}}{\sum P(\alpha)} = \overline{\cos\{\alpha - \alpha(\text{best})\}} \tag{7.40}$$

For reflections with a very well-defined phase angle both m and $\overline{\cos\{\alpha - \alpha(\text{best})\}} \cong 1$. In other words $\{\alpha - \alpha(\text{best})\}$ is very small for phase angles α, which have a high value of $P(\alpha)$. If the curve for $P(\alpha)$ shows a large spread, the value of $\overline{\cos\{\alpha - \alpha(\text{best})\}}$ and of m is much smaller than 1. We can interpret m as the weighted mean of the cosine of the error in the phase angle.

Let us now estimate the error in the electron density map. Suppose that for reflection hkl the true structure factor is $\mathbf{F}_{hkl}(\text{true})$. However, because of errors in our data set, it has been determined as $\mathbf{F}_{hkl}(\text{best})$. For $\bar{h}\bar{k}\bar{l}$ we have $\mathbf{F}^*(\text{best})$ instead of $\mathbf{F}^*(\text{true})$. The incorrect structure factors cause errors in the electron density map. The contribution to the error by reflections $h\,k\,l$ and $\bar{h}\,\bar{k}\,\bar{l}$ is, if we write $\mathbf{F_b}$ for $\mathbf{F}_{hkl}$(best) and $\mathbf{F_T}$ for $\mathbf{F}_{hkl}$(true):

$$\Delta\rho_{hkl} = \frac{1}{V}\{(\mathbf{F_b} - \mathbf{F_T})\,\exp[-2\pi i(hx + ky + lz)]$$

$$+ (\mathbf{F_b^*} - \mathbf{F_T^*})\,\exp[2\pi i(hx + ky + lz)]\}$$

$$(\Delta\rho_{hkl})^2 = \frac{1}{V^2}\{\cdots\}^2 \tag{7.41}$$

Equation (7.41) contains the following terms in the right part:

$$(\mathbf{F_b} - \mathbf{F_T})^2\,\exp[-4\pi i(hx + ky + lz)] = (\mathbf{F_b} - \mathbf{F_T})^2\{\cos[\cdots] - i\sin[\cdots]\}$$

$$(\mathbf{F_b^*} - \mathbf{F_T^*})^2\,\exp[4\pi i(hx + ky + lz)] = (\mathbf{F_b^*} - \mathbf{F_T^*})^2\{\cos[\cdots] + i\sin[\cdots]\}$$

$$2(\mathbf{F_b} - \mathbf{F_T})(\mathbf{F_b^*} - \mathbf{F_T^*}) = 2|\mathbf{F_b} - \mathbf{F_T}|^2$$

If the average is taken over all positions x, y, and z in the unit cell, the mean square value of $\Delta\rho_{hkl}$ is obtained: $\overline{(\Delta\rho_{hkl})^2}$. Averaging over x, y, and z makes all cosine and sin terms in Eq. (7.41) equal to 0 and the equation for $\overline{(\Delta\rho_{hkl})^2}$ simplifies to

$$\overline{(\Delta\rho_{hkl})^2} = \frac{2}{V^2}|\mathbf{F_b} - \mathbf{F_T}|^2 \tag{7.42}$$

$\mathbf{F_b} = \mathbf{F}$(best) has a fixed length and phase angle but $\mathbf{F_T}$ is unknown and, therefore, also $|\mathbf{F_b} - \mathbf{F_T}|^2$. The best value is the weighted average over all possible phase angles for $\mathbf{F_T}$:

$$\overline{(\Delta\rho_{hkl})^2} = \frac{2}{V^2} \times \frac{\displaystyle\int_{\alpha=0}^{2\pi} P_{hkl}(\alpha)[|\mathbf{F}_{\text{best}} - \mathbf{F}(\alpha)|^2]\,d\alpha}{\displaystyle\int_{\alpha=0}^{2\pi} P_{hkl}(\alpha)\,d\alpha} \tag{7.43}$$

Equation (7.43) tells us that the mean square error in the electron density is equal to $(2/V^2) \times$ the variance in $\mathbf{F}(\alpha)$ (see Section 5.1 on the Gauss error curve).

As before the integration is replaced by a summation and using

$$\left.\begin{aligned}\mathbf{F}(\text{best}) &= |F| \times \mathbf{m}\\ \mathbf{F}(\alpha) &= |F| \times \mathbf{r}\\ |r| &= 1\end{aligned}\right\}$$

$$\overline{(\Delta\rho_{hkl})^2} = \frac{2|F|^2}{V^2} \times \frac{\sum\limits_{\alpha} P(\alpha)|\mathbf{m} - \mathbf{r}|^2}{\sum\limits_{\alpha} P(\alpha)}$$

$$= \frac{2|F|^2}{V^2} \times \frac{\sum\limits_{\alpha} P(\alpha)|s|^2}{\sum\limits_{\alpha} P(\alpha)} \qquad (7.44)$$

Applying the cosine rule gives for $|s|^2$ (see Figure 7.19):

$$|s|^2 = |m|^2 + 1 - 2|m| \cos\{\alpha(\text{best}) - \alpha\}$$

Writing m for $|m|$ and substituting $|s|^2$ in Eq. (7.44):

$$\overline{(\Delta\rho_{hkl})^2} = \frac{2|F|^2}{V^2} \times \left\{ \frac{\sum\limits_{\alpha} P(\alpha)}{\sum\limits_{\alpha} P(\alpha)}(m^2 + 1) - \frac{2m\sum\limits_{\alpha} P(\alpha) \cos\{\alpha(\text{best}) - \alpha\}}{\sum\limits_{\alpha} P(\alpha)} \right\}$$

$$= \frac{2|F|^2}{V^2} \times \{m^2 + 1 - 2m^2\} = \frac{2|F|^2}{V^2} \times (1 - m^2) \qquad (7.45)$$

This is the contribution to the mean square error in the electron density caused by errors in one reflection and its Bijvoet mate. Adding the contributions from all reflections:

$$\overline{(\Delta\rho)^2} = \frac{2}{V^2}\sum_h \sum_k \sum_l |F|^2_{hkl}(1 - m^2_{hkl}) \qquad (7.46)$$

where the summation is over the reflections in half of reciprocal space. It can be derived that for difference Fourier maps a similar equation is valid:

$$\overline{\{\Delta(\Delta\rho)\}^2} = \frac{2}{V^2}\sum_h \sum_k \sum_l (\Delta|F|_{hkl})^2(2 - m^2_{hkl}) \qquad (7.47)$$

The extra $(\Delta|F|_{hkl})^2$ is due to the intrinsic error in difference Fourier maps, which is due to the fact that the direction of $\mathbf{F}_H$ is not known.

Suppose the average m is 0.8 and $m^2 = 0.64$. For the normal Fourier map we obtain

$$\overline{(\Delta\rho)^2} = 0.36 \times \left[\frac{2}{V^2}\sum_h \sum_k \sum_l |F|^2_{hkl} \right]$$

and

$$[\overline{(\Delta\rho)^2}]^{1/2} = 0.6 \times \left[\frac{2}{V^2}\sum_h \sum_k \sum_l |F|^2_{hkl} \right]^{1/2}$$

and for the difference Fourier map:

$$\overline{(\{\Delta(\Delta\rho)\}^2)} = 1.36 \times \left[\frac{2}{V^2}\sum_h\sum_k\sum_l(\Delta|F|_{hkl})^2\right]$$

If $\Delta|F|$ is of the order of $0.1 \times |F|$, and, therefore,

$$(\Delta|F|)^2 = 0.01 \times |F|^2$$

the error in the difference Fourier map is

$$\overline{\{\Delta(\Delta\rho)\}^2} = 1.36 \times 0.01 \times \left[\frac{2}{V^2}\sum_h\sum_k\sum_l(|F|_{hkl})^2\right]$$

$$= 0.0136 \times \left[\frac{2}{V^2}\sum_h\sum_k\sum_l(|F|_{hkl})^2\right]$$

$$\overline{(\{\Delta(\Delta\rho)\}^2)}^{1/2} = 0.12 \times \left[\frac{2}{V^2}\sum_h\sum_k\sum_l(|F|_{hkl})^2\right]^{1/2}$$

This comparison between the errors in a Fourier and a difference Fourier map shows that the errors in a difference Fourier map are appreciably smaller than in a normal Fourier map. This is the reason that reasonably accurate data can be derived from a difference Fourier map.

7.14. The Single Isomorphous Replacement (SIR) Method

A single heavy atom derivative gives two equally possible protein phase angles corresponding with the structure factors $\mathbf{F}_P(1)$ and $\mathbf{F}_P(2)$ (Figure 7.16). Only one of them is the correct phase angle. With a second heavy atom derivative the choice can be made, but suppose only one derivative is available. If the protein electron density map is then calculated with $\mathbf{F}_P(1) + \mathbf{F}_P(2)$, one can expect that the correct structure factors will lead to an acceptable map, whereas the incorrect ones will lead to noise in the map. The use of $\mathbf{F}_P(1) + \mathbf{F}_P(2)$ in the single isomorphous replacement method is in fact the method of calculating the best electron density map. $\mathbf{F}_P(1)$ and $\mathbf{F}_P(2)$ each has a probability of 0.5 of being the correct structure factor. Therefore,

$$\mathbf{F}_P(\text{best}) = \tfrac{1}{2}[\mathbf{F}_P(1) + \mathbf{F}_P(2)]$$

This best structure factor, $\mathbf{F}_P(\text{best})$, points along $\mathbf{F}_H$ (Figure 7.16) in the same or the opposite direction of $\mathbf{F}_H$. It has either the phase angle α_H of $\mathbf{F}_H$ or $\alpha_H + \pi$, with $\mathbf{F}_P(\text{best})$ within the smallest angle between $\mathbf{F}_P(1)$ and $\mathbf{F}_P(2)$.

The SIR method can work quite satisfactorily by producing sufficiently accurate protein phase angles for calculating an acceptable first electron density map. Its main problem is in a possible pseudocentrosymmetric relationship between the heavy atom positions. If the heavy atom arrangement has an exact center of symmetry, then for each reflection the phase angles of $\mathbf{F}_P(1)$ and $\mathbf{F}_P(2)$ are symmetric to each other with respect to the real axis in the Argand diagram. A set of structure factors, all having a phase angle opposite to the correct one, will lead to an electron density map that is centrosymmetric to the correct map. Instead of having the correct map with noise superimposed, we have the correct map with the centrosymmetric map superimposed. As a result, the interpretation of the map is impossible. Although protein structures themselves have no center of symmetry, the set of heavy atoms can have a pseudocenter of symmetry. This causes the appearance of the (undesirable) centrosymmetric protein structure in the electron density map. The result of the SIR method is appreciably improved if anomalous scattering by the heavy atoms is also taken into account in the phase determination of the protein (Section 9.2).

Summary

1. The X-ray intensities of the native and the derivative structures should be measured as accurately as possible, because the method depends on relatively small differences between these intensities, especially if anomalous differences are used.

2. The scaling of the data sets is critical. If there are absorption differences due to the nonspherical shape of the crystals, or the solvent around the crystal, or the capillary, a correction for absorption must be applied.

3. It is an advantage to have a low number of sites with a high occupancy, because it simplifies the interpretation of the difference Patterson map. Additional sites with low occupancy can be added later, after their location is determined from difference Fourier maps.

4. After determining the position of the heavy atom sites, it should be checked whether the interatomic vectors correspond with peaks in the Patterson map. Calculation of an anomalous Patterson map and comparison with the isomorphous map give an indication of the significance of the anomalous signal.

5. If it is known that noncrystallographic symmetry is present (such as a 5-fold axis of symmetry within one molecule), this symmetry may also be present between the heavy atom sites. If it is, a vector search in the difference Patterson map should reveal it.

Chapter 8
Phase Improvement

8.1. Introduction

After a first set of protein phases is obtained with the isomorphous replacement method, the molecular replacement method, or the multiple wavelength anomalous dispersion method and an electron density map calculated, the next step is the interpretation of the map in terms of the polypeptide chain. If this is successful and the major part of the chain can indeed be followed in the electron density map, refinement of the structure can begin. However, insufficient quality of the electron density map might hamper a complete and unambiguous tracing of the polypeptide chain, increasing the risk of introducing errors in the model, which cannot be easily removed during refinement. In such a case refinement should be preceded by a process to improve the quality of the map through improvement of the protein phase angles (Podjarny et al., 1987). During phase improvement, all available information on the structure should be used (Brünger and Nilges, 1993). This information may be in one of the following forms:

1. The structure is partially known.

2. The protein molecules distinguish themselves as relatively high regions of electron density and their boundaries can be estimated. The electron density between them is then set to a constant value or adjusted otherwise.

3. Noncrystallographic symmetry within the asymmetric unit is present and is known from molecular replacement. As in 2, molecular boundaries must then be determined and the solvent region modified. Moreover, the density of all molecules (or subunits of a molecule) related by noncrystallographic symmetry is averaged.

4. Correct protein electron density maps have a characteristic frequency distribution for the values of the electron density (histogram matching).

Methods 2 and 3 are examples of density modification methods. Usually application of these methods improves the electron density map to such an extent that the interpretation is no longer a problem. The crystallographer is not necessarily restricted to the resolution limits set by the isomorphous or molecular replacement method. If the X-ray pattern of the native protein crystals allows, the data limit can be moved very gradually to higher resolution. We shall now discuss these phase improvement methods in some detail.

8.2. The OMIT Map with and Without Sim Weighting

Frequently the interpretation of part of an electron density map is somewhat doubtful. For instance, a loop at the surface of the molecule cannot be traced satisfactorily. In such cases it is useful to calculate an OMIT map, that is, an electron density map with the observed structure factor amplitudes ($|F|$) and with phase angles α_K, calculated only for the part of the structure that is known correctly. This part should then show up in full height in the map, whereas the missing part is expected to show up at only about half of the actual height (Bhat and Cohen, 1984; Bhat, 1988). Because phase angles dominate electron density maps more than the amplitudes, such a map is biased toward the correctly known part of the structure (Read, 1997). The picture of the troublesome part can be improved by introducing a suitable weighting factor. A low weight should be given to the amplitudes of reflections for which the phase angles α_K can be expected to differ appreciably from the correct phase angles, and more weight given to reflections for which this difference can be expected to be small. It is assumed that the best weights are those that minimize the mean square error in electron density due to the errors in the phase angles. It was shown by Sim (1959, 1960) that the best weights are

$$w = \frac{I_1(X)}{I_0(X)}$$

for noncentric reflections and

$$\tanh\left(\frac{X}{2}\right)$$

for centric reflections, where

$$X = \frac{2|F| \times |F_K|}{\sum\limits_{1}^{n} f_i^2} \tag{8.1}$$

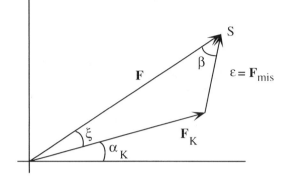

Figure 8.1. The structure factors involved in the calculation of an Omit map. **F** is the total structure factor for the reflection with an observed structure amplitude $|F|$ and a phase angle $\alpha_K + \xi$. $\mathbf{F_K}$ is the structure factor for the known part of the structure with phase angle α_K. The structure factor for the missing part is $\varepsilon = \mathbf{F}_{mis}$.

$I_0(X)$ and $I_1(X)$ are modified Bessel functions of order zero and one, respectively. $|F|$ is the observed structure factor amplitude. $|F_K|$ is the amplitude for the known part of the structure (Figure 8.1). If f_is are atomic scattering factors for n missing atoms, then $\Sigma_1^n f_i^2$ can be estimated from the structure factor amplitudes $|F|$ and $|F_K|$. Bricogne (1976) suggested $||F|^2 - |F_K|^2|$, whereas Read (1986) proposed $n(|F| - |F_K|)^2$ with $n = 1$ for centric and $n = 2$ for noncentric reflections. Bricogne's suggestion can be understood by applying the cosine rule in the triangle in Figure 8.1. $|F_{mis}|^2 = |F|^2 + |F_K|^2 - 2|F| \times |F_K| \cos \xi$. With the assumption that $|F| \cos \xi = |F_K|$ we obtain $\overline{|F_{mis}|^2} = \sum_1^n f_i^2 = \overline{(|F| - |F_K|)^2}$. Read's proposal can be easily understood for centric reflections, for which $||F| - |F_K|| = |F_{mis}|$. For noncentric reflections the assumption is made that $|F_{mis}|\cos \beta = |F| - |F_K|$ (Figure 8.1). Then $|F_{mis}|^2 = (|F| - |F_K|)^2 \times \dfrac{1}{\cos^2 \beta}$. In averaging over all reflections $\overline{\cos^2 \beta} = \dfrac{1}{2}$ and therefore $\overline{|F_{mis}|^2} = 2 \times \left(\overline{|F|^2 - |F_K|^2} \right)$.

The weighting factor w for the noncentric reflections can be derived in the following way. For one reflection, let the structure factor of the known part have an amplitude $|F_K|$ and a phase angle α_K. The correct structure factor for the entire structure has an amplitude $|F|$ and a phase angle $\alpha = \alpha_K + \xi$. $\varepsilon = \mathbf{F}_{mis}$ is the structure factor of the missing part. A single reflection $(h\,k\,l)$ and its Friedel mate $(\bar{h}\,\bar{k}\,\bar{l})$ contribute to the correct electron density ρ at position x, y, z with

$$\rho_{hkl}(x\;y\;z) = \frac{2}{V}|F|\cos[2\pi(hx + ky + lz) - \alpha]$$

$$= \frac{2|F|}{V}\cos\Psi \tag{8.2}$$

If electron density is calculated with correct $|F|$ but erroneous phase angle α_K, then α is replaced by $\alpha_K = \alpha - \xi$ and Ψ by $\Psi + \xi$ in Eq. (8.2). In this case, if a weighting factor w is applied to the amplitude of reflection $(h\,k\,l)$, the error in the electron density is

$$\Delta\rho_{hkl}(x\;y\;z) = \frac{2}{V}|F|[\cos\Psi - w\cos(\Psi + \xi)]$$

$$= \frac{2}{V}|F|[\cos\Psi - w\cos\Psi\cos\xi + w\sin\Psi\sin\xi]$$

$$= \frac{2}{V}|F|[\cos\Psi(1 - w\cos\xi) + w\sin\Psi\sin\xi]$$

$$\{\Delta\rho_{hkl}(x\;y\;z)\}^2 = \frac{4|F|^2}{V^2}[\cos^2\Psi(1 - w\cos\xi)^2 + w^2\sin^2\Psi\sin^2\xi$$

$$+ 2\cos\Psi\sin\Psi(1 - w\cos\xi)w\sin\xi]$$

The mean value for $\{\Delta\rho_{hkl}(x\;y\;z)\}^2$ over all x, y, z is obtained by averaging over Ψ:

$$\overline{\{\Delta\rho_{hkl}(x\;y\;z)\}^2} = \frac{2}{V^2}|F|^2[(1 - w\cos\xi)^2 + w^2\sin^2\xi]$$

$$= \frac{2}{V^2}|F|^2(1 - 2w\cos\xi + w^2)$$

The minimum of $\overline{\{\Delta\rho_{hkl}(x\;y\;z)\}^2}$ with respect to w is found by differentiation:

$$\frac{d\overline{\{\Delta\rho_{hkl}(x\;y\;z)\}^2}}{dw} = \frac{4}{V^2}|F|^2(w - \cos\xi) = 0 \tag{8.3}$$

It follows from Eq.(8.3) that the best value of w is $w = \cos\xi$. But the phase error ξ is unknown and the best that can be done is to use its average value, given by

$$w = \int_0^{2\pi} p(\xi)\cos\xi\;d\xi \tag{8.4}$$

where $p(\xi)$ is the probability of finding the phase angle error between ξ and $\xi + d\xi$. Note that w is equal to the figure of merit m as originally defined by Blow and Crick in their calculation of the "best" Fourier map. In Section 7.13 it was shown that m is the weighted mean of the cosine of the error in the phase angle, just as w is.

$p(\xi)$ can be derived from Figure 8.1 in the following way. In Section 5.3 (Eq. 5.4) it was found that

$$p(\mathbf{F})d(\mathbf{F}) = \frac{1}{\pi \times \sum\limits_{j=1}^{n} f_j^2} \exp\left[-\frac{|F|^2}{\sum\limits_{j=1}^{n} f_j^2} \right] d(\mathbf{F}) \qquad (8.5)$$

If the f_js are the atomic scattering factors for the n missing atoms, $\Sigma_{j=1}^{n} f_j^2$ is the average value expected for $|F_{\text{mis}}|^2$ of the missing atoms and the distribution function for the noncentrosymmetric reflections [Eq.(8.5)] is

$$p(\varepsilon)d(\varepsilon) = \frac{1}{\pi \times \sum\limits_{j=1}^{n} f_j^2} \exp\left[-\frac{|\varepsilon|^2}{\sum\limits_{j=1}^{n} f_j^2} \right] d(\varepsilon) \qquad (8.6)$$

With $d(\varepsilon) = |F|d\xi d|F|$ (Figure 8.2):

$$p(\varepsilon)d(\varepsilon) = p(\varepsilon)|F|d\xi d|F| = \frac{|F|}{\pi \times \sum\limits_{j=1}^{n} f_j^2} \exp\left[-\frac{|\varepsilon|^2}{\sum\limits_{j=1}^{n} f_j^2} \right] d|F|d\xi$$

$p(\varepsilon)$ is equal to the probability of finding the measured $|F|$ with a certain ξ. Application of the cosine rule in Figure 8.1 gives the conditional probability:

$$p(\xi; |F|)\, d\xi = \frac{|F|}{\pi \times \sum\limits_{j=1}^{n} f_j^2} \exp\left[-\frac{(|F|^2 + |F_{\text{K}}|^2)}{\sum\limits_{i=1}^{n} f_i^2} \right] \exp\left[\frac{2|F| \times |F_{\text{K}}| \cos \xi}{\sum\limits_{i=1}^{n} f_i^2} \right] d\xi$$

$$(8.6\text{a})$$

Normalization requires

$$\int\limits_{\xi=0}^{2\pi} N \times p(\xi; |F|)\, d\xi = 1$$

where N is a normalization constant.

$$2 \times N \times \frac{|F|}{\pi \times \sum\limits_{j=1}^{n} f_j^2} \exp\left[-\frac{(|F|^2 + |F_{\text{K}}|^2)}{\sum\limits_{i=1}^{n} f_i^2} \right] \times \int\limits_{0}^{\pi} \exp\left[\frac{2|F| \times |F_{\text{K}}| \cos \xi}{\sum\limits_{i=1}^{n} f_i^2} \right] d\xi = 1$$

$$(8.7)$$

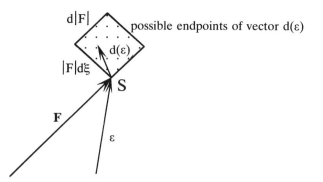

Figure 8.2. The possible endpoints of vector $d(\varepsilon)$ are found in a rectangle bounded by $|F|d\xi$ and $d|F|$.

The integral is a modified Bessel function of order 0. The general form for a modified Bessel function of order n is

$$I_n(x) = \frac{1}{\pi} \int_{\xi=0}^{\pi} \exp[x \cos \xi] \cos n\xi \, d\xi$$

$$I_0(x) = \frac{1}{\pi} \int_{\xi=0}^{\pi} \exp[x \cos \xi] \, d\xi$$

Therefore, N can be written as

$$N = \left\{ 2 \times \frac{|F|}{\sum_{j=1}^{n} f_j^2} \exp\left[-\frac{(|F|^2 + |F_K|^2)}{\sum_{i=1}^{n} f_i^2} \right] \times I_0\left(\frac{2|F| \times |F_K|}{\sum_{i=1}^{n} f_i^2} \right) \right\}^{-1}$$

and the normalized distribution function becomes

$$p(\xi; |F|) \, d\xi = \left\{ \exp\left[\frac{2|F| \times |F_K| \cos \xi}{\sum_{i=1}^{n} f_i^2} \right] d\xi \right\} \Big/ \left\{ 2\pi I_0\left[\frac{2|F| \times |F_K|}{\sum_{i=1}^{n} f_i^2} \right] \right\}$$

$$= \frac{\exp[X \cos \xi] \, d\xi}{2\pi I_0(X)} \qquad (8.8)$$

where

$$X = \frac{2|F| \times |F_K|}{\sum_{1}^{n} f_i^2}$$

Or, because $\alpha_K = \alpha - \xi$ and α_K is a constant,

$$P_{\text{par}}(\alpha; |F|)\, d\alpha = \frac{1}{2\pi I_0(X)} \exp[X \cos(\alpha - \alpha_K)]\, d\alpha \qquad (8.9)$$

The weighting factor w [Eq. (8.4)] becomes

$$w = \int_0^{2\pi} p(\xi) \cos\xi\, d\xi = \frac{\displaystyle\int_0^{2\pi} \exp[X\cos\xi]\cos\xi\, d\xi}{2\pi I_0(X)} = \frac{I_1(X)}{I_0(X)}$$

For centric reflections a different expression for w must be used:

$$w = \tanh\left(\frac{X}{2}\right)$$

This can be derived as follows.

For centric reflections the probability distribution of the structure factors is (Section 5.3):

$$p(\varepsilon)d(\varepsilon) = \frac{1}{\sqrt{2\pi\sum_i f_i^2}} \exp\left[-\frac{\varepsilon^2}{2\sum_i f_i^2}\right] d(\varepsilon)$$

There are two possibilities:

1. F in the same direction as F_K, implying $\xi_1 = 0$ and $\cos\xi_1 = 1$.

$$(\varepsilon)^2 = (|F| - |F_K|)^2 = |F^2| + |F_K^2| - 2|F||F_K|$$

and the probability for obtaining the required ε is

$$p_1(|\varepsilon|; |F|) = \frac{1}{\sqrt{2\pi\sum_i f_i^2}} \exp\left[-\frac{|F|^2 + |F_K|^2}{2\sum_i f_i^2}\right] \exp\left[\frac{2|F||F_K|}{2\sum_i f_i^2}\right]$$

$$= C \times \exp\left[\frac{2|F||F_K|}{2\sum_i f_i^2}\right]$$

2. F and F_K in opposite directions, implying $\xi_2 = \pi$ and $\cos\xi_2 = -1$.

$$(\varepsilon)^2 = (|F| + |F_K|)^2 = |F|^2 + |F_K|^2 + 2|F||F_K|$$

and $p_2(|\varepsilon|; |F|) = C \times \exp\left[-\dfrac{2|F||F_K|}{2\sum_i f_i^2}\right]$

According to eq. 8.4 the weight w of the reflection is given by the average value of $\cos \xi$. Thus,

$$w = \frac{p_1 \cos \xi_1 + p_2 \cos \xi_2}{p_1 + p_2} = \frac{p_1 - p_2}{p_1 + p_2}$$

Writing X for $\dfrac{2|F||F_K|}{\displaystyle\sum_i f_i^2}$, this becomes

$$w = \frac{\exp\left[\dfrac{X}{2}\right] - \exp\left[-\dfrac{X}{2}\right]}{\exp\left[\dfrac{X}{2}\right] + \exp\left[-\dfrac{X}{2}\right]} = \tanh\left(\frac{X}{2}\right)$$

As mentioned before w is equal to the figure of merit $m = \overline{\cos \xi}$, where ξ is the deviation from the phase angle α_K. In the calculation of w it was assumed that the known part of the structure is exactly known and has no errors in the parameters of the atoms. In that case,

$$X = \frac{2|F| \times |F_K|}{\displaystyle\sum_1^n f_i^2}$$

However, in practice, the known part of the structure does have errors. X must then be taken (Srinivasan, 1966) as

$$X = \frac{2\sigma_A |E| \times |E_K|}{1 - \sigma_A^2}$$

$|E|$ and $|E_K|$ are normalized structure factors and σ_A will be defined in Section 15.6. The result of the OMIT map procedure can be improved further by previous refinement of the partial structure. It appears that this can best be done with simulated annealing, discussed in Section 13.4.5 (Hodel et al., 1992). It often occurs that phase information from different sources must be combined in a joint probability curve. In Chapter 14 a convenient method for doing this will be presented. The contribution from partial structure information is then $p_{par}(\alpha)$ [Eq. (8.9)].

$I_0(X)$ depends on known quantities only: $|F|$, $|F_K|$, and $\Sigma_1^n f_i^2$, and can be put into a normalizing constant: $p(\xi) = N^* \times \exp[X \cos \xi]$ and

$$p_{par}(\alpha) = N^* \times \exp[X \cos(\alpha - \alpha_K)] \tag{8.9a}$$

In Chapter 14 it will be shown how this probability curve is combined with probability curves from other sources.

8.3. Solvent Flattening

The principle of this method is fairly simple. From highly refined protein crystal structures it is known that the electron density map is rather flat in the solvent region between the protein molecules. This is due to the liquid character of the solvent molecules in those regions. This does not mean that no solvent molecules can be observed. However, the more static solvent molecules are found only internally in the protein molecules, or as a monolayer or double layer at their surface. The rest of the solvent has a dynamic nature and its time-averaged electron density has a low constant value. If the region occupied by the protein molecules can be identified, a nonoptimal electron density map shows noise peaks in the solvent region and they are removed, simply by setting the electron density in this region to a low constant value.

The simplest method for defining a molecular envelope around the protein molecules is by visual inspection of the preliminary electron density map. However, this is rather subjective for a noisy map. An automated method proposed by Wang (1985) and modified by Leslie (1987) is more objective and much easier to apply. In the Wang method the noisy electron density map is smoothed in the following way:

A three-dimensional grid is superimposed on the unit cell. At each grid point j, the electron density is replaced by a new value that is proportional to the weighted sum of the densities within a sphere of radius R with the center in that grid point. R is typically on the order of $10\,\text{Å}$.

$$\rho'_j = K \sum_i^R w_i \rho_i$$

with $w_i = 0$ for $\rho_i < 0$ and $w_i = 1 - (r_{ij}/R)$ for $\rho_i > 0$. The summation is over the grid points i within the sphere. r_{ij} is the distance between the grid points i and j. K is an arbitrary constant. A grid spacing of 1/3 of the resolution is adequate. The molecular boundary is revealed in the new map by tracing a threshold density level. In the beginning of the process the threshold is usually chosen such that the volume of solvent in the map is a little smaller (e.g., 10–15%) than the known or estimated volume fraction of the solvent (estimated using the formula given in Section 3.9). Also, only low resolution data should be used because the phase angles of the high resolution data are still very poor. If necessary, the envelope can be polished, e.g., by removing internal voids. During the next cycles of solvent flattening the solvent fraction can gradually be increased up to, e.g., 5% below the estimated value. With the slightly smaller solvent region there are fewer chances that outer loops of the protein molecule are cut off. Moreover, the envelope should be updated in these cycles, because the electron density map improves and this allows us to trace a better envelope. In the solvent region (outside the envelope), the average value of the solvent density is assigned to each grid point. Structure factor amplitudes and phases are calculated for this new map (a process known as map inversion), using the fast Fourier transform technique (Section 13.3).

In the next step an electron density map is calculated with *observed* structure factor amplitudes and with phase angles either from the solvent flattening procedure alone or by combining them with phase angles from isomorphous or molecular replacement or any other phase information. The choice depends on the quality of the initial phases, the resolution limit and the solvent content. If, at the present resolution, no further improvement is obtained, data at higher resolution can be added in small steps.

Abrahams (1997) noted that phase improvement converges prematurely because a large contribution to the calculated structure factors is from the original structure. This is not new information. In other words, the improved phase angles are to a large extent biased by the present electron density. This causes a problem in the recombination step of the density modification cycle, where independence between the recombined data is assumed. Abrahams proposed to remove or diminish this bias by inversion of the solvent features. They are not flattened to a low constant value but their sign is reversed (solvent flipping). After this correction, the modified structure factors can be treated as statistically independent, allowing a straightforward recombination. A similar treatment is valid for other types of density modification.

Explanation of solvent flipping

The electron density map calculated without $\mathbf{F}(000)$ has negative values at some grid points and positive values at others, with average overall density at the zero level. However, it is convenient for the explanation if the mean density in the volume occupied by the solvent, instead of the overall density, is zero. This can be achieved by adding a constant value to every grid point. This has no effect on the structure factors, except on $\mathbf{F}(000)$.

Proof of this statement, which we call here **Property 1**, is

$$\mathbf{F}(\mathbf{h}) = V \int_{\text{cell}} \rho(\mathbf{x}) \exp[2\pi i \mathbf{h}.\mathbf{x}] \, dv_{\text{real}} \qquad (8.10)$$

If $\rho(\mathbf{x})$ has a constant value c:

$$\mathbf{F}(\mathbf{h}) = c.V \int_{\text{cell}} \exp[2\pi i \mathbf{h}.\mathbf{x}] \, dv_{\text{real}}$$

and this is zero, unless $\mathbf{h} = 0$ where $\mathbf{F}(\mathbf{h} = 0) = c.V$

The reverse is also true (**Property 2**).

Property 2:

If a function in reciprocal space is everywhere zero, except at the origin ($\mathbf{h} = 0$) its transform in real space is a constant.

Let us call the uncorrected electron density $\rho(\mathbf{x})$:

$$\rho(\mathbf{x}) = \frac{1}{V} \int_{\mathbf{h}} \mathbf{F}(\mathbf{h}) \exp[-2\pi i\, \mathbf{h}.\mathbf{x}]\, dv_{\text{rec.}} \qquad (8.11)$$

[for convenience we use the integral over $\mathbf{h}$ instead of the summation]

Now that the mean solvent density $\rho(\mathbf{x})$ is zero, solvent flattening is equal to multiplying the electron density $\rho(\mathbf{x})$ with a modifying function $g(\mathbf{x})$, which has the value 1 for grid points within the envelope and zero outside it. Its Fourier transform is $\mathbf{G}(\mathbf{h})$.

In real space: $\rho(\mathbf{x})$ and $g(\mathbf{x})$
In reciprocal space: $\mathbf{F}(\mathbf{h})$ and $\mathbf{G}(\mathbf{h})$

The solvent flattened map has density $g(\mathbf{x}) \times \rho(\mathbf{x})$. The reciprocal space equivalent of this multiplication is found by means of the convolution theorem (7.13) and the following property:

Property 3:

The Fourier transform of a convolution of two functions is equal to the product of the individual Fourier transforms of each of the two functions (Section 7.2)

$$\text{Tr}\,[\mathbf{G}(\mathbf{h}) * \mathbf{F}(\mathbf{h})] = g(\mathbf{x}) \times \rho(\mathbf{x})$$

Tr means Fourier transform and $*$ convolution

$\mathbf{G}(\mathbf{h})$ can be split into two parts: $\mathbf{G}(\mathbf{h} \neq 0) + \mathbf{G}(\mathbf{h} = 0)$, and we redefine solvent flattening in reciprocal space accordingly:

$$\text{Tr}\,[\mathbf{G}(\mathbf{h}) * \mathbf{F}(\mathbf{h})] = \text{Tr}\,[\mathbf{G}(\mathbf{h} \neq 0) * \mathbf{F}(\mathbf{h})] + \text{Tr}\,[\mathbf{G}(\mathbf{h} = 0) * \mathbf{F}(\mathbf{h})]$$

Application of Property 3 results in:

$$g(\mathbf{x}) \times \rho(\mathbf{x}) = g_1(\mathbf{x}) \times \rho(\mathbf{x}) + g_2(\mathbf{x}) \times \rho(\mathbf{x}) \qquad (8.12)$$

where $g_1(\mathbf{x})$ is the Fourier transform of $\mathbf{G}(\mathbf{h} \neq 0)$ and $g_2(\mathbf{x})$ of $\mathbf{G}(\mathbf{h} = 0)$

From property 2, it follows that $g_2(\mathbf{x})$ is a constant, independent of $\mathbf{x}$. Abrahams (1997) calls it γ. It is clear that the second term in the right hand part of (8.12) is the bias component. It is nothing more than a scaled down version of what was already known. Its value is obtained as follows: Its transform $\mathbf{G}(\mathbf{h} = 0)$ is equal to the number of gridpoints with a value 1 (the gridpoints inside the envelope), just as $\mathbf{F}(\mathbf{h} = 0)$ is equal to the total number of electrons in the unit cell (Section 4.10). According to property 1, $G(h = 0) = g_2 \times V$ with V here equal to the total number of gridpoints in the cell.

$$\gamma = \frac{\text{sum of gridpoints in envelope}}{\text{total number of gridpoints in the cell}} = \frac{V_P}{V}$$

with V_P the volume of the envelope.

Conclusion: Solvent flattening corresponds to multiplication of $\rho(x)$ with $g(x)$, but $g(x)$ contains a term γ which leaves the original electron density unchanged. The seriousness of this bias depends on $\frac{V_P}{V}$. The larger the protein content the more serious it is. The bias is removed by using $g_1(x)$ and not $g(x)$ as the modifying function:

$$g_1(x) = g(x) - \gamma \tag{8.13}$$

Example: Suppose $\frac{V_P}{V} = 0.5$. Application of (8.13) results in values for $g_1(x) = 0.5$ within the envelope and -0.5 in the solvent region. Rescaling gives 1 within the envelope and -1 for the solvent region. In other words the protein density is not affected but the features in the solvent region change sign.

For the protein phase angles a probability curve $p_{SF}(\alpha)$ is chosen similar to $p_{par}(\alpha)$ in the Sim weighting procedure [Eq. (8.9a)]. The argument is that the solvent flattened structure can be regarded as the "known" part of the structure in the Sim conception.

$$p_{SF}(\alpha) = N \exp[X' \cos(\alpha_P - \alpha_{calc})] \tag{8.14}$$

where N is a normalizing constant, α_P is the protein phase angle, α_{calc} is the phase angle calculated for the map with the flattened solvent density, and $X' = 2|F_{obs}| \times |F_{calc}|/\overline{|I_{obs} - I_{calc}|}$.

$\overline{|I_{obs} - I_{calc}|}$ is the mean intensity contributed by the unknown part of the structure. It replaces $\Sigma_{j=1}^n f_j^2$ in the Sim procedure. The closer the solvent flattened structure is to the true structure, the smaller $\overline{|I_{obs} - I_{calc}|}$ is and the stronger the phase indication. Appropriate scaling of the calculated to the observed structure factors is of course required. $\overline{|I_{obs} - I_{calc}|}$ is calculated in shells of increasing resolution.

From the phase probability curve—either the solvent flattened alone or a combined one—the "best" phases and figures of merit can be derived and used in the calculation of a "best" electron density map. The entire method can be repeated until no further improvement of the map is obtained. For a schematic representation of the method see Scheme 8.1

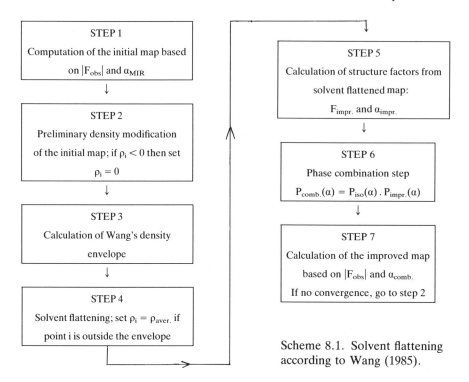

Scheme 8.1. Solvent flattening according to Wang (1985).

and for an example Figure 8.3. Leslie has given a reciprocal space method for calculating the smoothed electron density map, which is computationally much faster than Wang's real space method. The density for the smoothed map is

$$\rho'(i) = \sum_{r=0}^{R} \rho(i + r) \times \left(1 - \frac{r}{R}\right) \tag{8.15}$$

with $r \leq R$. Equation (8.15) is a convolution of the function ρ with the function $[1 - (r/R)]$ [see Eq. (7.13)]. Therefore, the transform of the function $\rho'(i)$ is equal to the product of the transform of $\rho(i)$ and the transform of $[1 - (r/R)]$. The transform of $\rho(i)$ is easily calculated with a fast Fourier program. The transform of $[1 - (r/R)]$ is the transform of 1 minus the transform of r/R. The transform of a function that has a constant value of 1 in the region $r \leq R$ and is 0 outside that region is the transform of a sphere with radius R. We shall meet this transform (the G-function) again in discussing the rotation function (Section 10.2; Fig. 10.3). It has the form

$$G(x) = \frac{3(\sin x - x \cos x)}{x^3} \quad \text{with} \quad x = \frac{4\pi R \sin \theta}{\lambda}$$

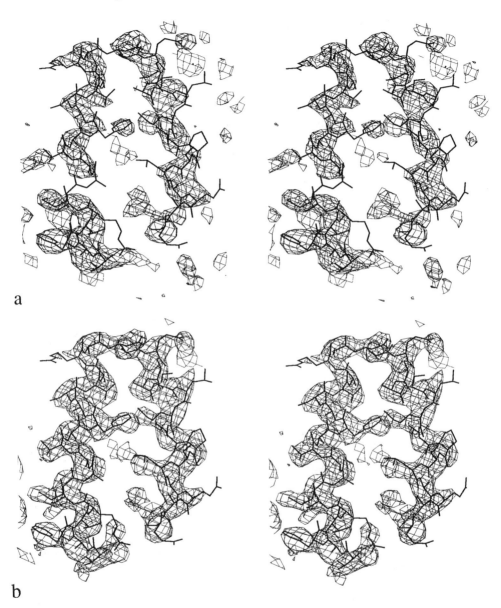

a

b

Figure 8.3. Stereo pairs of part of the electron density map of the *E. coli* enzyme soluble lytic transglycosylase at 3.3 Å resolution. The crystal contained 60% solvent. (a) The map as obtained with the multiple isomorphous replacement method, using two derivatives and including anomalous scattering by the heavy atoms. (b) The solvent flattened map. Disconnected parts of density before solvent flattening are nicely connected in (b). (Source: Dr. A.M.W.H. Thunnissen.)

The transform of the function which is r/R in the region $r \le R$ and 0 for $r > R$, is

$$L(x) = \frac{3}{x^4}\{2x \sin x - (x^2 - 2) \cos x - 2\}$$

The total transform of $[1 - (r/R)]$ is then $G(x) - L(x)$.

The following steps for calculating the smoothed map should be performed:

1. Calculate the structure factors for the original electron density $\rho(r)$, including all low angle reflections, because they depend much more on the shape of the solvent region than the higher resolution reflections.
2. Multiply them by $G(x) - L(x)$.
3. Use these modified structure factors for calculating the smoothed electron density map ρ'.

See Scheme 8.2 for a schematic representation. Solvent flattening is most powerful for crystals with a high solvent content. Xiang et al. (1993) introduced entropy maximization into the process of map improvement with phase extension. The constraining is by the molecular envelope and a basis set of reflections with reliable phase angles (high figures of merit). The parameters of the model are then adjusted to values consistent with the constraints. This process is continued until the extrapolated amplitudes (those of the reflections outside the basis set) match the observed amplitudes.

8.4. Noncrystallographic Symmetry and Molecular Averaging

From the first characterization of a protein crystal it usually becomes clear how many protein molecules or subunits are contained in the asymmetric unit. It is fairly common to find more than one molecule, related by one or more symmetry operators. This is noncrystallographic symmetry (NCS) and it is active only within each asymmetric unit, not through the entire crystal. The orientation and nature of a noncrystallographic axis follow from the self-rotation function (Chapter 10). A possible translation component along the axis (screw axis) is not detected in this way. The actual location of the axis can easily be found if a sufficient number of heavy-atom positions is available from isomorphous replacement. If a sufficient number is not available, NCS axes can be located by maximizing the correlation coefficient (Appendix 2) between electron density regions in the asymmetric unit or by calculating the translation function (Chapter 10).

The electron density in the molecules (subunits), related by this noncrystallographic symmetry, is essentially equal, although the difference in

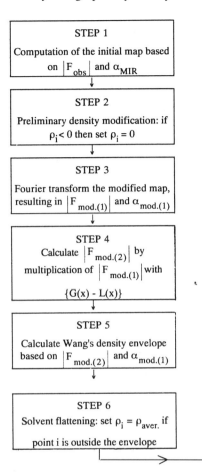

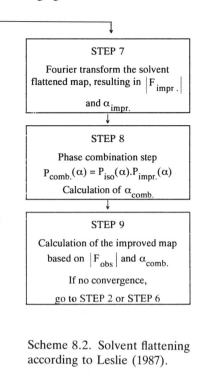

Scheme 8.2. Solvent flattening according to Leslie (1987).

the contact with neighbors may cause some deviation from exact equality. The equal density in the molecules related by the noncrystallographic symmetry imposes a constraint on the protein structure factors and, therefore, on the protein phase angles. Reciprocal space methods to derive these phase relationships were proposed by Rossmann and Blow (1963), Main (1967), and Crowther (1969). They were not very successful, but a real space approach, developed by Bricogne (1974), has found wide application. It consists of the following steps:

1. Determine the NCS operator(s).

2. The envelope of the molecules in the asymmetric unit must be defined in the electron density map, calculated with the available phase information. The envelope should contain as much protein density as possible, but it is sometimes chosen slightly smaller than the actually observed one, because of the possible difference between the molecules in the contact with their neighbors. One must be careful that no overlap

with symmetrically related envelopes occurs. A special case occurs if a noncrystallographic axis is parallel to a crystallographic axis. Then the Patterson map can supply useful information (Section 10.3 and Figure 10.7).

3. The electron density of the molecules in the asymmetric unit is averaged, the solvent region flattened, and its density set equal to the average electron density in that region; then the asymmetric unit is reconstituted.

4. Phase angles for this new model are calculated by back-transforming the electron density map.

5. If necessary, this phase information is combined with previously known phase information and "best" phases and figures of merit are obtained.

6. A new and improved electron density map is calculated with observed structure factor amplitudes and phase information from step 4.

7. The procedure is repeated starting in step 1 with refinement of the NCS operator(s).

Bricogne (1976) developed a complete set of programs for the application of this method. Another program has been described by Johnson (1978). More recently, Rossmann et al. (1992) presented a new program for the execution of the method, based on previous experience and using improved computer technology. Rossmann and co-workers applied averaging and phase extension with great success for the structure determination of several spherical viruses (Rossmann, 1990 and 1995; McKenna et al., 1992). Rayment (1983) studied the use of noncrystallographic symmetry and solvent flattening as a phase constraint on model structure factors of icosahedral particles. He found that the phase angles can be successfully refined against low resolution data. He also pointed out that it is important to include calculated values for unrecorded data in the refinement.

The phase information is derived from this averaging procedure in the same way as in the solvent flattening method and is based on Sim's phase probability function. The average structure is regarded as the known part in Sim's conception and $p_{\text{average}}(\alpha_P)$ has the same form as $P_{\text{SF}}(\alpha_P)$ [Eq. (8.14)]:

$$P_{\text{average}}(\alpha_P) = N \exp[X' \cos(\alpha_P - \alpha_{\text{calc}})] \qquad (8.16)$$

with $X' = 2|F_{\text{obs}}| \times |F_{\text{calc}}|/\overline{|I_{\text{obs}} - I_{\text{calc}}|}$.

The method of averaging is most powerful in cases of high noncrystallographic symmetry, such as viruses, because averaging improves the signal to noise ratio of the order of $\sqrt{N}$ with N the number of independent copies. But it can also give excellent improvement of a density map at lower noncrystallographic symmetry (Jacobson et al., 1994); an example is also shown in Figure 8.4. Moreover, the averaging method is not restricted to

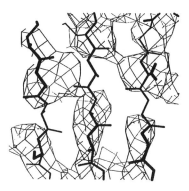

a

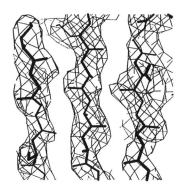

b

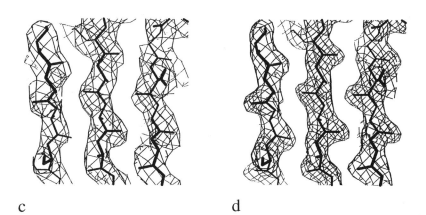

c   d

Figure 8.4. Example of the effect of density modification by NCS averaging and phase extension in the structure determination of the epoxide hydrolase from *Agrobacterium radiobacter* AD1; the asymmetric unit contains four molecules: (a) Electron density map from single isomorphous replacement supplemented by anomalous signal information (SIRAS) at 3.7 Å resolution; (b) electron density map after averaging at 3.7 Å resolution; (c) electron density map after averaging and phase extension from 3.7 Å to 2.6 Å resolution; (d) final {2|F(obs)| − |F(calc)|} map at 2.1 Å resolution. (Source: Dr. Marco Nardini).

noncrystallographic symmetry within one crystal, but can be used equally well if proteins crystallize in more than one crystal form. A detailed discussion of noncrystallographic symmetry averaging in phase refinement and extension can be found in Vellieux and Read (1997) and in (Kleywegt and Read, 1997).

8.5. Histogram Matching

Like solvent flattening, the method of histogram matching rectifies a distorted image. Histogram matching is usually applied in combination with solvent flattening. It does not deal with the electron density map as such, like solvent flattening does, but it uses the frequency distribution of electron density values (Figure 8.5). It appears that these distributions as a function of ρ are fairly independent of the protein, at least at the same resolution. The temperature factor also has an effect, but this can be eliminated by using temperature factor corrected structure factors. Histogram matching is one of many techniques applied in the general field of image processing. It requires the availability of another image, preferably a high quality image of the same kind. The assumption is that high quality images of a particular kind of object have the same frequency distribution of gray levels, the standard distribution. A poor image has a different distribution and this is then scaled to the standard distribution and this process results in an improved image. It was pointed out by Lunin in 1988 and in subsequent papers (Lunin, 1988; Lunin et al., 1990; Lunin & Skovoroda, 1991; Lunin & Vernoslova, 1991) that histogram matching could be useful for improving electron density maps. Zhang and Main (1990a) presented it in a simplified form and incorporated it in the program SQUASH (Zhang and Main, 1990b); DM in the CCP4 library (Cowtan and Main, 1996; CCP4, 1994). The frequency distribution of electron density levels calculated at grid points is plotted as a function of ρ for the poor electron density map and compared with the standard plot. Next, the plots are divided into bins containing an equal number of grid points. The bin boundaries for the poor map are ρ_i, ρ_{i+1}, with $i = 1 \rightarrow n$ and n of the order of 100, and for the standard map ρ'_i, ρ'_{i+1}. If the two maps are identical $\rho_i = \rho'_i$, $\rho_{i+1} = \rho'_{i+1}$, etc. In general they are not. By scaling with a

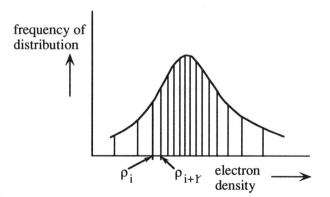

Figure 8.5. The frequency of electron density occurrence at grid points

factor $a = \dfrac{\rho'_{i+1} - \rho'_i}{\rho_{i+1} - \rho_i}$ the bin width from ρ_i to ρ_{i+1} is given the correct value. The bin is then moved over a distance b such that ρ_i moves to ρ'_i and ρ_{i+1} to ρ'_{i+1}. From $\rho'_i = a \times \rho_i + b$ it can easily be derived that $b = \dfrac{\rho_{i+1}\rho'_i - \rho_i\rho'_{i+1}}{\rho_{i+1} - \rho_i}$.

The electron densities in each bin are corrected with the appropriate a and b. This results in an improved electron density map from which new phase angles can be calculated.

According to the experience of Zhang et al. (1997) with a particular protein, solvent flattening and averaging improve the phase angles of the low resolution reflections, but not of the high resolution reflections. Combined with histogram matching, the phasing of the higher resolution reflections also improved considerably.

8.6. wARP: Weighted Averaging of Multiple-Refined Dummy Atomic Models

If an electron density map is still uninterpretable after solvent flattening, averaging, and histogram matching, but the resolution is at least 2.5 Å, wARP can appreciably improve the protein phase angles (and the electron density map) by an automatic procedure (Lamzin and Wilson, 1993; Perrakis et al., 1997; van Asselt et al., 1998). The condition is not only a reasonably high resolution but also that the protein region in the asymmetric unit can be identified.

A number of atoms, usually oxygen atoms, are placed in regions of significantly high electron density. This preliminary "seed model" of approximately 100 atoms is expanded fully automatically by adding more atoms at gridpoints with a density of at least 2σ above the mean density and within 5 Å from atoms in the "seed model." The grid should have small spacings, of approximately 0.25 Å. The preliminary "seed model" is removed and the resulting "model" is further expanded by gradually adding atoms in significant electron density at distances of 1.1 to 1.8 Å from existing atoms. After this expansion in several cycles, the threshold density is lowered in steps of 0.1σ from 2.0 to 0.9σ and additional atoms are placed in these weaker electron density regions. The result is a "model" with many more atoms than the expected number of nonhydrogen protein atoms.

The efficacy of the method is greatly improved by producing more "models" and averaging them later. The extra "models" are built, for instance, by starting from intermediates that led to the first "model" or they are created by random positional shifts of 0.5 Å. Usually six "models" are produced. Before proceeding further, the number of free atoms in these "pseudoprotein models" is reduced to 110% of the number of nonhydrogen atoms in the asymmetric unit by removing atoms in weak density.

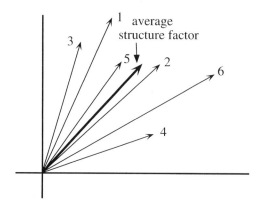

Figure 8.6. Argand diagram with the structure factors from the six "models" produced by wARP, and the average structure factor, for one reflection.

In the next step a refinement procedure starts. In this refinement (Section 13.1), the atoms are moved freely until the calculated structure factor amplitudes match the observed ones as close as possible. All six models are independently refined, preferably by a maximum likelihood procedure (Section 13.2.2). After a refinement cycle, electron density maps with calculated phase angles are subjected to an automatic correction procedure: it removes atoms from low density regions in a $(2m|F_o| - D|F_c|)$ map and adds them to high density in a $(m|F_o| - D|F_c|)$ difference map (Section 7.7), taking into account distance criteria (too close to, or too far from existing atoms). The procedure of "model" refinement + correction is repeated in an iterative process until convergence is reached.

In the final step the result from the six "models" is averaged to maximize the quality of the final phase set. In this averaging step, the calculated structure factor amplitudes are first scaled to the observed ones. Then the structure factors are averaged in the Argand diagram, assigning a weight to each vector depending on its distance from the average vector (Figure 8.6). The phase angles of the averaged structure factors are combined with the observed amplitudes to calculate the final electron density map. The advantage of wARP is that it is a fast procedure to go from an uninterpretable to an interpretable map. Its main limitation is the restriction to electron density maps with a resolution better than 2.5 Å.

8.7. Further Considerations Concerning Density Modification

Density modification procedures start with a poor electron density map that has been calculated with structure factors having the correct amplitude $|F_{obs}|$, but a wrong phase angle α_{wr}. This structure factor can be thought of as composed of a contribution by the protein part of the crystal structure [$\mathbf{F}_{wr}(pr)$] and a contribution by the solvent part [$\mathbf{F}_{wr}(s)$]. The designation $\mathbf{F}_{wr}$ means that the protein and also the solvent contribution are both incorrect in this stage. Flattening the solvent part in the structure means that $\mathbf{F}_{wr}(s)$ is replaced by an improved $\mathbf{F}_{impr}(s)$ (Figure 8.7). A new and improved electron density map is then calculated with the structure factor $\mathbf{F}_{impr}$ (total) having the amplitude $|F_{obs}|$ and the improved phase angle α_{impr}. From this new map an improved envelope can be derived and the process repeated. Usually, several cycles of density modification are required to shift the protein phase angles close enough to their correct value to allow an interpretation of the electron density map in terms of the polypeptide chain. If in the process, the solvent area becomes flatter, the amplitudes of the $\mathbf{F}(s)$ contributions become smaller, except for the low order reflections (see Section 13.1). The shape of the envelope is mainly determined by these reflections and they should be incorporated as much as possible in solvent flattening. If the solvent content of the crystal is high, the contribution of $\mathbf{F}(s)$ to the total structure factor is also relatively high and solvent flattening is more powerful.

In solvent flattening the driving force is the gradual improvement of $\mathbf{F}(s)$. In the averaging procedure, two density modifications are applied in each step: (1) averaging the electron densities of the noncrystallographically related molecules and (2) solvent flattening. Therefore, both

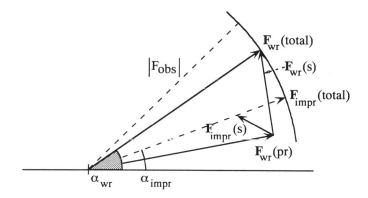

Figure 8.7. Structure factors in solvent flattening. The electron density maps are always calculated with observed structure factor amplitudes $|F_{obs}|$. For the starting map the poor protein phase angles α_{wr} are used for the structure factors $\mathbf{F}_{wr}$(total). This structure factor is composed of a protein contribution, $\mathbf{F}_{wr}(pr)$, and a solvent contribution, $\mathbf{F}_{wr}(s)$. The flattening does not change the protein contribution, but improves the solvent contribution, replacing $\mathbf{F}_{wr}(s)$ by $\mathbf{F}_{impr}(s)$. The next electron density map is then calculated with structure factors having the same amplitudes as before, $|F_{obs}|$, but improved protein phase angles α_{impr}.

$F_{wr}(s)$ and $F_{wr}(pr)$ are replaced by improved structure factor contributions: $F_{impr}(s)$ and $F_{impr}(pr)$. This speeds up the process considerably compared with solvent flattening alone. As a result it is much more powerful, particularly if the number of symmetry-related molecules is higher.

Summary

The interpretation of an electron density map in terms of the polypeptide chain is based on the chemical structure of the protein. Additional information is the higher electron density in the protein compared with the rather flat electron density in the solvent region between the protein molecules and noncrystallographic symmetry, if present. The application of this additional information (density modification) can make all the difference for the interpretation of a hitherto uninterpretable map. Moreover, the resolution can often be increased in small steps. Usually several cycles of density modification and model building are required, using calculated model phases combined with previous phase information. For this procedure the fast Fourier transform algorithm is indispensible. If only part of the electron density map, e.g., rather mobile loops, cannot satisfactorily be interpreted, OMIT maps, preferably with Sim weighting, should be calculated. If the protein region in the asymmetric unit can be identified and the resolution is at least 2.5 Å, the protein phase angles (and the electron density map) can be appreciably improved with the automatic wARP procedure.

Chapter 9
Anomalous Scattering in the Determination of the Protein Phase Angles and the Absolute Configuration

9.1. Introduction

Anomalous scattering is not a new subject. It was already introduced in Chapter 7. There, you learned that anomalous scattering by an atom is due to the fact that its electrons cannot be regarded as completely free electrons. This effect depends on the wavelength, but it is in general stronger for the heavier atoms than for the light atoms in the upper rows of the periodic system. If heavy atoms are present in a protein structure, the consequence of their anomalous scattering is that the intensities of a reflection $h\,k\,l$ and its Bijvoet mate $\bar{h}\,\bar{k}\,\bar{l}$ are no longer equal. In Chapter 7 this effect was used in combination with the isomorphous replacement differences in the search for the heavy atom positions and in the refinement of these positions. In this chapter it will be shown how anomalous scattering information can help to determine the phase angle of the protein reflections and the absolute configuration of the protein structure. Moreover, it will be discussed how anomalous scattering is exploited for protein phase angle determination by the multiple wavelength anomalous dispersion (MAD) method.

9.2. Protein Phase Angle Determination with Anomalous Scattering

In principle the anomalous scattering by heavy atoms contributes to the determination of the protein phase angles as much as the isomorphous replacement does. This can best be explained in Figure 9.1. Three circles are drawn in that figure, with radii F_P, $F_{PH}(+)$, and $F_{PH}(-)$; the $(+)$ and the (−) indicate a Bijvoet pair of reflections. The F_P circle has its center at O. For the $F_{PH}(+)$ circle the center is at the end of the vector $-\mathbf{F}_H(+)$ and for the $F_{PH}(-)$ circle at the end of the vector $-\mathbf{F}_H(-)$. The two intersections of the F_P and $F_{PH}(+)$ circles at α_1 and α_2 indicate two possible protein phase angles. Two other possibilities are found at the two intersections of the circles F_P and $F_{PH}(-)$: α_1' and α_2'. Since the reflections $(h\,k\,l)$ and $(\bar{h}\,\bar{k}\,\bar{l})$ of the native protein crystal have opposite phase angles (Section 4.11) the correct choice is for the phase angles α_1 for $(h\,k\,l)$ and α_1' for $(\bar{h}\,\bar{k}\,\bar{l})$. This is illustrated in a simpler way in Figure 9.2. Here the vector $-\mathbf{F}_H(-)$ is drawn with the opposite phase angle (mirror image with respect to the horizontal axis). Now the correct phase angle is found at the intersection of the three circles F_P, F_{PH} (+) and $F_{PH}(-)$, assuming that the data are error-free. The conclusion is that

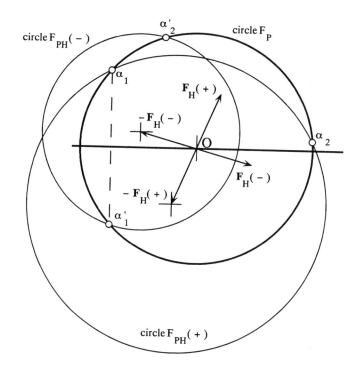

Figure 9.1. The Harker diagram for protein phase angle determination by anomalous scattering. $|F_P|$ is the structure factor amplitude for the native protein and $|F_{PH}(+)|$ and $|F_{PH}(-)|$ for the Friedel mates of the heavy atom derivative. The contribution to the structure factor by the heavy atom is $\mathbf{F}_H(+)$ for one member of the Friedel pair and $\mathbf{F}_H(-)$ for the other member. These two structure factors are not symmetric with respect to the horizontal axis because of an anomalous component. The positions of the intersection points α_1 and α_1' do have a position symmetric with respect to the horizontal axis because the structure factor of the native protein has no anomalous component.

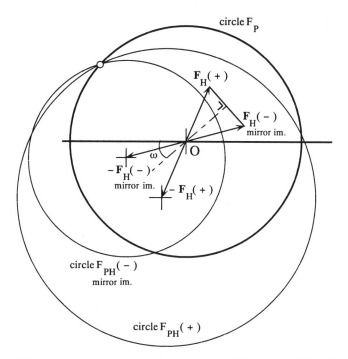

Figure 9.2. This figure gives the same information as Figure 9.1. The difference is that the vector $-\mathbf{F}_H(-)$ is now drawn with the opposite phase angle (mirror image with respect to the horizontal axis). The consequence is a different position for the $F_{PH}(-)$ circle. The advantage of this drawing is that the three circles have one common point of intersection. The dashed line indicates the direction of the nonanomalous scattering part of the heavy atoms.

in principle, the protein phase angle problem can be solved with one isomorphous heavy atom derivative if anomalous scattering is incorporated (SIRAS).

9.3. Improvement of Protein Phase Angles with Anomalous Scattering

From the isomorphous replacement method a probability curve for the protein phase angle is obtained for each reflection: $P_{iso}(\alpha)$. The information from the anomalous scattering data could easily be combined with the $P_{iso}(\alpha)$ curve if it could also be expressed in a probability curve: $P_{ano}(\alpha)$. The combined probability would then be

$$P(\alpha) = P_{iso}(\alpha) \times P_{ano}(\alpha)$$

This can be done in the following way:

$$|F_{PH}(+)| - |F_{PH}(-)| = \Delta PH_{obs}$$

In Section 7.9 it was derived [Eq. (7.28)] that

$$|F_{PH}(+)| - |F_{PH}(-)| = \frac{2f''}{f'}|F_H| \sin(\alpha_{PH} - \alpha_H)$$

or

$$|F_{PH}(+)| - |F_{PH}(-)| = 2\frac{|F_H|}{k} \sin(\alpha_{PH} - \alpha_H) \qquad (9.1)$$

where

$$k = \frac{f'}{f''} = \frac{|F_H|}{|F''_H|}$$

ΔPH_{calc} must be expressed as a function of the protein phase angle α_P.
In triangle ABE in Figure 7.12 the sine rule gives

$$\frac{\sin - (\alpha_{PH} - \alpha_H)}{|F_P|} = \frac{\sin(\alpha_H - \alpha_P)}{|F_{PH}|}$$

or

$$\sin - (\alpha_{PH} - \alpha_H) = \frac{|F_P|}{|F_{PH}|} \sin(\alpha_H - \alpha_P)$$

$$\Delta PH_{calc} = \frac{2|F_H|}{k} \sin(\alpha_{PH} - \alpha_H) = -\frac{2|F_P| \times |F_H|}{k|F_{PH}|} \sin(\alpha_H - \alpha_P)$$

In the ideal case $\Delta PH_{calc} = \Delta PH_{obs}$. In practice for each reflection a value
$\varepsilon_{ano}(\alpha) = \Delta PH_{obs} - \Delta PH_{calc}$ is found depending on the phase angle α_P of
the protein. This is comparable with the lack of closure error ε in the
isomorphous replacement phase triangle of the protein. Here also a
Gauss probability distribution is assumed:

$$P_{ano}(\alpha) = N' \exp\left[-\frac{\varepsilon_{ano}^2(\alpha)}{2(E')^2}\right] \qquad (9.2)$$

N' is a normalization constant and $(E')^2$ is the mean square value of ε.
Because anomalous scattering data are taken from the same crystal,
lack of isomorphism does not cause errors in the values of $|F_{PH}(+)| - |F_{PH}(-)|$. Therefore, although these differences are small, the errors in
$|F_{PH}(+)| - |F_{PH}(-)|$ are inherently smaller than errors in $|F_{PH}| - |F_P|$,
and E' is smaller than E and can be taken as, e.g., $\frac{1}{3}E$. Equation (9.2) can
now be combined with $P_{iso}(\alpha)$ [Eq. (7.34)]. One should be careful to
combine the anomalous data with the correct set of isomorphous data,
that is, the set that gives the electron density of the protein in the

absolute configuration (see next section) and not to combine it with the wrong set.

If the multiple isomorphous replacement method includes anomalous scattering information, it is called the MIRAS method. It should be stressed that in collecting anomalous scattering data, great care should be taken, because the difference in intensity between the Bijvoet pairs is very small. One generally prefers to collect Bijvoet pairs close in time to avoid experimental errors.

9.4. The Determination of the Absolute Configuration

Without anomalous scattering the isomorphous replacement method results in either the correct protein structure or its enantiomorph (mirror image). If the resolution in the electron density map is sufficiently high and the configuration at the $C(\alpha)$ position in the amino acid residues can be observed, it can easily be checked whether the configuration is correct with the amino acid residues having the L-configuration. And if α-helices appear in the map, they should be right-handed for the correct configuration of the protein. However, the absolute configuration of the protein can be derived more straightforwardly from the intensity differences between the two members of the Bijvoet pairs. This will be discussed below.

In Figures 9.1 and 9.2 the situation was presented with the correct set of heavy atom positions and $|F_{PH}(+)| > |F_{PH}(-)|$. In Figure 9.3 the situation is drawn with a correctly chosen set of axes, $|F_{PH}(+)| > |F_{PH}(-)|$, but the choice of the heavy atom positions was incorrect, because from the difference Patterson map the wrong set of the two equally possible centrosymmetrically related sets of positions was chosen (see Section 7.6). The consequence is that the entire set of vectors $\mathbf{F}_H(+)$, $-\mathbf{F}_H(+)$, $\mathbf{F}_H(-)$, and $-\mathbf{F}_H(-)$ is reflected with respect to the horizontal axis. This causes a rotation of the circles $F_{PH}(+)$ and $F_{PH}(-)$ by angle 2ω around the center O of the F_P circle. As a result an incorrect value for the phase angle of $\mathbf{F}_P$ will be found, which is different from the correct one by the value 2ω. If the anomalous information is combined with each of the two possibilities from the isomorphous data, the correct combination will give an electron density map of the protein that is superior to the map calculated with the incorrect combination.

An alternative method to find the absolute configuration is the following. Use the single isomorphous replacement method with anomalous scattering (SIRAS), as just described, for the calculation of two sets of phase angles, corresponding to the two centrosymmetrically related sets of heavy atoms. With these two sets of protein phase angles two dif-

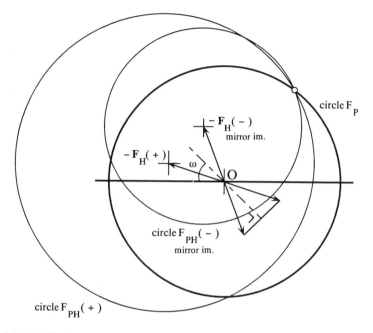

Figure 9.3. In this figure the crystallographic axes are chosen as in Figure 9.2, but the wrong (the centrosymmetric) set of heavy atom positions is chosen. The dashed line is the mirror image of the dashed line in Figure 9.2 and with respect to this dashed line the vectors $\mathbf{F}_H(+)$ and $\mathbf{F}_H(-)$ (mirror im.) are drawn.

ference Fourier maps are calculated for a second heavy atom derivative, PH(2):

1. amplitudes $|F_{PH(2)}| - |F_P|$ and one set of SIRAS phase angles and
2. amplitudes $|F_{PH(2)}| - |F_P|$ and the alternative set of SIRAS phase angles.

The difference Fourier map calculated with the correct set of phase angles (and the correct set of heavy atom positions) will show the highest peaks. This fixes the absolute configuration.

A somewhat simpler way to derive the absolute configuration of the protein from the anomalous scattering is as follows. According to Eq. (9.1), $|F_H|$ can be expressed as

$$|F_H| = \frac{k}{2 \sin(\alpha_{PH} - \alpha_H)} [|F_{PH}(+)| - |F_{PH}(-)|] \qquad (9.3)$$

$|F_{PH}(+)|$ and $|F_{PH}(-)|$ are observed values and $(\alpha_{PH} - \alpha_H)$ can be derived from isomorphous data without anomalous scattering. $|F_H|$ should now be calculated with Eq. (9.3) as a positive number. If the inverted image of the heavy atom set was chosen α_H and α_{PH} as well as $(\alpha_{PH} - \alpha_H)$

would have a value opposite to their correct value. $|F_H|$ would then be calculated as a negative number. If for a number of reflections, for which the heavy atom contribution is relatively strong, $|F_H|$ is calculated, the choice can easily be made. It can also be done by using all reflections and calculating a Fourier map with the amplitudes $|F_H|$ obtained with Eq. (9.3) and phase angles α_H as found from the isomorphous replacement. With the correct choice of the heavy atom positions, positive peaks will appear at the correct heavy atom positions. If the inverted set of heavy atom positions had been chosen, negative peaks would appear at these inverted positions. The coefficients used in the calculation of this Fourier map are

$$|F_H|\ \exp[i\alpha_H] = \frac{k}{2\ \sin(\alpha_{PH} - \alpha_H)}[|F_{PH}(+)| - |F_{PH}(-)|]\ \exp[i\alpha_H]$$

In another method the following coefficients are used:

$$\tfrac{1}{2}(|F_{PH}(+)| - |F_{PH}(-)|)\ \exp[i\alpha_{PH}]$$

In this expression,

$$\frac{1}{2}(|F_{PH}(+)| - |F_{PH}(-)|) = \frac{1}{k}|F_H|\ \sin(\alpha_{PH} - \alpha_H) \qquad (9.1)$$

Here also the correct or incorrect sign of $(\alpha_{PH} - \alpha_H)$ determines whether the peaks in the Fourier map will be positive or negative at the positions assumed for the heavy atoms.

9.5. Multiple Wavelength Anomalous Dispersion (MAD)

If the protein has anomalous scatterers in its molecule, the difference in intensity between the Bijvoet pairs, $|F_h(+)|^2$ and $|F_h(-)|^2$, can profitably be exploited for the protein phase angle determination. In the multiple wavelength method the wavelength dependence of the anomalous scattering is used. The principle of this method is rather old but it was the introduction of the tunable synchrotron radiation sources that made it a technically feasible method for protein structure determination. Hendrickson and colleagues (Hendrickson et al., 1988; Krishna Murthy et al., 1988) were the first to take advantage of this method and to use it for solving the structure of a protein (see also Guss et al., 1988). Of course, the protein should contain an element that gives a sufficiently strong anomalous signal. Therefore, the elements in the upper rows of the periodic system are not suitable. Hendrickson showed that the presence of one Se atom (atomic number 34) in a protein of not more than approximately 150 amino acid residues is sufficient for a successful application of MAD (Hendrickson et al., 1990; Leahy et al., 1992); however, this depends very much on the quality of the data. With more Se atoms the size of the protein can, of course, be larger. One way to introduce Se into a protein is by growing a microorganism on a Se-methionine substrate instead of a methionine-containing substrate. Condition for application of the method is that the wavelengths are carefully chosen to optimize the difference in intensity between Bijvoet pairs and between the diffraction at the selected wavelengths. In Figure 9.4 the anomalous scattering

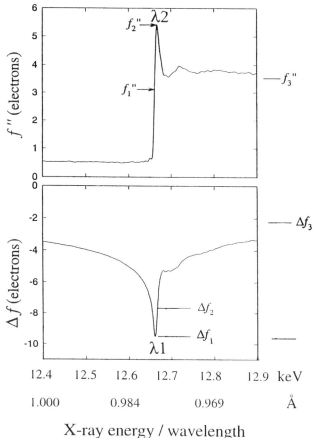

X-ray energy / wavelength

Figure 9.4. The curves for the real (Δf) and imaginary (f'') components of the anomalous scattering around the K-edge of selenium as a function of the wavelength. The minimum of Δf is at $\lambda 1$; f'' has the value f_1'' at this wavelength. The maximum of f'' is at $\lambda 2$; Δf has here the value Δf_2. The third wavelength, $\lambda 3$, is far to the right at 0.93Å. At this remote wavelength Δf has the value Δf_3, and f'' has the value f_3''. Copied in a slightly modified form from: Ramakrishnan, V. and Biou, V. in Methods in Enzymology, Vol. 276, published by Academic Press Inc. Permission obtained.

contributions Δf (the real part) and f'' (the imaginary part) for Se are presented at three different wavelengths.[1] The position of the extremes in the curves is slightly affected by the chemical environment. To obtain optimal results the curves must be determined for the protein crystal under investigation. See Note 1 in Section 9.5.3.

Usually diffraction data are collected at three different wavelengths (Figure 9.4):

- $\lambda 1$ (also called the "edge"), where Δf has its minimum.
- $\lambda 2$ (also called the "white line") where f'' has its maximum and where the Bijvoet difference (between $|F(\mathbf{h})|$ and $|F(-\mathbf{h})|$) is largest.
- $\lambda 3$, remote from the edge, on the left or right, where Δf and f'' are small.

The position of the minimum of Δf (at $\lambda 1$) corresponds to the position of the inflection point (f_1'') in the f''-curve. The dispersion difference is large between $\lambda 1$ and $\lambda 3$. The dispersion difference is defined as

$$\Delta F_{\Delta\lambda}(\mathbf{h}) = \overline{|F^{\lambda 1}(\mathbf{h})|} - \overline{|F^{\lambda 2}(\mathbf{h})|}$$

where

$$\overline{|F^{\lambda}(\mathbf{h})|} = \frac{|F^{\lambda}(\mathbf{h})| + |F^{\lambda}(-\mathbf{h})|}{2}$$

9.5.1. The Algebraic Processing of MAD Data

The most frequently occurring situation in which there is one type of anomalously scattering atoms will be discussed here. It was developed by Hendrickson and coworkers, and is based on Karle's treatment of the problem (Karle, 1980), which has the advantage that the nonanomalous scattering of all atoms in the structure is separated from the wavelength-dependent part. Each anomalously scattering atom has an atomic scattering factor of $f = f_0 + \Delta f + if''$ (Section 7.8). In Figure 9.5 $\mathbf{F_B}$ is the contribution to the structure factor by the nonanomalously scattering atoms, $\mathbf{F_A}$ is the nonanomalous contribution of the anomalously scattering atoms, and the complete nonanomalous part is $\mathbf{F_{BA}} = \mathbf{F_B} + \mathbf{F_A}$. The anomalous scattering contribution is

$$\frac{\Delta f}{f_0} \times \mathbf{F_A} + i\frac{f''}{f_0} \times \mathbf{F_A} = \mathbf{a}$$

ϕ_{BA} is the phase angle of $\mathbf{F_{BA}}$, ϕ_A of vector $\mathbf{F_A}$, and ϕ_a of vector $\mathbf{a}$. $\Delta\phi = \phi_{BA} + (180° - \phi_a) = 180° + (\phi_{BA} - \phi_a)$.

---

[1] Note that we indicate the real part of the anomalous signal by Δf, whereas in the literature it is often called f'.

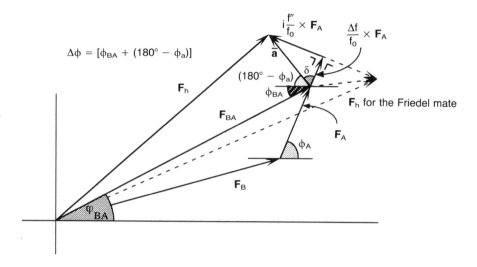

Figure 9.5. The structure factor diagram for the reflection of a protein crystal that contains one kind of anomalously scattering atoms. F_B is the contribution to the structure factor by the nonanomalously scattering atoms, F_A is the nonanomalous contribution of the anomalously scattering atoms, and the complete nonanomalous part is $F_{BA} = F_B + F_A$. $\Delta f/f_0 \times F_A + i(f''/f_0) \times F_A = a$ is the anomalous contribution of the anomalously scattering atoms. The anomalous component is exaggerated in this figure. The dotted lines are for the mirror image of the Friedel mate.

Writing $|F|$ for $|F(h)|$ and applying the cosine rule:

$$|F|^2 = |F_{BA}|^2 + |a|^2 - 2|F_{BA}| \times |a| \times \cos \Delta \phi$$

with

$$|a|^2 = \frac{(\Delta f)^2 + (f'')^2}{f_0^2} \times |F_A|^2$$

$$|F|^2 = |F_{BA}|^2 + \frac{(\Delta f)^2 + (f'')^2}{f_0^2} \times |F_A|^2$$

$$+ 2\frac{\Delta f}{f_0} \cos \delta |F_{BA}| \times |F_A| \cos(\phi_{BA} - \phi_a)$$

$$+ 2\frac{f''}{f_0} \sin \delta |F_{BA}| \times |F_A| \cos(\phi_{BA} - \phi_a)$$

$$\phi_a = \phi_A + \delta; \quad \cos(\phi_{BA} - \phi_a) = \cos(\phi_{BA} - \phi_A - \delta)$$

Because ϕ_{BA} and ϕ_A are independent of the wavelength λ, whereas δ depends on λ, the cosine will be split into

$$\cos(\phi_{BA} - \phi_A) \cos \delta + \sin(\phi_{BA} - \phi_A) \sin \delta$$

$$|F|^2 = |F_{BA}|^2 + \frac{(\Delta f)^2 + (f'')^2}{f_0^2} \times |F_A|^2$$

$$+ 2\frac{\Delta f}{f_0}|F_{BA}| \times |F_A| \times \{\cos^2 \delta \cos(\phi_{BA} - \phi_A)$$

$$+ \sin \delta \cos \delta \sin(\phi_{BA} - \phi_A)\}$$

$$+ 2\frac{f''}{f_0}|F_{BA}| \times |F_A| \times \{\sin \delta \cos \delta \cos(\phi_{BA} - \phi_A)$$

$$+ \sin^2 \delta \sin(\phi_{BA} - \phi_A)\}$$

Grouping together the $\cos(\phi_{BA} - \phi_A)$ terms and also the $\sin(\phi_{BA} - \phi_A)$ terms:

$$|F|^2 = |F_{BA}|^2 + \frac{(\Delta f)^2 + (f'')^2}{f_0^2} \times |F_A|^2$$

$$+ |F_{BA}| \times |F_A| \times \left\{2\frac{\Delta f}{f_0} \cos^2 \delta + 2\frac{f''}{f_0} \sin \delta \cos \delta\right\} \cos(\phi_{BA} - \phi_A)$$

$$+ |F_{BA}| \times |F_A| \times \left\{2\frac{\Delta f}{f_0} \sin \delta \cos \delta + 2\frac{f''}{f_0} \sin^2 \delta\right\} \sin(\phi_{BA} - \phi_A)$$

Because

$$\left(\frac{\Delta f}{f_0} \cos \delta + \frac{f''}{f_0} \sin \delta\right) \cos \delta = \frac{\Delta f}{f_0}$$

and

$$\left(\frac{\Delta f}{f_0} \cos \delta + \frac{f''}{f_0} \sin \delta\right) \sin \delta = \frac{f''}{f_0}$$

we finally obtain

$$|F|^2 = |F_{BA}|^2 + p|F_A|^2 + |F_{BA}| \times |F_A|$$

$$\times [q \cos(\phi_{DA} - \phi_A) + r \sin(\phi_{DA} - \phi_A)]$$

with

$$p = \frac{(\Delta f)^2 + (f'')^2}{f_0^2}, \quad q = 2\frac{\Delta f}{f_0}, \quad \text{and} \quad r = 2\frac{f''}{f_0}$$

p, q, and r are functions of λ and can be derived from the atomic absorption coefficient. The $|F|^2$ values are different for the Friedel mates but they can be determined experimentally. The unknown quantities are $|F_{BA}|$, $|F_A|$, and $(\phi_{BA} - \phi_A)$, all three independent of λ and equal for Friedel mates, except for the sign of $(\phi_{BA} - \phi_A)$. Therefore, a data set for one value of λ gives two sets of equations for these three unknowns and in principle measurements at two different wavelengths are sufficient to find $|F_{BA}|$, $|F_A|$, and $(\phi_{BA} - \phi_A)$ for each reflection. To calculate the electron density map of the protein, ϕ_{BA} is needed. This is obtained by solving the A-structure, that is, locating the anomalously scattering atoms from a Patterson map with coefficients $|F_A|^2$ or by direct methods. From the A-structure ϕ_A can be calculated and then ϕ_{BA} from the known value of $(\phi_{BA} - \phi_A)$. Terwilliger (1994a) pointed out that the estimates of $|F_A|$ may be unrealistically high. He introduces probability functions for the expected values of $|F_A|$ and, in combination with the MAD information, he obtains weighted-average values as improved values for $|F_A|$. Moreover, he treats the dispersive differences $\Delta F = \overline{|F(\lambda_i)|} - \overline{|F(\lambda_j)|}$ as isomorphous replacement information (Terwilliger, 1994b).

Because no anomalous scattering is taken into account for the calculation of the A-structure, the real structure or its enantiomorph is obtained. The solution of this problem is to calculate ϕ_A angles for both structures. This gives two sets of ϕ_{BA} angles and two protein electron density maps from which the best one must be selected.

9.5.2. The "Isomorphous Replacement" Method for Processing MAD Data

The physical effect of anomalous scattering is not different from isomorphous replacement. In the latter method a change in scattering is produced by introducing a sufficiently heavy atom and in MAD by changing the scattering of an existing atom. The real and imaginary contributions to atomic scattering are wavelength dependent, and this effect is exploited in MAD (Figure 9.6). A typical strategy for data collection and processing is presented here. We shall call the real contribution to the anomalous scattering by all anomalous scatterers in the crystal $\Delta \mathbf{F}$. For the data collected with $\lambda 1$ this is $\Delta \mathbf{F}1$, and for the other data sets $\Delta \mathbf{F}2$ and $\Delta \mathbf{F}3$. The imaginary contribution is $\mathbf{F}''1$ for the $\lambda 1$ data set, $\mathbf{F}''2$ for the $\lambda 2$ data set, and $\mathbf{F}''3$ for the $\lambda 3$ data set. The wavelengths $\lambda 1$, $\lambda 2$ and $\lambda 3$ are chosen as in Figure 9.4. The $\lambda 3$ data set is chosen as the parent; the other two sets are scaled to the parent set. We distinguish two processes: isomorphous or dispersive[2] MAD (Figure 9.6a) and anomalous MAD (Figure 9.6b).

In isomorphous MAD, the large difference between $\Delta \mathbf{F}3$ and $\Delta \mathbf{F}1$ is exploited as the major change in the structure factor by the anomalous scatterers. To remove the relatively small $\mathbf{F}''$ contribution, the Bijvoet pairs in the $\lambda 1$ and $\lambda 3$ sets are averaged. In fact, $\mathbf{F}_{parent}$ is taken as $\mathbf{F}_B + \mathbf{F}_A + \Delta \mathbf{F}3$ and triangle ABC in Figure 9.6a is comparable to triangle ABC in Figure 7.8. The processing is exactly as in isomorphous replacement. It

---

[2] Dispersive means wavelength dependent.

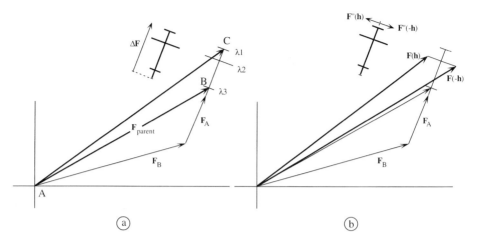

Figure 9.6. MAD as isomorphous replacement. The vectors F_A and F_B have the same meaning as in Figure 9.5. The parent structure factor $F_{parent} = F_B + F_A + \Delta F3$. The anomalous scattering contributions ΔF and F'' have been exaggerated. Moreover, the sign of ΔF is taken as positive for easier drawing. (a) The processing of the data as isomorphous (dispersive) information. $\Delta F1 - \Delta F3$ is comparable with the heavy-atom contribution in real isomorphous replacement. (b) The processing of the data from anomalous information by exploiting the difference between $|F(h)|$ and $|F(-h)|$. As usual, the structure factor $F(-h)$ in the figure is the mirror image of the actual $F(-h)$ with respect to the horizontal axis in the Argand diagram.

requires calculation of a difference Patterson—called a dispersive difference Patterson—for the localization of the anomalous scatterers followed by the determination of the preliminary phase angles by SIR. The $\lambda2$ set has relatively strong differences between the Bijvoet pairs because of the large $F''2$ (Figure 9.6b). With an anomalous difference Patterson, the anomalous scatterers can be localized as in Section 7.9 and preliminary phase angles determined by SIR. The combined information from anomalous and isomorphous MAD is the input into a refinement procedure such as MLPHARE (Otwinowski, 1991), PHASIT (Furey and Swaminathan, 1997), SHARP (de La Fortelle and Bricogne, 1997), or WHALESQ (Bella and Rossmann, 1998).

9.5.3. Notes

1. Anomalous contributions to the structure factor are small and, therefore, it is important to select the three wavelengths near an absorption edge of the anomalous scatterer. More precisely, $\lambda1$ for the dispersive difference exactly at the dip in the Δf curve and $\lambda3$ for the anomalous difference on the sharp peak in the f'' curve. Moreover, the synchrotron beam must be highly stable.

The anomalous contribution to the atomic scattering factor is a function of the atomic absorption coefficient for the anomalously scattering element and can be derived from the experimental values of this coefficient. The absorption coefficient must be measured at the absorption edge of the element and at some distance from the edge. Because the precise position of the absorption edge depends on the chemical environment of the element, the spectrum of the atomic absorption coefficient as a function of the X-ray wavelength (or photon energy) should be measured on the crystal itself. This can conveniently be done by measuring the fluorescence from the element when radiated by an incident beam. Fluorescence is a product of the absorption, because in absorption an electron is removed from its atomic orbital and fluorescent radiation is emitted when the empty position is filled up by another electron. The wavelength of the fluorescent radiation is characteristic for the irradiated element. In converting the fluorescence spectrum to the atomic absorption spectrum, background and scaling corrections are made in such a way that the experimental values for the atomic absorption coefficient fit their theoretical values. The latter cannot be calculated accurately inside the edge region, but it can be done outside the edge region.

$$\mu_a = s \times R - (a + b\Delta + c\Delta^2)$$

μ_a is the atomic absorption coefficient, R is the fluorescence ratio $I_{fl}(E)/I_0(E)$, in which $I_{fl}(E)$ is the fluorescent intensity, and $I_0(E)$ is the intensity of the incident beam. E is the energy of the incident photons. $\Delta = E - E_0$, where E_0 corresponds with the photon energy at the absorption edge. s is a scale factor that is assumed to be independent of E. s, a, b, and c are chosen such that the experimental and theoretical curves fit as closely as possible to each other.

The atomic absorption curve (μ_a as a function of λ) is directly related to the wavelength dependence of the imaginary part of the anomalous scattering, and the position of the peak in the absorption curve corresponds exactly with the position of the peak in the f'' curve (for an explanation see the end of this Section). The real part, Δf, can be derived from the imaginary part by the Kramers–Kronig transformation (Hendrickson et al., 1988). The mathematical relationships are derived in James (1965). It turns out that the dip in the Δf curve corresponds with the inflection point in the f'' curve, as indicated in Figure 9.4.

2. Bijvoet differences are most accurately measured if the corresponding reflections, $h\,k\,\ell$ and $\bar{h}\,\bar{k}\,\bar{\ell}$, or a symmetry related one, are measured close together in time, for instance by the inverse beam geometry: measure $h\,k\,\ell$, rotate the crystal 180°, and measure $\bar{h}\,\bar{k}\,\bar{\ell}$ (see Figure 4.20). By measuring them close together differences because of beam instability, radiation damage of the crystal, or absorption by ice growing on the specimen in a cryo experiment, can be avoided.

3. f'' is always positive. Δf can be either positive or negative.

4. The great advantage of MAD is ideal isomorphism. Data collection for all wavelengths is on the same crystal, or the same type of crystal.

5. Anomalous scattering is the result of tightly bound electrons close to the atomic nucleus. Because this inner region is small in size, the anomalous effect stays rather constant up to the resolution edge for proteins. It means that the relative contribution by the anomalous scattering increases with resolution.

The Relation Between Absorption and the Imaginary Part of Anomalous Scattering

The direct relationship between absorption and the imaginary part of anomalous scattering can easily be understood with the Argand diagram in Figure 9.7. The phase of the incident beam is indicated by a vector on the real axis, pointing to the right. A free electron scatters the X-ray beam with a phase angle of π with respect to the incident beam; its vector is pointing to the left, opposite to the incident beam vector (Section 4.14). In Section 4.15 we have seen that a plane of free electrons (or atoms with free elec-

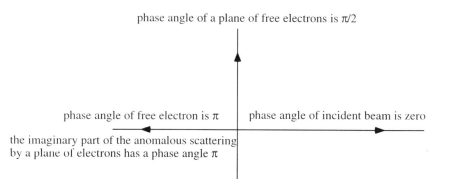

Figure 9.7. Vectors indicating phase angles in the Argand diagram. The vector for the incident beam is on the real axis, pointing to the right, because it is given phase angle zero. A free electron scatters the X-ray beam with a phase angle of π with respect to the incident beam; its vector is pointing to the left. A plane of free electrons (or atoms with free electrons) scatters $\frac{\pi}{2}$ behind a free electron and the arrow is pointing upward. If the electrons are not completely free, the imaginary part of anomalous scattering by a plane of nonfree electrons has a phase angle $\frac{\pi}{2}$ ahead of the scattering by free electrons, and its vector is pointing along the real axis to the left. The consequence is that a plane of anomalously scattering electrons has a phase angle just opposite to the incident beam and diminishes its intensity; in other words: absorption!

trons) scatters $\frac{\pi}{2}$ behind a free electron. If the electrons are not completely free, there is an anomalous contribution to the scattering. The imaginary part of it has a phase angle $\frac{\pi}{2}$ ahead of the scattering by free electrons; this is also true for a plane of electrons. The consequence is that a plane of anomalously scattering electrons has a phase angle just opposite to the incident beam. This expresses itself as absorption.

Summary

Tightly bound electrons in an atom cause measurable anomalous scattering of X-rays. For light atoms (C, N, and O) the effect is negligible but for heavier atoms (from S onward for the commonly used X-ray wavelengths) a measurable effect does occur and this causes a difference in intensity between Friedel pairs of reflection (anomalous effect). Moreover, anomalous scattering is wavelength dependent (dispersive effect). Both effects are exploited in MAD. Because the contribution by anomalous scattering to the structure factors is small, intensities must be measured with extreme accuracy.

From anomalous scattering data two kinds of information can be derived:

1. the choice between a structure and its enantiomorph and
2. phase angle information: this can be additional to the isomorphous replacement information, or by collecting data at suitably chosen X-ray wavelengths, it can supply phase information for a native protein crystal if this does contain an anomalously scattering atom.

Chapter 10
Molecular Replacement

10.1. Introduction

With the isomorphous replacement method a preliminary set of protein phases and a first model of the protein structure can be obtained. As we shall see in Chapter 13 such a model can be refined by minimizing the difference between the observed $|F|$-values and the $|F|$-values calculated from the model. An easier way to obtain a first model can be followed if the structure of a protein with a homologous amino acid sequence has already been established. The structure of this homologous protein is—as it were—borrowed by the protein for which the structure must be determined and serves as a very first model that can subsequently be refined. This procedure is based on the observation that proteins, homologous in their amino acid sequence, have a very similar folding of their polypeptide chain. Also, if for another reason two structures can be expected to be similar, one known and the other unknown, the procedure can be applied.

The problem is to transfer the known protein molecular structure from its crystalline arrangement to the crystal of the protein for which the structure is not yet known. The solution is the molecular replacement method, which was initiated in pioneering studies by Rossmann and Blow (1962). Placement of the molecule in the target unit cell requires its proper orientation and precise position. In short, it involves two steps: rotation and translation. In the rotation step the spatial orientation of the known and unknown molecule with respect to each other is determined while in the next step the translation needed to superimpose the now correctly oriented molecule onto the other molecule is calculated. The molecular replacement method can also serve another purpose: If a crystal structure has more than one protein molecule or a number of equal subunits in the asymmetric unit, then their relative position can be determined. This noncrystallographic symmetry is useful information in the process of improving protein phase angles by molecular averaging (Section 8.4).

The basic principle of the molecular replacement method can be understood by regarding the Patterson function of a protein crystal structure. The Patterson map is a vector map: vectors between atoms in the real structure show up as vectors from the origin to maxima in the Patterson map. If the pairs of atoms belong to the same molecule, then the corresponding vectors are relatively short and their end-points are found not too far from the origin in the Patterson map; they are called *self-Patterson vectors*. If there were no intermolecular vectors (*cross-Patterson vectors*), this inner region of the Patterson map would be equal for the same molecule in different crystal structures, apart from a rotation difference. For homologous molecules it is not exactly equal but very similar. Therefore, the self-Patterson vectors can supply us with the rotational relationship between the known and the unknown molecular structures. From the cross-Patterson vectors the translation required for moving the molecules to their correct position can be derived. The principle of separating the Patterson vectors into these two groups and using them for orientation and translation determination was given by Hoppe (1957).

10.2. The Rotation Function

We shall first consider how the angular relationship between identical units within one asymmetric unit (*self-rotation function*) or between equal or closely related molecules in two different crystal forms (*cross-rotation function*) can be derived from the X-ray data. This will be discussed following the original Rossmann and Blow procedure. It is true that in many of the software packages available to calculate the rotational orientation the conventional Rossmann and Blow procedure is replaced by the mathematically more elegant Crowther's fast rotation function (Crowther, 1972). However, the principle of the method can be best understood with the Rossmann and Blow procedure.

The self-Patterson peaks all lie in a volume around the origin with a radius equal to the dimension of the molecule (or subunit). If a number of identical molecules (or subunits) lie within one asymmetric unit, the self-Patterson vector distribution is exactly the same for all these molecules, except for a rotation that is the same as their noncrystallographic rotational symmetry in real space. Therefore, if the Patterson function is superimposed on a correctly rotated version, maximum overlap between the two Patterson maps will occur. Similarly, for two different lattices, the two different Patterson maps must be superposed to maximum overlap by a rotation of one of the two maps.

We assume that the crystal system has orthogonal axes. An atom in one system is located at position $\mathbf{x} = x_1\mathbf{a}_1 + x_2\mathbf{a}_2 + x_3\mathbf{a}_3$ in a crystallographic lattice with axes $\mathbf{a}_1$, $\mathbf{a}_2$, and $\mathbf{a}_3$. Rotation of the axial system, keeping the same origin, leads to a new set of axes: $\mathbf{a}_{r,1}$, $\mathbf{a}_{r,2}$ and $\cdot\mathbf{a}_{r,3}$. With respect to the new axes the position of the particular atom in position $\mathbf{x}$ is $\mathbf{x}_r = x_{r,1}\mathbf{a}_{r,1} + x_{r,2}\mathbf{a}_{r,2} + x_{r,3}\mathbf{a}_{r,3}$ and the relationship between the two sets of coordinates is

$$\left.\begin{array}{l} x_{r,1} = c_{11}x_1 + c_{12}x_2 + c_{13}x_3 \\ x_{r,2} = c_{21}x_1 + c_{22}x_2 + c_{23}x_3 \\ x_{r,3} = c_{31}x_1 + c_{32}x_2 + c_{33}x_3 \end{array}\right\} \quad \text{or in matrix notation } \mathbf{x}_r = [C]\mathbf{x}$$

A rotation of the axes has the same effect as a rotation of the structure in the opposite direction. If the structure rotates, its Patterson map rotates in the same way. Applying the rotation $[C]$ to the Patterson function $P(\mathbf{u})$ gives the rotated Patterson function $P_r(\mathbf{u}_r)$. An overlap function R of $P(\mathbf{u})$ with the rotated version, $P_r(\mathbf{u}_r)$, of the same crystal lattice (self-rotation function) or a different crystal lattice (cross-rotation function) is defined as

$$R(\alpha, \beta, \gamma) = \int_U P(\mathbf{u}) \times P_r(\mathbf{u}_r)\, d\mathbf{u} \qquad (10.1)$$

U is the volume in the Patterson map where the self-Patterson peaks are located. The product function R depends on the rotation angles (related to $[C]$) and will have a maximum value for correct overlap. $P(\mathbf{u})$ can be expanded in a Fourier series:

$$P(\mathbf{u}) = \frac{1}{V}\sum_{\mathbf{h}}|F(\mathbf{h})|^2 \exp[-2\pi i\mathbf{h}\mathbf{u}]$$

For $P_r(\mathbf{u}_r)$ can be written

$$P_r(\mathbf{u}_r) = \frac{1}{V}\sum_{\mathbf{h}'}|F(\mathbf{h}')|^2 \exp[-2\pi i\mathbf{h}'\mathbf{u}_r]$$

Because $\mathbf{u}_r = [C]\mathbf{u}$

$$P_r(\mathbf{u}_r) = \frac{1}{V}\sum_{\mathbf{h}'}|F(\mathbf{h}')|^2 \exp[-2\pi i\mathbf{h}'[C]\mathbf{u}]$$

$\mathbf{h}'[C]$ is equal to $[C^{-1}]\mathbf{h}'$, and therefore

$$P_r(\mathbf{u}_r) = \frac{1}{V}\sum_{\mathbf{h}'}|F(\mathbf{h}')|^2 \exp[-2\pi i[C^{-1}]\mathbf{h}'\mathbf{u}]$$

which is equal to

$$\frac{1}{V}\sum_{\mathbf{h}'}|F([C]\mathbf{h}')|^2\exp[-2\pi i\mathbf{h}'\mathbf{u}]$$

$P(\mathbf{u})$ and $P_r(\mathbf{u}_r)$ must now be superimposed and $P(\mathbf{u}) \times P_r(\mathbf{u}_r)$ calculated for every position $\mathbf{u}$ within U and then the integral must be taken to obtain $R(\alpha, \beta, \gamma)$:

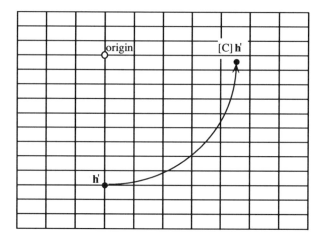

Figure 10.1. The effect of a rotation $[C]$ in reciprocal space is that an integral lattice point $\mathbf{h}'$ ends up at a nonintegral lattice position $([C]\mathbf{h}')$. This is illustrated here for a two-dimensional lattice.

$$R(\alpha, \beta, \gamma) = \frac{1}{V^2}\sum_{\mathbf{h}}\sum_{\mathbf{h}'}|F(\mathbf{h})|^2|F([C]\mathbf{h}')|^2$$

$$\times \int_U \exp[-2\pi i(\mathbf{h} + \mathbf{h}')\mathbf{u}]\,du \qquad (10.2)$$

A problem arises here: for the calculation of $P_r(\mathbf{u}_r)$ the coefficients must be sampled at positions $[C]\mathbf{h}'$ in reciprocal space, which is impossible, because, in general, $[C]\mathbf{h}'$ is at nonintegral reciprocal lattice positions (Figure 10.1). This problem can be solved together with another problem, namely that an enormous number of terms must be calculated, because a multiplication is required for each $\mathbf{h}$ with all $\mathbf{h}'$. However, it is not as difficult as it looks like, because of the weighting term

$$\int_U \exp[-2\pi i(\mathbf{h} + \mathbf{h}')\mathbf{u}]\,du$$

It limits the summation over $\mathbf{h}'$ to only a restricted number of $\mathbf{h}'$. This can be understood in the following way. Suppose a crystalline lattice has a very special structure: it contains in each unit cell a body with the shape and the volume of U. The center of U is at the origin of the unit cell. The electron density inside U is flat $[\rho(\mathbf{x}) = \rho]$, and outside U the unit cell is empty: $\rho(\mathbf{x}) = 0$ (Figure 10.2). The structure factor of this special structure at reciprocal lattice position $-(\mathbf{h} + \mathbf{h}')$ is

$$\mathbf{F}[-(\mathbf{h} + \mathbf{h}')] = V\int_V \rho(x)\,\exp[-2\pi i(\mathbf{h} + \mathbf{h}')x]\,dx$$

Because of the special electron density distribution

$$\mathbf{F}[-(\mathbf{h} + \mathbf{h}')] = V\rho \int_U \exp[-2\pi i(\mathbf{h} + \mathbf{h}')\mathbf{x}]\,d\mathbf{x} \qquad (10.3)$$

If the Fourier transform of the body with the shape and volume of U and *unit* electron density is $\mathbf{G}$, then the value of the transform for the body with uniform electron density ρ at reciprocal lattice position $-(\mathbf{h} + \mathbf{h}')$ is

$$U \times \rho \times \mathbf{G}[-(\mathbf{h} + \mathbf{h}')] \qquad (10.4)$$

Comparing Eqs. (10.3) and (10.4) gives

$$\int_U \exp[-2\pi i(\mathbf{h} + \mathbf{h}')\mathbf{x}]\,d\mathbf{x} = \frac{U}{V} \times \mathbf{G}[-(\mathbf{h} + \mathbf{h}')] \qquad (10.5)$$

The rotation function can thus be written as

$$R(\alpha, \beta, \gamma) = \frac{U}{V^3}\sum_{\mathbf{h}}\sum_{\mathbf{h}'}|F(\mathbf{h})|^2|F([C]\mathbf{h}')|^2 \times \mathbf{G}[-(\mathbf{h} + \mathbf{h}')] \qquad (10.6)$$

The properties of $\mathbf{G}[-(\mathbf{h} + \mathbf{h}')]$ allow us to solve the two problems mentioned above. Usually U is assumed to be spherical and the transform of a sphere with radius $\mathbf{r}$ at the origin of a unit cell is

$$G = \frac{3(\sin 2\pi x - 2\pi x \cos 2\pi x)}{(2\pi x)^3}$$

x is in our case equal to $(\mathbf{h} + \mathbf{h}')\cdot\mathbf{r}$. The graphic representation of the function is shown in Figure 10.3. $\mathbf{G}$ has its maximum value for $\mathbf{h}' = -\mathbf{h}$ and falls off very rapidly for values of $\mathbf{h}'$ differing from $-\mathbf{h}$. Therefore the summation in Eq. (10.2) can be performed for every $\mathbf{h}$ with only a limited number of $\mathbf{h}'$ terms, namely only those for which $\mathbf{h}'$ is close to $-\mathbf{h}$. This solves the second problem.

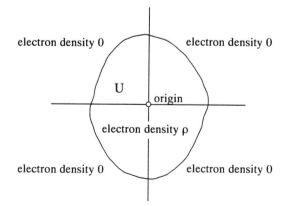

Figure 10.2. The integral in Eq. (10.2) is developed by using a special structure that has a flat electron density inside the three-dimensional body $U[\rho(\mathbf{x}) = \rho]$, and no electron density outside $U[\rho(\mathbf{x}) = 0]$. The center of U is in the origin of the unit cell.

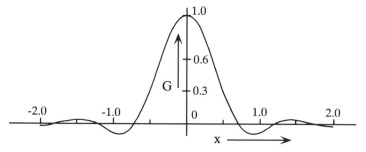

Figure 10.3. The function $G = [3(\sin 2\pi x - 2\pi x \cos 2\pi x)]/(2\pi x)^3$ is plotted as a function of x. Note the rapid fading away for increasing x.

The fall off of G can be illustrated with an example. Let the cell dimension be $80\,\text{Å}$ and the radius of the sphere $r = 20\,\text{Å}$, then we obtain the following values for G:

Distance from $-\mathbf{h}$ in Reciprocal Lattice Units	x	G
1	0.25	0.78
2	0.50	0.30
3	0.75	−0.029

To solve the first problem, $|F([C]\mathbf{h}')|^2$ is calculated for a number of integral lattice points around $([C]\mathbf{h}')$ and the $|F|^2$-value at $([C]\mathbf{h}')$ is obtained by interpolation, giving a weight $G[(\mathbf{h} + \mathbf{h}')]$ to the $|F|^2$-values at the integral lattice points around $([C]\mathbf{h}')$.

In the application of the rotation method it is important that all strong reflections are present because the calculation of the rotation function basically depends on the rotation of a Patterson map, which is mainly determined by the strong reflections. Another point to consider is the resolution range of the data used in the calculation of the rotation function. Low resolution data can be excluded because they are rather insensitive to rotation; moreover they are determined to an appreciable extent by the solvent region. High resolution data are more discriminating but are also more sensitive for the model. The best range is often found between 3 and 5 Å. Also, because of computational limitations, the integration is extended to a rather modest resolution.

Other parameters to choose are the shape and the size of the region U. For a matter of convenience the region is assumed to be spherical. Its radius can be chosen equal to, or somewhat less than the diameter of the molecule. In the calculation of the rotation function, the shorter intermolecular vectors confuse the situation. This can be improved if in the calculation of the rotation function the known molecule is put into a large artificial unit cell having no crystallographic symmetry (space group $P1$). The dimensions of the cell should be such that all cross-vectors are longer than the diameter of the molecule. Instead of working with the X-ray data from the crystal structure of the known molecule, the calculated structure factors of the artificial lattice are used. It is not always easy to find the optimal model structure and different models must be tried if the first results are unsatisfactory. For instance, the original model can be truncated by deleting side chains, doubtful parts, using one monomer if the original model was an oligomer, and using just one domain of the model molecule.

The magnitude of the rotation function is plotted in a three-dimensional space with the three angular rotations as the coordinates. Several alternatives and conventions for the directions, names, signs, and origins of the rotation angles exist. This can cause a great deal of confusion and it is extremely important to know the procedure in the available software package. The usual system works with Eulerian angles because then the symmetry of the rotation function shows up clearly. The system used by Rossmann and Blow in their original paper (Rossmann and Blow, 1962) applies first a rotation by the angle α around z of an orthogonal coordinate system, then around the new x-axis by an angle β, and finally a rotation by γ around the new z axis. However, another convention is now generally used (Machin, 1985): Rotation by the angle α around the z-axis, next a rotation by the angle β around the new y-axis, and finally a rotation by the angle γ around the new z-axis (Figure 10.4a). The sign for

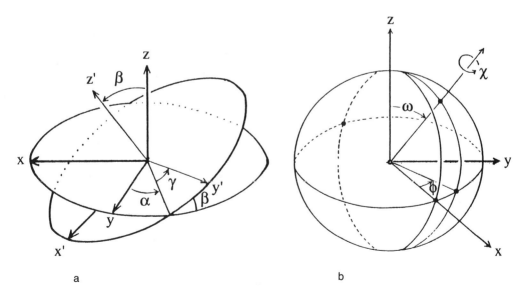

Figure 10.4. (a) Eulerian angles α, β, and γ; (b) Polar angles χ, ω, and ϕ. (Reproduced from the Proceedings of the Daresbury Study Weekend on Molecular Replacement, 15–16 February, 1985, with permission from the Daresbury Laboratory, Daresbury, Warrington, U.K.)

the rotation of the axes is positive for a clockwise rotation when looking from the origin along the positive rotation axis.

If one is searching for noncrystallographic symmetry in an asymmetric unit (self-rotation function) a zero rotation results in a high value for R, because it simply superposes the Patterson map on itself. An odd feature of R is the ridge in the section $\beta = 0$. It represents the set of equivalent zero rotations $(\alpha, 0, -\alpha)$. This stems from the fact that for $\beta = 0$ the rotations α and γ have the same effect on the rotation and, therefore, all rotations with the same $\alpha + \gamma$ are identical. The same is also true for $\beta = 180°$, where $(\alpha - \gamma)$ rotations are equivalent. Equivalent effects can also be obtained for different combinations of the three angular coordinates, which causes symmetry in the rotation function. It depends on symmetry in the Patterson functions that are rotated with respect to each other and on the choice of the system of variables chosen for the rotation. Because of the symmetry in the rotation function it need to be calculated only for its asymmetric unit. Methods for deriving the asymmetric unit in the rotation function have been given by Tollin and Rossmann (1966), Narasinga Rao et al. (1980), and Moss (1985).

When searching for noncrystallographic symmetry, it is convenient to work with spherical polar rotation (Figure 10.4b). ω and ϕ determine the position of the rotation axis and χ is the rotation around this axis. The advantage of this system derives from the fact that rotations of $\chi = 180°$ and $120°$ are very common and the search can be restricted to a fixed value for χ. In Figure 10.4b, Z is the polar axis. However, when working in a monoclinic space group it is convenient to have the unique y-axis as the polar axis.

From Eq. (10.1) it is clear that the strong reflections will dominate the calculation of R. This is true to such an extent that it suffices to incorporate only a fraction of all reflections in the calculation, for instance, 10% with the highest intensities (Tollin and Rossmann, 1966). To obtain an impression of the quality of the rotation function it is advisable to give the ratio of the highest to the next highest peak in the rotation map.

An, in principle, simpler way than the Rossmann and Blow method to solve the rotation problem was proposed by Huber and his colleagues (Huber, 1985). In their method the Patterson map is also rotated and the rotated map is superimposed on the original map. However, now the operation is performed, not in reciprocal, but in direct space. A number of high peaks (a few hundred to a few thousand) in the Patterson map are selected for a rough rotational search, which is done in steps of 10°. These peaks lie within a sphere around the origin where most of the self-Patterson peaks are located. However, the innermost region of this sphere, close to the high origin peak of the Patterson map, is neglected. As fit criterion the product of the map and its rotated version at corresponding grid points is used, as in the Rossmann and Blow procedure. After a highest or a few high peaks are found in the product function, the search can be continued in finer steps around these peaks.

10.2.1. The Locked Rotation Function

If the cross-rotation function is applied with a model molecule to be oriented in an unknown crystal structure, several solutions will be found if the crystal structure has crystallographic symmetry, but also because of noncrystallographic symmetry operators. If the latter do exist and are known from the self-rotation function, the solutions of the cross-rotation function are not independent but related through the noncrystallographic symmetry. This knowledge can be used as a constraint in the calculation of a "locked" rotation function, which is the average of n independent rotation functions and, consequently, has an improved peak-to-noise ratio (Tong and Rossmann, 1990; Hiremath et al., 1990). By the same token the self-rotation function can be improved if an assumed point group symmetry is imposed on the unknown molecule or combination of molecules.

10.2.1.1. The Locked Self-Rotation Function

For a locked self-rotation function an assumed point group in a standard orientation is rotated into the unknown unit cell by matrix $[E]$. Within this point group are n symmetry elements $[I_i]$ with $i = 1, \ldots, n$. Their application leaves the point group unchanged. A position vector $\mathbf{U}_1$ in the point group is moved to $\mathbf{U}_i$ by the application of the symmetry element $[I_i]$:

$$\mathbf{U}_i = [I_i] \times \mathbf{U}_1 \quad \text{with} \quad i = 1, \ldots, n$$

By application of $[E]$ the vector $\mathbf{U}_1$ in the point group is moved to $\mathbf{X}_1$ in the crystal:

$$\mathbf{X}_1 = [E] \times \mathbf{U}_1$$
$$\mathbf{X}_i = [E] \times \mathbf{U}_i$$

For the point group as well as for the crystal, an orthogonal lattice is defined. The relationship between each $\mathbf{X}_i$ and $\mathbf{X}_1$ is then.

$$\mathbf{X}_i = [E] \times [I_i] \times \mathbf{U}_1 = [E] \times [I_i] \times [E^{-1}] \times \mathbf{X}_1$$

From the self-rotation function we know that

$$\mathbf{X}_i = [\rho_i] \times \mathbf{X}_1$$
$$[\rho_i] = [E] \times [I_i] \times [E^{-1}] \tag{10.7}$$

In self-rotation the Patterson function is rotated and superimposed on itself. The rotation is by matrix $[C]$ in a real (not necessarily orthogonal) lattice.

$$[C] = [\alpha] \times [\rho] \times [\beta] \qquad (10.8)$$

where matrix $[\beta]$ orthogonalizes the real lattice and $[\alpha]$ deorthogonalizes it. Combining Eqs. (10.7) and (10.8):

$$[C_i] = [\alpha] \times [E] \times [I_i] \times [E^{-1}] \times [\beta] \qquad (10.9)$$

$[C_i]$ tests the ith symmetry element and this must be done for each of the n symmetry elements. Each $[C_i]$ corresponds with a self-rotation function R_i. By combining the results, the value of the locked rotation function is obtained as R_L:

$$R_L = \frac{\sum_{i=1}^{n} R_i}{n} \qquad (10.10)$$

The noise of this function will be $\sqrt{n}$ smaller than for the normal self-rotation function.

10.2.1.2. The Locked Cross-Rotation Function

In the calculation of this function one model molecule is oriented in the unknown crystal structure, which contains one or more similar molecules. In case of more molecules, the cross-rotation function gives several peaks corresponding with the superposition of the model molecule on each of these separate molecules. From the self-rotation function the relative orientation of these separate molecules within an asymmetric unit is usually known and can be used as a constraint in the calculation of the cross-rotation function. If $[E]$ is the matrix that relates the known model molecule with the first molecule in the unknown cell (cell with unknown molecular structure), we have

$$\mathbf{X}_1 = [E] \times \mathbf{U}$$

where $\mathbf{U}$ is a position vector in the model molecule. If $[\beta_u]$ is the orthogonalization matrix in the known cell and $[\alpha_x]$ the deorthogonalization matrix in the unknown cell

$$\mathbf{x}_1 = [\alpha_x] \times [E] \times [\beta_u] \times \mathbf{u} \qquad (10.11)$$

with $\mathbf{x}$ and $\mathbf{u}$ fractional coordinates in the unknown and known cell, respectively. $[I_i]$, with $i = 1, \ldots, n$, is a matrix that moves position vector $\mathbf{X}_1$ in the unknown structure to $\mathbf{X}_i$ in an orthogonal coordinate system; $\mathbf{X}_1$ and $\mathbf{X}_i$ are related by the noncrystallographic symmetry. In fractional coordinates

$$\mathbf{x}_i = [\alpha_x] \times [I_i] \times [\beta_x] \times \mathbf{x}_1 \qquad (10.12)$$

Combining Eqs. (10.11) and (10.12), it follows that

$$\mathbf{x}_i = [\alpha_x] \times [I_i] \times [\beta_x] \times [\alpha_x] \times [E] \times [\beta_u] \times \mathbf{u}$$

or

$$\mathbf{x}_i = [\alpha_x] \times [I_i] \times [E] \times [\beta_u] \times \mathbf{u}$$

The cross-rotation function relates $\mathbf{x}_i$ to $\mathbf{u}$ by means of the matrix $[C_i]$.

$$[C_i] = [\alpha_x] \times [I_i] \times [E] \times [\beta_u]$$

The locked cross-rotation function is now calculated as the average of n cross-rotation functions, each calculated with a different $[C_i]$.

$$R_L = \frac{\sum\limits_{i=1}^{n} R_i}{n} \tag{10.13}$$

Because n rotation functions instead of 1 must be calculated, the calculation time would be n times longer. However, it can be shown that the plot of the locked cross-rotation function has an n times higher symmetry and, therefore, the time of calculation for a normal and a locked cross-rotation function is not much different. The locked rotation function has the advantage of a better peak-to-noise ratio and the interpretation of its result is simpler because it involves the location of fewer peaks. A further advantage is that only the very strong reflections—even fewer than for the normal rotation function—need to be used, which speeds up the calculation appreciably.

10.3. The Translation Function

The rotation function is based on the rotation of a Patterson function around an axis through its origin. A translation is not incorporated. However, for the final solution of the molecular replacement method the translation required to overlap one molecule (or subunit) onto the other in real space must be determined, after it has been oriented in the correct way with the rotation function. The simplest way to do this is by trial and error. The known molecule is moved through the asymmetric unit and structure factors are calculated—F(calc)—and compared with the observed structure factor by calculating an R-factor or the correlation coefficient as a function of the molecular position.

$$R = \frac{\sum\limits_{hkl} ||F(\text{obs})| - k|F(\text{calc})||}{\sum\limits_{hkl} |F(\text{obs})|}$$

The standard linear correlation coefficient C is

$$C = \frac{\sum\limits_{hkl}(|F(\text{obs})|^2 - \overline{|F(\text{obs})|^2}) \times (|F(\text{calc})|^2 - \overline{|F(\text{calc})|^2})}{\left[\sum\limits_{hkl}(|F(\text{obs})|^2 - \overline{|F(\text{obs})|^2})^2\sum\limits_{hkl}(|F(\text{calc})|^2 - \overline{|F(\text{calc})|^2})^2\right]^{1/2}}$$

The advantage of this correlation coefficient over the R-factor is that it is scaling insensitive; replacement of $|F(\text{obs})|^2$ by $k|F(\text{obs})|^2$ + a constant (k is the scale factor for the intensities) gives the same value. During R-factor search calculations, one need not start calculating the $F(\text{calc})$ values from atomic coordinates every time the molecule is shifted, but instead calculate phase shifts for each set of the molecules related by crystallographic symmetry. This saves considerable computer time. The result from the rotation function can be refined by changing the rotation parameters in small steps. With increased computing power in the future, it might be possible, instead of calculating the rotation and translation function, to carry out a six-dimensional search with the three rotation angles and the three translation components as parameters. This has already been applied successfully in the structure determination of oligonucleotides and nucleic acids at low resolution (Rabinovich and Shakked, 1984). It can also be helpful to exploit packing analysis; the protein molecules in the unit cell cannot penetrate each other and this limits the possible positions of the molecules. Although not necessarily giving a unique solution, it excludes parts of the unit cell and in this way reduces the computer time (Hendrickson and Ward, 1976; Harada et al., 1981).

In a more straightforward method than the trial and error search, a translation function is calculated that gives the correlation between a set of cross-Patterson vectors for a model structure and the observed Patterson function. Cross-Patterson vectors in this context mean vectors in the Patterson map derived from vectors between atoms in two molecules in the model structure related by a crystallographic symmetry operation $[C]$ + $\mathbf{d}$. In space group $P1$, in which no crystallographic symmetry exists, the origin can be chosen everywhere and this has no influence on the absolute value of the structure factor. The calculation of the translation function is then not necessary.

10.3.1. The Crowther and Blow translation function

Here, the translation function will be presented as derived by Crowther and Blow (1967) and we shall do this for an unknown structure with crystals belonging to space group $P222$ (Figure 10.5). They have one molecule in the asymmetric unit and therefore four molecules (1–4) in the unit cell. We choose one pair of molecules, e.g., 1 and 2. Their orientation is known from the rotation function, but not their position in

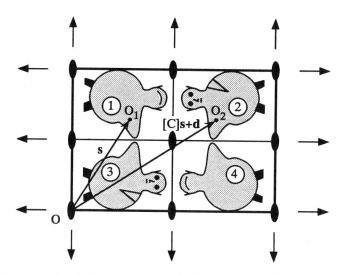

Figure 10.5. A unit cell for space group $P222$. It contains four molecules, 1–4. The origin of the unit cell is in O. Molecule 1 has its local origin in O_1 at position s, and the origin of molecule 2 is in O_2 at position $[C]s + d$, where matrix $[C]$ transfers molecule 1 to molecule 2; see also Figure 10.6.

the unit cell. With the translation function we can determine the position of molecule 1 with respect to the symmetry related molecule 2, and subsequently for any other pair of symmetry related molecules.

The local origin of molecule 1 is in O_1 and of molecule 2 in O_2. The origin of the unit cell is in O. Cross-Patterson vectors between the two molecules can then be calculated with (see Section 7.10).

$$P_{1,2}(\mathbf{u}) = \int_V \rho_1(\mathbf{x}) \times \rho_2(\mathbf{x} + \mathbf{u}) \, d\mathbf{x} \qquad (10.14)$$

If the electron density expressed with respect to the local origin of the first molecule, the model molecule M, is ρ_M, then (Figure 10.6)

$$\rho_1(\mathbf{x}) - \rho_M(\mathbf{x} - \mathbf{s})$$

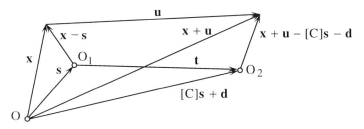

Figure 10.6. The position $(\mathbf{x})$ in molecule 1 and $(\mathbf{x} + \mathbf{u})$ in molecule 2. O_1 and O_2 are local origins.

and

$$\rho_2(x + u) = \rho(x + u - [C]s - d)$$

for the local origin in O_2. This is equal to the electron density in the model molecule (molecule 1) at the symmetry related position $[C^{-1}](x + u - [C]s - d)$ and, therefore

$$\rho_2(x + u) = \rho_M\{[C^{-1}](x + u - [C]s - d)\}$$

Equation (10.14) becomes

$$P_{1,2}(u, s) = \int_V \rho_M(x - s) \times \rho_M\{[C^{-1}](x + u - [C]s - d)\}\, dx$$

If ρ_M is written as a Fourier series in terms of the structure factors F_M, we have

$$P_{1,2}(u, s) = \int_V \sum_h F_M(h) \exp[-2\pi ih \cdot (x - s)]$$

$$\times \sum_p F_M(p) \exp[-2\pi ip[C^{-1}](x + u - [C]s - d)]\, dx$$

$$= \sum_h \sum_p F_M(h) \times F_M(p) \exp[2\pi i(h \cdot s + p[C^{-1}]([C]s + d))]$$

$$\times \exp[-2\pi ip[C^{-1}]u] \int_V \exp[-2\pi i(h + p[C^{-1}])x]\, dx$$

Because $[C]$ is a crystallographic rotation, $p[C^{-1}]$ is a reciprocal lattice point and, therefore, $(h + p[C^{-1}])$ is an integral number and as a consequence the integral vanishes, unless $(h + p[C^{-1}]) = 0$ (see Figure 4.15). Then the integral is equal to V with $p = -h[C]$. Deleting the constant V

$$P_{1,2}(u, s) = \sum_h F_M(h) \cdot F_M(-h[C]) \exp[2\pi ih(s - [C]s - d)] \exp[2\pi ih \cdot u]$$

If the intermolecular vector t is between O_1 and O_2:

$$t = -s + [C]s + d$$

Because $F_M(-h[C]) = F_M^*(h[C])$ with F_M^* the complex conjugate of F_M:

$$P_{1,2}(u, t) = \sum_h F_M(h) \cdot F_M^*(h[C]) \exp[-2\pi ih \cdot t] \exp[2\pi ih \cdot u] \qquad (10.15)$$

This is the cross-Patterson function of the model structure in which two molecules are related by crystallographic symmetry. This should now be compared with the observed Patterson function $P(u)$. To this end the translation function $T(t)$ is calculated:

$$T(t) = \int_V P_{1,2}(u, t) \times P(u)\, du \qquad (10.16)$$

When the intermolecular vector **t** is equal to the true intermolecular vector $\mathbf{t}_0$, the function $T(\mathbf{t})$ will reach a maximum value, because then the computed Patterson $P_{1,2}(\mathbf{u}, \mathbf{t})$ fits correctly on the observed Patterson function $P(\mathbf{u})$. If $P_{1,2}(\mathbf{u}, \mathbf{t})$ in Eq. (10.16) is replaced by the right part of Eq. (10.15) and $P(\mathbf{u})$ by $\Sigma_{\mathbf{p}}|F_{obs}(\mathbf{p})|^2 \exp[-2\pi i\mathbf{p} \cdot \mathbf{u}]\, d\mathbf{u}$ the translation function becomes:

$$T(\mathbf{t}) = \int_V \sum_{\mathbf{h}} \mathbf{F}_M(\mathbf{h}) \cdot \mathbf{F}_M^*(\mathbf{h}[C]) \exp[-2\pi i\mathbf{h} \cdot \mathbf{t}] \exp[2\pi i\mathbf{h} \cdot \mathbf{u}]$$

$$\times \sum_{\mathbf{p}} |F_{obs}(\mathbf{p})|^2 \exp[-2\pi i\mathbf{p} \cdot \mathbf{u}]\, d\mathbf{u}$$

Because the integration is over **u** we can take all terms with **u** together under the integral

$$T(\mathbf{t}) = \sum_{\mathbf{h}} \mathbf{F}_M(\mathbf{h}) \cdot \mathbf{F}_M^*(\mathbf{h}[C]) \exp[-2\pi i\mathbf{h} \cdot \mathbf{t}]$$

$$\times \sum_{\mathbf{p}} |F_{obs}(\mathbf{p})|^2 \int_V \exp[2\pi i(\mathbf{h} - \mathbf{p}) \cdot \mathbf{u}]\, d\mathbf{u}$$

For the same reason as before, the integral vanishes unless $\mathbf{h} - \mathbf{p} = 0$ or $\mathbf{p} = \mathbf{h}$ and, therefore, apart from the constant V,

$$T(\mathbf{t}) = \sum_{\mathbf{h}} |F_{obs}(\mathbf{h})|^2 \cdot \mathbf{F}_M(\mathbf{h}) \cdot \mathbf{F}_M^*(\mathbf{h} \cdot [C]) \exp[-2\pi i\mathbf{h} \cdot \mathbf{t}] \quad (10.17)$$

This final form of the translation function is a Fourier summation in which the coefficients are known. Therefore, it can easily be calculated. Unwanted background in the translation function is caused by self-Patterson vectors (vectors between atoms within one molecule) and by ignoring the Patterson vectors between molecules other than those under consideration. Self-Patterson vectors can easily be eliminated from the observed Patterson map, assuming that the known and unknown molecules have the same structure and, therefore, the same self-Patterson vectors. As an example we use again crystals with $P222$ symmetry. The orientation of the four molecules in the unit cell is known from the rotation function. Therefore, self-Patterson functions for each of the molecules can be calculated with coefficients $|F_{M(n)}(\mathbf{h})|^2$, where $n = 1 - 4$. The self-Patterson-corrected translation function is then for the example of Figure 10.5:

$$\mathbf{T}_1(\mathbf{t}) = \sum_{\mathbf{h}} \left\{ |F_{obs}(\mathbf{h})|^2 - \sum_{n=1}^{4} |F_{M(n)}(\mathbf{h})|^2 \right\} \cdot \mathbf{F}_M(\mathbf{h}) \cdot \mathbf{F}_M^*(\mathbf{h} \cdot [C])$$

$$\times \exp[-2\pi i\mathbf{h} \cdot \mathbf{t}] \quad (10.18)$$

In addition, the intermolecular vectors for pairs of molecules that have already been solved could be subtracted.

The translation functions $T(\mathbf{t})$ and $T_1(\mathbf{t})$ are two-dimensional with the intermolecular vectors **t** perpendicular to the symmetry axis under consideration. For a complete solution of the unknown structure, the various two-dimensional translation functions must be combined. This is straightforward for the *P*222 example discussed above, but more complicated in high-symmetry space groups. This is the reason that $T_1(\mathbf{t})$ is no longer a very popular translation function. An improved translation function has been derived by Crowther and Blow (1967); it results in peaks corresponding to all possible intermolecular vectors in the unknown structure. The model structure now contains not a pair of molecules, but the same number of molecules as the unknown crystal structure has in its unit cell.

Because of the large number of intermolecular vectors, it is now more convenient to use the position vector **m** of the model molecule M as the variable. The positions and orientations of the other molecules ($j = 2 \dots n$) in the unit cell are derived from the model molecule by the operations $[C_j]\mathbf{m} + \mathbf{d}_j$. We must now derive an expression for the calculated Patterson function.

$$\mathbf{F}_{\text{calc}}(\mathbf{h}, \mathbf{m}) = \sum_{j=1}^{n} \mathbf{F}_{\text{M}}(\mathbf{h}[C_j]) \exp[-2\pi i \mathbf{h}([C_j]\mathbf{m} + \mathbf{d}_j)]$$

where $\mathbf{F}_{\text{M}}(\mathbf{h}[C_j])$ is the contribution to the structure factor of reflection **h** by the *j*th molecule with respect to its local origin in the orientation defined by $[C_j]$, and $\exp[-2\pi i \mathbf{h}([C_j]\mathbf{m} + \mathbf{d}_j)]$ takes care of the fact that the origin of the molecule is at $[C_j]\mathbf{m} + \mathbf{d}_j$.

$$\mathbf{F}_{\text{calc}}(\mathbf{h}, \mathbf{m}) = \sum_{j=1}^{n} \mathbf{F}_{\text{M}}(\mathbf{h}[C_j]) \exp[-2\pi i \mathbf{h} \mathbf{d}_j] \exp[-2\pi i \mathbf{h}[C_j]\mathbf{m}]$$

Because $|\mathbf{F}_{\text{calc}}(\mathbf{h}, \mathbf{m})|^2 = \mathbf{F}_{\text{calc}}(\mathbf{h}, \mathbf{m}) \cdot \mathbf{F}^*_{\text{calc}}(\mathbf{h}, \mathbf{m})$

$$|\mathbf{F}_{\text{calc}}(\mathbf{h}, \mathbf{m})|^2 = \sum_{j=1}^{n} \sum_{k=1}^{n} \mathbf{F}_{\text{M}}(\mathbf{h}[C_j]) \mathbf{F}^*_{\text{M}}(\mathbf{h}[C_k]) \times \exp[-2\pi i \mathbf{h}(\mathbf{d}_j - \mathbf{d}_k)]$$
$$\times \exp[-2\pi i \mathbf{h}([C_j] - [C_k])\mathbf{m}] = Q$$

The calculated Patterson function is then $\mathbf{P}_{\text{calc}}(\mathbf{m}, \mathbf{u}) = \sum_{\mathbf{h}} Q \exp[-2\pi i \mathbf{h} \cdot \mathbf{u}]$, and the new translation function

$$\mathbf{T}_2(\mathbf{m}) = \int_V P_{\text{obs}}(\mathbf{u}) \times P_{\text{calc}}(\mathbf{m}, \mathbf{u}) \, d\mathbf{u}$$

As before, this equation contains the integral

$$\int_V \exp[2\pi i(\mathbf{h} - \mathbf{p}) \cdot \mathbf{u}] \, d\mathbf{u}$$

which vanishes unless $\mathbf{h} - \mathbf{p} = 0$, and therefore,

$$T_2(\mathbf{m}) = \sum_{\mathbf{h}} |\mathbf{F}_{\text{obs}}(\mathbf{h})|^2 \sum_{j=1}^{n} \sum_{k=1}^{n} \mathbf{F}_{\text{M}}(\mathbf{h}[C_j]) \mathbf{F}^*_{\text{M}}(\mathbf{h}[C_k])$$
$$\times \exp[-2\pi i \mathbf{h}(\mathbf{d}_j - \mathbf{d}_k)] \times \exp[-2\pi i \mathbf{h}([C_j] - [C_k])\mathbf{m}]$$

This can be calculated as a Fourier summation if, instead of the normal index $\mathbf{h}$, the index is taken as $\mathbf{h}([C_j] - [C_k])$.

The name $T_2(\mathbf{m})$ is confusing because it suggests a two-dimensional function, whereas, in fact, it is a three-dimensional function, since it utilizes all symmetry operators simultaneously. So far we have assumed that only one molecule is present in the asymmetric unit. Driessen et al. (1991) expanded this to asymmetric units containing more than one molecule (subunit) by associating each subunit in the asymmetric unit with an independent translation vector. An improvement of the signal-to-noise ratio can be obtained if from both the observed and the calculated Pattersons, the self-Patterson vectors are subtracted. For further improvement it is recommended that normalized structure factor amplitudes be used (Section 6.3) because this sharpens the Patterson map. Alternatively, a negative temperature factor parameter B can be applied.

In the two-step procedure, first the calculation of the rotation function and subsequently the translation function, the latter usually causes more problems than the first. For example, the search model differs too much from the unknown structure, or the highest peak in the rotation function does not correspond to the correct orientation. Several methods have been proposed to improve this situation:

- Combine the translation search with a limited systematic variation of the orientation parameters using the R-factor or the correlation coefficient as criterion.
- Refine the search model and its orientation after the rotation, but before the translation search (Patterson correlation refinement).
- Combine any existing phase information from isomorphous replacement with the translation function (phased translation function).
- Zhang and Matthews (1994) propose a modified translation function if part of the structure is known (addition and subtraction strategy).

The first method has been described and successfully used by Fujinaga and Read (1987). Inspired by previous work of Harada et al. (1981), they were the first to introduce the standard linear correlation coefficient as a criterion for solving the translation problem:

$$C = \frac{\sum \left(|F_o|^2 - \overline{|F_o|^2} \right) \times \left(|F_c|^2 - \overline{|F_c|^2} \right)}{\left[\sum \left(|F_o|^2 - \overline{|F_o|^2} \right)^2 \times \sum \left(|F_c|^2 - \overline{|F_c|^2} \right)^2 \right]^{\frac{1}{2}}}$$

This criterion was later also applied in the second improvement method by Brünger (Brünger, 1990; Brünger and Nilges, 1993; Brünger, 1997a) but now in a form with normalized structure factors:

$$C_{tr} = \frac{\overline{|E_h(obs)|^2 \times |E_h(calc)|^2} - \overline{|E_h(obs)|^2} \times \overline{|E_h(calc)|^2}}{[\{\overline{|E_h(obs)|^4} - (\overline{|E_h(obs)|^2})^2\} \times \{\overline{|E_h(calc)|^4} - (\overline{|E_h(calc)|^2})^2\}]^{1/2}}$$

In fact Brünger introduced two techniques for increasing the success in finding the correct translation of the search model. First he calculates a normal rotation function; peaks close together are clustered to a single peak. He selects not only the highest peak in the rotation function but a number of peaks, e.g., 200, and calculates the standard linear correlation coefficient, C_{tr}, between the squares of the normalized observed $(E_{obs})^2$ and calculated $(E_{calc})^2$ structure factors, as a function of the coordinates of the center of gravity of the search model and its symmetry mates. This three-dimensional search for the optimal coordinates is done for each of the selected orientations and the orientation that gave the highest value for the correlation coefficient is then chosen for the translation search.

If this search is unsuccessful, the highest peaks of the rotation function are again selected, but now, before a translation search is carried out, the search model and its orientation are adjusted by minimizing $E_{total}(\mathbf{r})$ in Eq. (10.19). A single copy of this search model is put into a triclinic cell, identical in geometry to the crystal unit cell and in an orientation derived from the rotation function. The translation is then of no relevance because a change of the molecule's position with respect to the origin changes only the phases and not the magnitudes of the structure factors. The target function to be minimized in the refinement is the energy term:

$$E_{total}(\mathbf{r}) = E_{PC}(\mathbf{r}) + E_{emp}(\mathbf{r}) \qquad (10.19)$$

$E_{emp}(\mathbf{r})$ is an empirical energy function (Section 13.4.4). $E_{PC}(\mathbf{r})$ is a pseudoenergy term, related to the standard linear correlation coefficient, $C(\mathbf{r}, \Omega)$, which measures the correlation between the squares of the normalized observed $(E_{obs})^2$ and calculated $(E_{calc})^2$ structure factors:

$$E_{PC}(\mathbf{r}) = W_{PC}\{1 - C(\mathbf{r}, \Omega)\}$$

Ω is the rotation matrix defining the orientation. C is a function of E_{calc}, and E_{calc} is a function of Ω and of the positional parameters of the atoms.

If the correlation is at its maximum, $E_{PC}(\mathbf{r})$ has a minimum value. W_{PC} is a suitably chosen weighting factor that determines the relative weight of the empirical and the pseudoenergy term. In many applications the minimization is carried out exclusively with $E_{PC}(\mathbf{r})$.

The refinement is particularly important if the protein molecule has flexible parts, which have different relative positions in the crystal structure to be solved, compared with the search model. This was, e.g., true for the structure determination of the antigen-binding fragment of an immunoglobulin molecule (Brünger, 1991a). The orientation and translation of the individual domains of the search model (a homologous antigen binding fragment) were appreciably modified in the refinement. The combination of the adjusted search model and its orientation produced an unambiguous solution for the translation search.

For the third method, the phased translation function, any bit of phase information, even from a poor heavy atom derivative, can facilitate the resolution of the translation problem. The method is based on overlapping the electron density map computed with the prior phase information, with the electron density map of one copy of the search model correctly oriented in the unit cell (Read and Schierbeek, 1988; Cygler and Desrochers, 1989; Bentley and Houdusse, 1992; Zhang and Matthews, 1994). The variable is the translation vector **t**, which moves the model away from its arbitrary position in the unit cell to its correct position. The criterion is the standard linear correlation coefficient Corr.(t).

$$\text{Corr.}(t) = \frac{\int\limits_V [\rho_P(\mathbf{x}) - \overline{\rho_P}] \times [\rho_M(\mathbf{x} - \mathbf{t}) - \overline{\rho_M}]\, d\mathbf{x}}{\left\{\int\limits_V [\rho_P(\mathbf{x}) - \overline{\rho_P}]^2 d\mathbf{x} \times \int\limits_V [\rho_M(\mathbf{x} - \mathbf{t}) - \overline{\rho_M}]^2 d\mathbf{x}\right\}^{1/2}} \quad (10.20)$$

Alternatively the symmetry operations in the unit cell can be applied to the search model and this would improve the signal somewhat. However, it has been shown that the simple "one copy of search model" technique works satisfactorily.

Equation (10.20) can be simplified because the average electron densities $\overline{\rho_P}$ and $\overline{\rho_M}$ are zero if $F(000)$ is omitted from the Fourier summation. Moreover the integral $\int_V [\rho_M(\mathbf{x} - \mathbf{t})]^2 d\mathbf{x}$ (with a single copy of the search model in the unit cell) is independent of the actual position of the model in the unit cell and is, therefore, independent of **t**; the integral is equal to $\int_V [\rho_M(\mathbf{x})]^2 d\mathbf{x}$.

$$\rho_P(\mathbf{x}) = \frac{1}{V}\sum_{\mathbf{h}} m_P |F_0(\mathbf{h})|\, \exp[i\alpha_P]\, \exp[-2\pi i(\mathbf{h} \cdot \mathbf{x})]$$

$$\rho_M(\mathbf{x}) = \frac{1}{V}\sum_{\mathbf{h}'} \mathbf{F}_M(\mathbf{h}')\, \exp[-2\pi i(\mathbf{h}' \cdot \mathbf{x})]$$

$$= \frac{1}{V}\sum_{\mathbf{h}'} \mathbf{F}_M^*(\mathbf{h}')\, \exp[2\pi i(\mathbf{h}' \cdot \mathbf{x})]$$

because $\mathbf{F}_M(-\mathbf{h}') = \mathbf{F}_M^*(\mathbf{h}')$, the complex conjugate of $\mathbf{F}_M(\mathbf{h}')$; m_p is the figure of merit. In the integration in the nominator of Eq. (10.20) all terms cancel except if $\mathbf{h}' = \mathbf{h}$, and it can easily be verified that the result is

$$\int\limits_V \rho_P(\mathbf{x}) \times \rho_M(\mathbf{x} - \mathbf{t})\, d\mathbf{x} = \frac{1}{V}\sum_{\mathbf{h}} m_P |F_0(\mathbf{h})|\, \exp[i\alpha_P]\, \mathbf{F}_M^*(\mathbf{h})\, \exp[-2\pi i(\mathbf{h} \cdot \mathbf{t})]$$

It can be shown in a similar way for the denominator terms that

$$\int\limits_V \{\rho_P(\mathbf{x})\}^2 d\mathbf{x} = \frac{1}{V}\sum_{\mathbf{h}} \{m_P |F_0(\mathbf{h})|\}^2$$

and

$$\int_V \{\rho_M(\mathbf{x})\}^2 d\mathbf{x} = \frac{1}{V} \sum_{\mathbf{h}} \{|F_M(\mathbf{h})|\}^2$$

The final result for Corr.(**t**) is

$$\text{Corr.}(\mathbf{t}) = \frac{\dfrac{1}{V} \sum_{\mathbf{h}} m_P |F_0(\mathbf{h})| \exp[i\alpha_P] \, F_M^*(\mathbf{h}) \, \exp[-2\pi i(\mathbf{h} \cdot \mathbf{t})]}{\dfrac{1}{V}\left[\displaystyle\sum_{\mathbf{h}} \{m_P|F_0(\mathbf{h})|\}^2 \times \sum_{\mathbf{h}} \{|F_M(\mathbf{h})|\}^2\right]^{1/2}} \qquad (10.21)$$

Because $F_M^*(\mathbf{h}) = |F_M(\mathbf{h})| \exp[-i\alpha_M]$, the correlation function can now straightforwardly be calculated as a Fourier summation with amplitudes $m_P|F_0(\mathbf{h})| \times |F_M(\mathbf{h})|$ and phase angles $(\alpha_P - \alpha_M)$, multiplied with

$$V\left[\sum_{\mathbf{h}} \{m_P|F_0(\mathbf{h})|\}^2 \times \sum_{\mathbf{h}} \{|F_M(\mathbf{h})|\}^2\right]^{-1/2}.$$

The maximum of Corr.(**t**) should give the correct translation vector **t**.

The addition and subtraction strategy of Zhang and Matthews (1994) can improve the quality of the rotation function as well as the translation function if part of the structure is known and another part is as yet unknown. For solving the rotation function of the unknown part with the addition strategy, the Patterson function of the known part is added in its correct orientation to the Patterson function of the search model for the unknown part:

$$\{P_{\text{known}} + [C]P_{\text{search}}\}.$$

This modified Patterson is calculated as a function of $[C]$ and compared with the observed Patterson. In the subtraction method the Patterson function of the known part is subtracted from the observed Patterson and $[C]P_{\text{search}}$ is solved as a function of $[C]$ by comparing with $\{P_{\text{obs}} - P_{\text{known}}\}$. Of course, a normal rotation function must be calculated first. If then the orientation of the molecule or subunit or part of it is known, the rotation result can best be improved by incorporating the subtraction strategy, which will reduce noise in the map of the rotation function. Zhang and Matthews also propose a modified translation function in which the addition and subtraction strategies are applied simultaneously.

10.3.2. AMoRe

Navaza (1994) has written a molecular replacement program AMoRe which incorporates several features, such as the exploration of many potential solutions and already existing knowledge of the models. In fact it is a combination of programs with three core programs: ROTING for orientation, TRAING for translation, and FITING for refinement of the results (Navaza and Vernoslova, 1995; Navaza and Soludjian, 1997).

AMoRe has the following characteristics:

1. Fast computation of structure factors from the continuous Fourier transform of the model, produced by the program TABLING, in all stages of the process.
2. Many potential solutions are explored.
3. Correlation coefficients are the main criteria for selection.
4. Efficient links between the various programs allow a high degree of automation.

ROTING is a fast rotation function, defined as the overlap of observed and calculated Patterson functions (Eq. 10.1). The Patterson functions are expanded in spherical harmonics, as in Crowther's fast rotation function (Crowther, 1972) but with an increased accuracy. The resulting peaks in the rotation function are tested and selected with the correlation coefficient between calculated and observed structure factor amplitudes.

The output of ROTING is the multiple input for the translation function TRAING. However, this is not a single translation function but a collection of four different fast translation functions from which the user can choose:

1. A Patterson overlap function similar to T_2 of Crowther and Blow (1967).
2. A phased translation function (Section 10.3.1).
3. A function similar to the one proposed by Harada et al. (1981); it incorporates packing considerations in the translation function and uses a modified correlation coefficient in terms of intensities as target function. This function can be evaluated by Fast Fourier Techniques.
4. This fourth function uses the normal correlation coefficient in terms of intensities as the target function. The correlation coefficient in terms of intensities has the advantage that it can be rapidly evaluated using Fast Fourier Techniques contrary to the coefficient in terms of amplitudes (Navaza and Vernoslova, 1995).

The highest peaks in the translation function(s) are tested with the correlation coefficient in terms of amplitudes because it is the most discriminating criterion. On this basis, acceptable solutions are selected. Molecular replacement solutions are usually followed by rigid body refinement: the orientation and position of the search model in the target unit cell are refined (Section 13.4.1). In AMoRe, this is done using the program FITING (Castellano et al., 1992). The difference with other rigid body refinement strategies is that FITING presents the search model as an electron density map and not as an atomic model. It is similar, though numerically improved, to a method previously proposed by Huber and Schneider (1985).

Rigid Body Refinement in AMoRe

If $\rho(\mathbf{r})$ is the electron density in the unit cell of the target crystal structure (the structure to be refined), its structure factors are

$$\mathbf{F}_{obs}(\mathbf{h}) = \int_{\substack{\text{unit} \\ \text{cell}}} \rho(\mathbf{r}) \exp[2\pi i\mathbf{h} \cdot \mathbf{r}] \, d\mathbf{r} \tag{10.22}$$

With s symmetry operators, each characterized by a crystallographic rotation matrix $[C]_s$ and a translation $\mathbf{t}_s$, this is equal to

$$\sum_s \exp[2\pi i\mathbf{h}\mathbf{t}_s] \int_{\substack{\text{asym.} \\ \text{unit}}} \rho(\mathbf{r}) \exp[2\pi i\mathbf{h}[C]_s\mathbf{r}] \, d\mathbf{r} \tag{10.23}$$

The search model (the known structure) is positioned in a suitable model box. Its electron density is $\tilde{\rho}(\mathbf{u})$, where $\mathbf{u}$ are the coordinates in the box. For transferring this electron density $\tilde{\rho}(\mathbf{u})$ to the crystal structure, we must apply the rotation matrix $[C]_R$, the translation vector $\mathbf{x}$, and, in addition, two transformation matrices: $[O]$ for transforming fractional into orthogonal coordinates in the model box and $[D]$ to transform orthogonal into fractional coordinates in the crystal. After the model is transferred into the crystal structure its positional coordinates $\mathbf{r}$ are related to its coordinates $\mathbf{u}$ in the model box as $\mathbf{r} = [D][C]_R[O]\mathbf{u} + \mathbf{x}$.

The structure factors calculated for the model in the crystal structure are:

$$\mathbf{F}_{calc}(\mathbf{h}) = \sum_s \exp[2\pi i\mathbf{h}\mathbf{t}_s] \int_{\substack{\text{asym.} \\ \text{unit}}} \tilde{\rho}(\mathbf{u}) \exp[2\pi i\mathbf{h}[C]_s\mathbf{r}] \, d\mathbf{r}$$

$$= \sum_s \exp[2\pi i\mathbf{h}\mathbf{t}_s] \int_{\substack{\text{model} \\ \text{box}}} \tilde{\rho}(\mathbf{u}) \exp[2\pi i\mathbf{h}[C]_s([D][C]_R[O]\mathbf{u} + \mathbf{x})] \, d\mathbf{u}$$

$$= \sum_s \exp[2\pi i\mathbf{h}\mathbf{t}_s] \exp[2\pi i\mathbf{h}[C]_s\mathbf{x}] \times$$

$$\int_{\substack{\text{model} \\ \text{box}}} \tilde{\rho}(\mathbf{u}) \exp[2\pi i\mathbf{h}[C]_s[D][C]_R[O]\mathbf{u}] \, d\mathbf{u}$$

$$= \sum_s \exp[2\pi i\mathbf{h}\mathbf{t}_s] \exp[2\pi i\mathbf{h}[C]_s\mathbf{x}] \times \mathbf{f}(\mathbf{h}[C]_s[D][C]_R[O])$$

$$\tag{10.24}$$

where $\mathbf{f}(\mathbf{h}[C]_s[D][C]_R[O])$ is the Fourier transform of the model in its box at reciprocal lattice position $(\mathbf{h}[C]_s[D][C]_R\,[O])$. In a least-squares procedure the difference between the amplitudes of $\mathbf{F}_{obs}(\mathbf{h})$ (Eq. 10.23) and $\mathbf{F}_{calc}(\mathbf{h})$ (Eq. 10.24) is minimized with respect to $[C]_R$, $\mathbf{x}$ and the scalefactor λ:

$$Q([C]_R, \mathbf{x}, \lambda) = w(\mathbf{h})\sum_{\mathbf{h}}\{|F_{obs}(\mathbf{h})| - \lambda|F_{calc}(\mathbf{h})|\}^2$$

$w(\mathbf{h})$ is any convenient weighting function.

10.3.3. Notes

1. A special translation case should be mentioned. If a crystal has noncrystallographic symmetry and a local axis is parallel to another local axis or a crystallographic axis, this is expressed in the Patterson map as cross-vectors between the subunits (or molecules) (Stubbs et al., 1996). A special case is illustrated in Figure 10.7 (Eagles et al., 1969; Epp et al., 1971). If there is a noncrystallographic two-fold axis parallel to a crystallographic two-fold, one can easily find the position of the noncrystallographic axis with respect to the crystallographic one from the Patterson map. It is clear that the Patterson will show a high peak corresponding to the distance Δ between the atoms in molecule 1 and 2, and their counterparts in molecules 3 and 4; this is equal to the distance between the local two-fold axes.

2. An interesting application of the translation function has been reported by Antson et al. (1995). They expected to find 11 Br atoms in the heavy atom-derivative of *trp* RNA-binding attenuation protein but were unable to locate them from an isomorphous or anomalous Patterson map. Because of 11-fold symmetry in the structure it was expected that the 11 Br atoms would lie in one plane. The translation function (AMoRe) was successfully applied against the isomorphous and anomalous differences with a model structure consisting of a ring of 11 Br atoms. The radius of the ring was varied stepwise.

3. The cross-Patterson function $P_{1,2}(\mathbf{u},\mathbf{t})$ (Eq. 10.15) is a noncentrosymmetric function, just as the cross-Patterson function $P(\mathbf{u})$ in section 7.10. Therefore, the translation function is also noncentrosymmetric. This property can be helpful in determining the correct space group, for instance making the choice between P3$_1$21 and P3$_2$21. Apply the translation function in both space groups and choose the one which gives the highest peak. Alternatively, look in the relevant Harker section whether the molecules at position 1/3 c higher are rotated with respect to the previous one by 120° (P3$_1$21) or 240° (P3$_2$21).

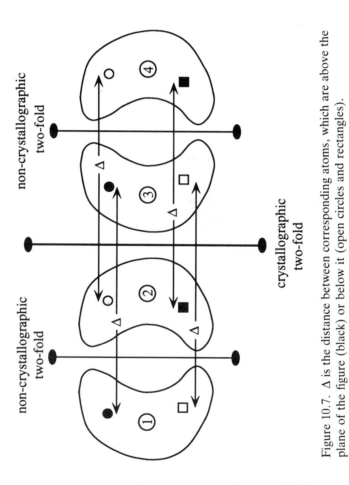

Figure 10.7. Δ is the distance between corresponding atoms, which are above the plane of the figure (black) or below it (open circles and rectangles).

Summary

For the molecular replacement technique a known structure is required, which serves as a model for the unknown structure. Homology in the amino acid sequence is an indication of whether a model is suitable. The solutions of the rotation and translation functions are not always found in a straightforward way. It can be necessary to modify the model, for instance, by ignoring side chains and deletions/additions in the model, and to play with the resolution range of the X-ray data. With the rapid increase in the number of successful protein structure determinations, molecular replacement has become an extremely useful technique for protein phase angle determination.

Chapter 11
Direct Methods

11.1. Introduction

The major problem in X-ray crystallography is to determine the phase angles of the X-ray reflections. In protein crystallography this problem is solved by the application of either isomorphous replacement or molecular replacement or multiple wavelength anomalous dispersion. In small molecule crystallography a completely different solution is applied. There, direct methods are the standard techniques for determining the phase angles of the structure factors. They use the principle that phase information is included in the intensities and this principle depends on the basic assumptions that the electron density is always positive [$F(000)$ included in the Fourier summation] and the crystal consists of discrete atoms that are sometimes even considered to be equal. Phase relations based on probability theory have been formulated and these phase relations are applied to suitably chosen clusters of reflections. Although these direct methods work perfectly well for small molecule crystals, it has thus far not been easy to extend them successfully to protein crystals (Karle, 1989; Brünger and Nilges, 1993).

11.2. Shake-and-Bake

This is one of the few *ab initio* or direct methods that has been rather successful in the determination of protein phase angles. It has a similar rational as the method proposed by Sheldrick and Gould (1995). The basis of Shake-and-Bake is the triplet relation between phase angles and in Section 11.2.1 this relationship is derived.

11.2.1. The Phase Triplet

Structure factors are not independent from each other, but related through structure. A particular relation exists for the reflections from lattice planes that belong to one zone. A group of lattice planes forms a zone if they are all parallel to one axis—the zone axis. The relationship is the triplet:

$$\phi(\mathbf{h}_1) + \phi(\mathbf{h}_2) + \phi(-\{\mathbf{h}_1 + \mathbf{h}_2\}) = 0 \qquad (11.1)$$

The ϕ's are the phase angles of the reflections $\mathbf{h}_1$, $\mathbf{h}_2$ and $-\{\mathbf{h}_1 + \mathbf{h}_2\}$. This triplet forms the basis of the Shake-and-Bake phase determination method and a simplified derivation shall be presented.

Section 4.7.2 mentioned that all atoms in a reflecting plane scatter in phase. Also, all member planes of a set of parallel reflecting planes scatter in phase. In the derivation, we assume that all atoms are identical and lie only on lattice planes, not in between. We shall focus on three sets of planes: 1, 2, and 3 (Figure 11.1a) and choose the origin O of the system in

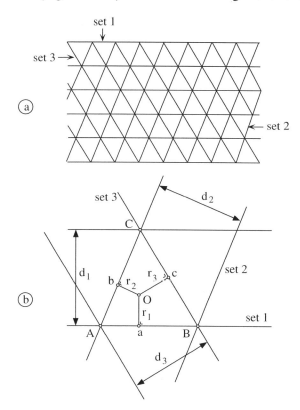

Figure 11.1. (a) Three sets of lattice planes belonging to one zone. The drawing is of a plane perpendicular to the zone axis. (b) A close-up of one of the triangles in (a). O is the origin of the system; d_1, d_2, and d_3 are interplanar spacings.

an arbitrary position. The phase angle of a reflection is determined by the distance of the origin O of the system to the nearest member of the set of reflecting planes. In Figure 11.1a and 11.1b, the plane perpendicular to the zone axis and through the origin O is drawn, and the phases of point a, b, and c with respect to O are required.

The individual phase angles depend on the position of O, but there is a fixed phase relation among ϕ_1, ϕ_2, and ϕ_3, where ϕ_1 is the phase angle of the beam reflected against set 1, ϕ_2 for set 2, and ϕ_3 for set 3. It is easy to prove, with our assumption of atoms on lattice planes, that

$$\phi_1 + \phi_2 + \phi_3 = 0 \tag{11.2}$$

Proof:

  Set 1: interplanar spacing d_1; distance of origin to nearest member r_1.
  Set 2: interplanar spacing d_2; distance of origin to nearest member r_2.
  Set 3: interplanar spacing d_3; distance of origin to nearest member r_3.

The position of a, b, and c in Figure 11.1b is comparable to the position of the atoms in the unit cell in Figure 4.12. Instead of atoms 1, 2, and 3 in that figure, we now have points a, b, and c, with the only difference that not all three points a, b, and c can be located in one unit cell. Suppose b and c are inside one unit cell and a is in another unit cell. This means that r_2 and r_3 have positive values, and r_1 has a negative value.

$$\phi_1 = \frac{r_1}{d_1} \times 2\pi; \quad \phi_2 = \frac{r_2}{d_2} \times 2\pi; \quad \phi_3 = \frac{r_3}{d_3} \times 2\pi \tag{11.3}$$

The area, S, of the triangle ABC in Figure 11.1b is:

$$S = \frac{1}{2}[-(r_1 \times AB) + (r_2 \times AC) + (r_3 \times BC)] \tag{11.4}$$

Combining equations 11.3 and 11.4 gives

$$\begin{aligned} 2 \times S \times 2\pi = &-(d_1 \times AB \times \phi_1) + (d_2 \times AC \times \phi_2) \\ &+ (d_3 \times BC \times \phi_3) \end{aligned} \tag{11.5}$$

$2 \times S$ is also equal to $d_1 \times AB = d_2 \times AC = d_3 \times BC$
  Divide both sides in eq. (11.5) by $2 \times S$:

$$2\pi = -\phi_1 + \phi_2 + \phi_3$$

The full phase circle of 2π can be subtracted and the result is that the sum of $(-\phi_1)$, ϕ_2, and ϕ_3 is equal to zero. This result is independent of the position of the origin O. The relation between the Miller indices of the lattice plane sets 1, 2, and 3 can be derived from the reciprocal lattice vectors of the reflections (Figure 11.2). Set 2 has indices $\{h_2, k_2\}$ and set 3 has $\{h_3, k_3\}$. From Figure 11.2, it is clear that set 1 has the indices $\{h_1, k_1\}$ = $\{(h_2 + h_3), (k_2 + k_3)\}$. It follows that

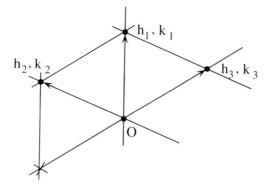

Figure 11.2. A reciprocal space representation of the sets of lattice planes 1, 2, and 3. Set 2 has indices $\{h_2, k_2\}$ and set 3 has $\{h_3, k_3\}$. It is clear that set 1 has the indices $\{h_1, k_1\} = \{(h_2 + h_3), (k_2 + k_3)\}$.

$$-\phi\{(h_2 + h_3), (k_2 + k_3)\} + \phi\{h_2, k_2\} + \phi\{h_3, k_3\} = 0$$

or, if $\{h_2, k_2\}$ is called $\mathbf{h}_1$ and $\{h_3, k_3\}$ is named $\mathbf{h}_2$:

$$-\phi(\{\mathbf{h}_1 + \mathbf{h}_2\}) + \phi(\mathbf{h}_1) + \phi(\mathbf{h}_2) = 0$$

or, because $-\phi(\{h_1 + h_2\}) = \phi(-\{h_1 + h_2\})$

$$\phi(-\{\mathbf{h}_1 + \mathbf{h}_2\}) + \phi(\mathbf{h}_1) + \phi(\mathbf{h}_2) = 0 \qquad (= \text{Eq. (11.1)})$$

This is the triplet relation between phase angles. In the derivation, it was assumed that all atoms were located on lattice planes. In fact, this is not true and Eq. (11.1) is only approximately valid. It is closer to the truth for strong reflections (large structure factor amplitudes), because then the lattice planes are heavily occupied with atoms, and there are not many atoms in between.

11.2.2. Execution of Shake-and-Bake

Eq. (11.1) is successfully used in the phase angle determination for crystal structures of small compounds by applying it to the strongest reflections only. For structures with more than 150 nonhydrogen atoms, the unit cell is so evenly filled with atoms that the method does not work. This is certainly true for proteins. Hauptman and coworkers have modified the method in such a way that it provides a fully automatic procedure for structures up to 1000 nonhydrogen atoms. A requirement is that $\leq 1.2\,\text{Å}$ resolution data are available (DeTitta et al., 1994; Weeks et al., 1994; Hauptman, 1997a and b). The technique is called Shake-and-Bake and has been applied successfully to a number of small protein structures (Ealick, 1997), as well as for the location of selenium atoms using anomalous difference data in the application of MAD (Section 9.5) (Turner et al., 1998). In the latter case, 3 Å data were sufficient to resolve the selenium atoms.

The principle of the modification is that the triplet sum (Eq. 11.1) is no longer set to zero, but to a value ϕ. Or, $\cos \phi$ is no longer equal to 1. The value they do give to $\cos \phi$ can be understood by comparison with the Sim weighting factor (Section 8.2). In the derivation of the Sim weighting factor it was found that the probability distribution for the unknown phase angle ξ, given the observed structure factor amplitude $|F|$ and the calculated amplitude $|F_K|$, depends on $|F|$ and $|F_K|$. The optimal weighting factor, $w = \cos \xi$, for noncentric reflections, turned out to be $w = \dfrac{I_1(X)}{I_0(X)}$, with $X = \dfrac{2|F| \times |F_K|}{\sum\limits_i f_i^2}$ and I_1 and I_0 are modified Bessel functions of order one and zero respectively. This result is now transferred to Shake-and-Bake: instead of putting $\cos \phi$ equal to 1 a procedure has been developed to have $\cos \phi$ approaching $\dfrac{I_1[X(\mathbf{h}_1, \mathbf{h}_2)]}{I_0[X(\mathbf{h}_1, \mathbf{h}_2)]}$ where $X(\mathbf{h}_1, \mathbf{h}_2)$ is now equal to $\dfrac{2}{\sqrt{N}} |E(\mathbf{h}_1)| \times |E(\mathbf{h}_2)| \times |E(\mathbf{h}_1 + \mathbf{h}_2)|$. The $|E|$s are the normalized structure factor amplitudes for the reflections $\mathbf{h}_1$, $\mathbf{h}_2$, and $-(\mathbf{h}_1 + \mathbf{h}_2)$. N is the number of identical atoms per unit cell. The actual function to be minimized with respect to ϕ is $\mathscr{R}(\phi)$ (Debaerdemaeker and Woolfson, 1983). It comprises a chosen set of the stronger reflections:

$$\mathscr{R}(\phi) = \frac{\sum\limits_{\mathbf{h}_1, \mathbf{h}_2} X(\mathbf{h}_1, \mathbf{h}_2) \left[\cos\{\phi(\mathbf{h}_1, \mathbf{h}_2)\} - \dfrac{I_1\{X(\mathbf{h}_1, \mathbf{h}_2)\}}{I_0\{X(\mathbf{h}_1, \mathbf{h}_2)\}} \right]^2}{\sum\limits_{\mathbf{h}_1, \mathbf{h}_2} X(\mathbf{h}_1, \mathbf{h}_2)}. \qquad (11.6)$$

$X(\mathbf{h}_1, \mathbf{h}_2)$ is a weighting function which is large for large values of the $|E|$s, giving more weight to $\left[\cos\{\phi(\mathbf{h}_1, \mathbf{h}_2)\} - \dfrac{I_1\{X(\mathbf{h}_1, \mathbf{h}_2)\}}{I_0\{X(\mathbf{h}_1, \mathbf{h}_2)\}} \right]^2$. The denominator $\sum\limits_{\mathbf{h}_1, \mathbf{h}_2} X(\mathbf{h}_1, \mathbf{h}_2)$ normalizes the function.

It appears that minimization of $\mathscr{R}(\phi)$ does not necessarily lead to the correct solution of the phase angles. It does so only if constraints are introduced and this is done in real space. This constraint is a simple density modification step: A number of the largest peaks are chosen, equal to the expected number of atoms in the asymmetric unit. They are all made equal in height. The rest of the electron density map is given zero density. A minimum atomic distance constraint of 1 Å is applied.

According to Weeks et al. (1994), the number of peaks chosen should be equal to or less than the expected number of atoms in the asymmetric unit. It is important to use fewer peaks for protein structures, which are likely to have many disordered atoms or atoms with high thermal motion. Such atoms are unlikely to be found during the preliminary stages of a solution, and inclusion of an unnecessarily large number of incorrect peaks does more harm than good. In fact, if oxygen is the heaviest element present, it is good to choose the number of peaks to be approximately 80% of the expected number of atoms. If several sulfurs are present, it is better to use even fewer peaks (40% of expected atoms). For atoms heavier than oxygen, an appropriate number of the highest peaks are weighted by the appropriate atomic numbers. The remainder (identified as C, N, or O) are weighted as nitrogens. Good weighting can have a significant impact on the success of the procedure.

The computer program for executing Shake-and-Bake is called SnB (Miller et al., 1994). The steps in the procedure are given in the flow chart in Figure 11.3.

Conclusion: Shaking (of the phase angles) occurs in reciprocal space. In real space, baking is applied by removing low density regions.

11.3. The Principle of Maximum Entropy

A nonclassical approach in direct phase angle determination for proteins, which is mathematically very demanding, has been pioneered by Bricogne (1993). We shall not go into details here. A full explanation can be found in Bricogne (1997). The process from structure factor amplitudes to phase angles is a stepwise process under the control of the concepts of Bayesian statistics. Probabilities play a major role, as they do in classical direct methods. However, macromolecules require a more subtle treatment than small molecules. This is provided by maximum entropy distributions instead of random distributions for unknown atomic positions.

The maximum extropy principle originated in information theory. Skilling (1988) proposed that the entropy $S(f, m)$ of an image f relative to a model m, is

$$S(f, m) = -\int f(x) \log \frac{f(x)}{m(x)} dx \qquad (11.7)$$

This is true for both $f(x)$ and $m(x)$ normalized. Equation (11.7) cannot be proven but it has been shown that it is the only one that gives correct results. In crystallography the image $f(x)$ is the normalized electron density distribution in the unit cell: $q(\mathbf{x}) = \rho(\mathbf{x})/F(000)$ with $F(000) = \int_V \rho(\mathbf{x})d\mathbf{x}$.

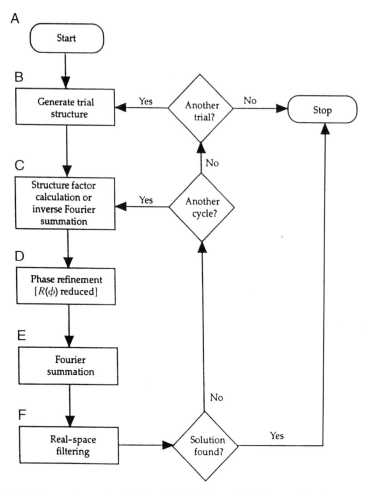

Figure 11.3. Flow chart for SnB. Copied in a slightly modified form from: Hauptman H.A. in Methods in Enzymology, Vol. 277, published by Academic Press Inc. Permission obtained.

The formula for the entropy is

$$S(q,m) = -\int q(\mathbf{x}) \log \frac{q(\mathbf{x})}{m(\mathbf{x})} d\mathbf{x}$$

or, for discrete grid points

$$S(q,m) = -\sum_j q_j \log \frac{q_j}{m_j}$$

The m_j distribution is the prior information that exists about the electron density distribution. If no other information (no structure factors) is available $q_j = m_j$ for all j and $S(q,m)$ has a maximum value equal to zero. If besides m_j, extra information is available, $S(q, m) < 0$, because the experimental data impose restrictions on q_j and, therefore, the entropy is reduced. q_j is always ≥ 0 and $q_j \geq m_j$.

In the various maximum entropy methods $S(q,m)$ is maximized with respect to q under the restraints of the extra information. More specifically, the maximization of $S(q,m)$ should occur under the conditions that

$$(|F_{obs}(h\ k\ l)| - |F_{calc}(h\ k\ l)|)^2$$

or

$$(I_{obs}(h\ k\ l) - I_{calc}(h\ k\ l))^2$$

and any other X-ray terms (for instance, related to phase information) are minimized. In Figure 11.4 the procedure is expressed in a pictorial way. Maximizing $S(q,m)$ is finding the shortest route (vector A in Figure 11.4) from the starting point to the region determined by the restraints. If in a next step more restraints can be added, the true solution can be approached in an iterative way. This is a formidable mathematical problem and the progress in the use of the maximum entropy formalism for the phase determination of a protein depends largely on solving these problems. A *de novo* structure determination of a protein by maximum entropy alone is not yet possible. However, a combination of maximum entropy with existing information from, for instance, MIR or MAD, which by itself gives an uninterpretable map, does result in some cases in a successful solution of the protein structure. Also knowledge of the molecular envelope could be useful.

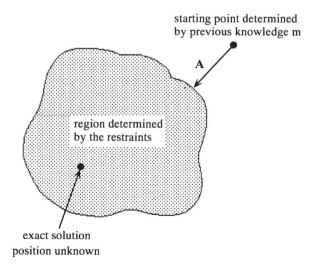

Figure 11.4. Simple schematic picture of the mathematical process in the application of the maximum entropy method.

Summary

So far, direct methods have been used in protein X-ray crystallography with only limited success and they have not yet been promoted to the level of standard techniques but are still in a state of development. However, much effort is put into improving them and in the not too distant future they may become of extreme importance in solving protein crystal structures.

The Shake-and-Bake method is rather successful, but so far its application is restricted to proteins with up to 600 nonhydrogen atoms. A completely different method has been developed and will be developed further by Bricogne. It is formulated on the basis of Bayesian statistics and maximum entropy.

Chapter 12
Laue Diffraction

12.1. Introduction

When a stationary crystal is illuminated with X-rays from a continuous range of wavelengths (polychromatic or "white" radiation), a Laue diffraction pattern is produced. The very first X-ray diffraction pictures of a crystal were in fact obtained in this way by Friedrich, Knipping, and Laue in 1912. However, since then, monochromatic beams were used nearly exclusively in X-ray crystal structure determinations. This is due to the fundamental problem that a single Laue diffraction spot can contain reflections from a set of parallel planes with different d/n, where d is the interplanar distance and n is an integer. These spots are multiples instead of singles. This is easily explained by Bragg's law:

$$2d \sin \theta = \lambda$$

$$2 \sin \theta = \frac{\lambda}{d} = \frac{\lambda/2}{d/2} = \frac{\lambda/3}{d/3} = \text{etc.}$$

The reflection conditions are satisfied, not only for the interplanar spacing d and wavelength λ, but also for $d/2$ and wavelength $\lambda/2$ and $d/3$ and wavelength $\lambda/3$, etc. Another problem with conventional X-ray sources is that their spectral properties with anode-specific lines are not very suitable for Laue diffraction.

The availability of synchrotrons as X-ray sources has changed this situation. Their fully polychromatic beam with a smooth spectral profile, having very high intensity and very small divergence, make them excellent sources for Laue diffraction of protein crystals. Moreover the harmonics (or multiple) problem turned out not to be as serious as previously thought and not to be a limiting obstacle (Cruickshank et al., 1987). The extremely high intensity of synchrotron X-ray sources, combined with their broad effective spectrum, allows X-ray diffraction pictures taken in times as short as 150 picoseconds (Srajer et al., 1996). This opens perspectives for time-resolved X-ray structure determinations (Moffat, 1989). Major contributions to the application of the Laue method in the area of protein X-ray crystallography have been made in the Daresbury Laboratory of the Science and Engineering Research Council in the U.K. and in laboratories in the USA. The Laue method is extensively discussed in the book by Cruickshank, Helliwell, and Johnson (1992); see also (Hajdu and Johnson, 1990; Pai, 1992; Sweet et al., 1993; Moffat, 1997).

12.2. The Accessible Region of Reciprocal Space

The range of reflections registered on a Laue photograph depends on the minimum and maximum values of the wavelength region ($\lambda_{max} - \lambda_{min}$). This is illustrated in Figure 12.1. Absorption and radiation damage increase with the wavelength and become very serious around 2 Å, which determines the useful maximum wavelength. These problems exist to a far lesser degree at shorter wavelength, but here the weaker diffraction intensity is limiting, because of its proportionality to λ^2 (see Section 4.14). The

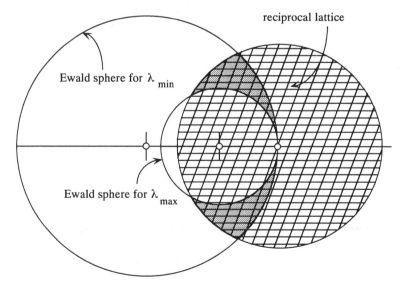

Figure 12.1. A reciprocal space representation of Laue diffraction by a stationary crystal (and thus a stationary reciprocal lattice). For the reciprocal lattice points in the shaded area the reflection conditions are satisfied, at least if not truncated by a maximum diffraction angle.

efficiency of the detector as a function of the wavelength also plays a role.

12.3. The Multiple Problem

A Laue spot can contain the reflections from a set of planes that are "harmonics" of each other, for instance, 3 1 7; 6 2 14; 9 3 21, etc., or in general a set of reflections $nh\ nk\ nl$, where n is an integer $\geqslant 1$ and $h\ k\ l$ the first harmonic, which has 1 as the greatest common divisor of its indices. In reciprocal space such a set of reflections forms a line through the origin and the nth harmonic reflects in the same direction as $h\ k\ l$, but with a spacing $d(nh\ nk\ nl) = (1/n) \times d(h\ k\ l)$ and for a wavelength $\lambda(nh\ nk\ nl) = (1/n)\lambda(h\ k\ l)$ (Figure 12.2). The maximum multiplicity of a Laue spot is $d(h\ k\ l)/d_{min}$, where d_{min} is the resolution limit. However, due to the finite wavelength region the multiplicity may be lower (Figure 12.2).

For the processing of the data one must know whether a spot is a single or a multiple. The intensities of the singles can be used directly, but the multiples must be unscrambled. Fortunately a large fraction of the observed spots will be singles as can be shown in the following way

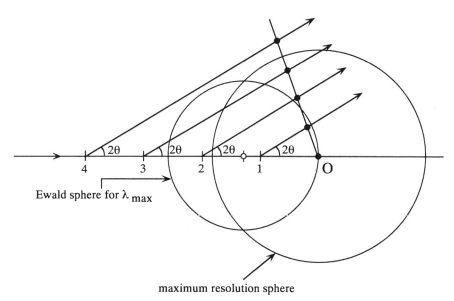

Figure 12.2. Reciprocal space construction for a double. In this example the first harmonic does not diffract, because it falls inside the Ewald sphere for λ_{max} with radius $1/\lambda_{max}$. The fourth and higher harmonics fall outside the resolution range. The origin of the reciprocal lattice is in O. The points 1, 2, 3, and 4 are the centers of the $\lambda(h\ k\ l)$, the $\lambda(2h\ 2k\ 2l)$, the $\lambda(3h\ 3k\ 3l)$, and the $\lambda(4h\ 4k\ 4l)$ spheres, respectively.

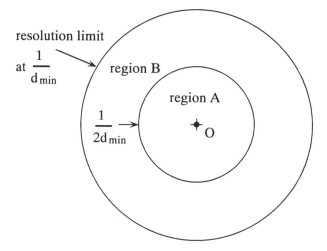

Figure 12.3. Region A in reciprocal space is within the sphere with radius $1/2d_{min}$ and region B is the shell between the two spheres. Of the reflections in region B 83% are measured as singles, but all reflections in region A are—in principle— members of a multiple.

(Cruickshank et al., 1987). If the common divisor is p, then every index (h or k or l) has the chance of $1/p$ to have that divisor. For the three indices together, it is $1/(p^3)$. The probability that p is not a common divisor is $\{1 - 1/(p^3)\}$; p has one of the prime numbers 2, 3, 5, 7, ..., and the probability that the point is the first on an harmonics line is

$$\{1 - 1/(2^3)\} \times \{1 - 1/(3^3)\} \times \{1 - 1/(5^3)\} \times \{1 - 1/(7^3)\} \times \cdots = 0.83$$

This gives the confidence that 83% of the reflections, corresponding with a spacing d, which is *smaller* than half the resolution limit (reflections in region B in Figure 12.3) are measured as singles. However, all reflections corresponding with a spacing *larger* than half the resolution limit (reflections in region A in Figure 12.3) are measured as multiples, unless all but one of the harmonics are excluded by the wavelength limiting Ewald spheres. Therefore, the low resolution reflections suffer most from harmonic overlaps. In addition, due to the Laue geometry, only relatively few low resolution reflections are in diffracting position (Figure 12.1).

12.4. Unscrambling of Multiple Intensities

The intensity of a multiple spot is the sum of the component intensities. For instance, in Figure 12.2 the reciprocal lattice points on the line radiating from the origin in reciprocal space form a double. Because of symmetry and Friedel's law, this spot is measured a number of times: K. Depending on the crystal orientation the spots i, with $i = 1$ to K, are either singles, doubles, triples, ... This forms the basis for the unscrambling procedure (Ren and Moffat, 1995). Each measurement i is the sum of the intensities of the contributing Bragg reflections:

$$I(\text{spot } i) = \sum_{j_{\min}}^{j_{\max}} I(\text{Bragg refl.} j) = \sum_{j_{\min}}^{j_{\max}} \frac{|F|^2 (\text{Bragg refl.} j)}{f(\text{Bragg refl.} j)} \qquad (12.1)$$

$j_{\min}$ and $j_{\max}$ are integers, indicating the lowest and the highest order of the reciprocal lattice spots on the radial line that forms one Laue spot. The function in the denominator "$f(\text{Bragg refl.} j)$" is a wavelength-dependent correction factor composed of many terms, such as Lorentz and polarization correction and wavelength-normalization factor $g(\lambda)$ (Section 12.6). There are K observations of the spots i, giving K equations of type 12.1. Most spots i are singles ($j_{\min} = j_{\max}$) (Section 12.3), and for those spots f-values can be obtained. The K equations can then be solved and in a least squares refinement procedure accurate values for the structure factor amplitudes of the separate Bragg reflections within each multiple can be obtained.

12.5. The Spatial Overlap Problem

Although the divergence of a synchrotron beam is small and the spots on the detector are sharp, there are so many spots on a typical protein Laue photograph (Figure 12.4) that spatial overlap of neighboring spots is a problem. To reduce this problem the crystal-to-detector distance can be increased; one should, however, realize the effect of this increase on the resolution. Profile fitting also reduces the problem. Standard profiles should be derived from nonoverlapping spots in subregions of reciprocal space and applied to partially overlapping spots in the same region. Profile fitting also improves the accuracy of weak intensities, assuming that strong and weak reflections share a common profile.

12.6. Wavelength Normalization

One of the complexities in the evaluation of Laue diffraction data is to compensate for wavelength-dependent parameters influencing the intensities of the spots. They are numerous, for instance:

1. The spectral characteristics of the white radiation at the sample.
2. Sample absorption, which is stronger for longer wavelength.
3. Wavelength-dependent detector response including absorption edges.
4. Anomalous scattering if present.

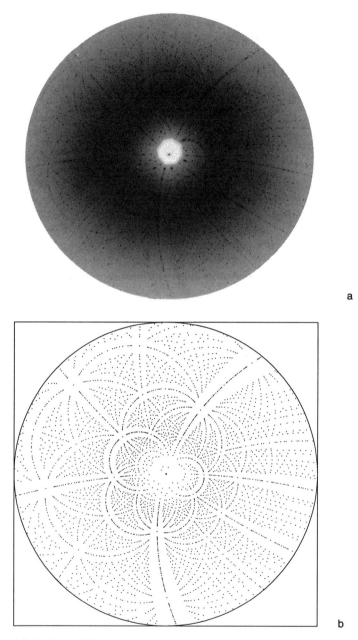

Figure 12.4. (a) A Laue diffraction picture of a crystal of the human oncogene product Ha-*ras*-p21 (molecular weight 21 kDa). The film was exposed for less than a second on the Wiggler beamline 9.5 at the SRS in Daresbury (U.K.). (b) The calculated diffraction picture for the same crystal setting. (Source: Dr. Axel J. Scheidig, Max-Planck-Institut für Medizinische Forschung, Heidelberg, Germany.)

5. The crystal scattering dependence on λ^2.
6. Wavelength-dependent source polarization.

All these factors are corrected for by multiplying each reflection intensity with a wavelength dependent factor $1/g(\lambda)$. This factor is determined empirically, either by using symmetry-related reflections, measured at different wavelengths within a single image, or by using the same reflection, measured at different wavelengths in different crystal orientations. In practice, $g(\lambda)$ is represented by a suitable mathematical function such as a simple polynomial. The parameters of the function are adjusted to obtain the best agreement between the reflection intensities recorded at different wavelengths (Ren and Moffat, 1995).

Summary

The advantages of the Laue method:

1. Extremely short data collection time with synchrotron radiation.
2. Consequently structural changes in the 150 psec–1 sec range can be observed.
3. One or only a few exposures at different angular settings cover a substantial portion of reciprocal lattice. The higher the symmetry in the crystals, the fewer exposures are needed.
4. Very small crystals can be used in principle but the mosaicity of the crystal and background scattering may be unfavorably high.

The main problems:

1. An unbalanced coverage of reciprocal space with relatively few low order reflections.
2. The crystal must withstand short exposures to extremely intense X-ray radiation.
3. Spatial overlap which requires a low mosaicity.

These problems can be addressed by careful experimental design (Moffat, 1997). However, because of the availability of extremely strong monochromatic radiation from synchrotron sources, which allow reduced exposure times, the interest in Laue diffraction as a routine, static data collection method has diminished. It remains the only technique that is appropriate to very rapid, time-resolved data collection on time scales shorter than a few msec. In such studies, it is important that the initiation of the structural and chemical reaction be synchronized in all unit cells, be uniform throughout the crystal, and be nondamaging. These conditions have been met in certain systems (Stoddard et al., 1996; Srajer et al., 1996) and have permitted the pulsed nature of the synchrotron beam to be utilized to reveal structural changes on the nsec time scale.

Chapter 13
Refinement of the Model Structure

13.1. Introduction

From isomorphous replacement or molecular replacement an approximate model of the protein structure can be obtained in which the broad features of the molecular architecture are apparent. Structure factors calculated on the basis of this model are generally in rather poor agreement with the observed structure factors. The agreement index between calculated and observed structure factors is usually represented by an R-factor as defined in Eq. (13.1). Thus an R-factor of 50% is not uncommon for the starting model, whereas for a random acentric structure it would be 59% (Wilson, 1950).

$$R = \frac{\sum\limits_{hkl} ||F_{obs}| - k|F_{calc}||}{\sum\limits_{hkl} |F_{obs}|} \times 100\% \qquad (13.1)$$

Refinement is the process of adjusting the model to find a closer agreement between the calculated and observed structure factors. Several methods have been developed and, if applied, they lower the R-factor substantially, reaching values in the 10 to 20% range or even lower. The adjustment of the model consists of changing the positional parameters and the temperature factors for all atoms in the structure, except the hydrogen atoms. Because hydrogen atoms have one electron only, their influence on X-ray scattering is low and they are normally disregarded in the structure determination. But still the number of nonhydrogen atoms is very high. In the refinement of the papain crystal structure at 1.65 Å resolution 25,000 independent X-ray reflections had been measured. The parameters to be refined were three positional parameters (x, y, and z) and one isotropic temperature factor parameter (B) for each of the 2000 nonhydrogen atoms, making a total of 8000 parameters. Therefore, the ratio of observations to parameters is only 3, and this is a poor over-determination. This is the reason why as many as possible additional "observations" are incorporated in the refinement process (Brünger and Nilges, 1993). They are in the first place the stereochemical data from small molecular structures. Their bond lengths and angles have been determined with high precision and it can be safely assumed that these data for amino acids and small peptides are also valid in proteins (Engh and Huber, 1991). A further "observation" is that the bulk solvent that fills the channels between protein molecules in their crystalline arrangement is not ordered and should appear as a flat region in the electron density map. This "solvent flattening"—which has already been discussed in Section 8.3—can, therefore, be imposed on the map. Finally noncrystallographic symmetry (Section 8.4)—if it does exist and is detected by molecular replacement—makes an important contribution to the refinement of the protein structure by imposing the equality in structure of the noncrystallographically related molecules or subunits. One can apply the stereochemical information on bond lengths, bond angles, etc. in two different ways:

• They are taken as rigid and only dihedral angles can be varied. In this case the geometry and the refinement are called *constrained*. This effectively reduces the number of parameters to be refined. In the application of this method it is difficult to move small parts of the structure to a "best fit" position because many angular motions are involved.

• If, on the other hand, the stereochemical parameters are allowed to vary around a standard value, controlled by an energy term, the refinement is called *restrained*. The atomic coordinates are the variables and the restraints are on bond lengths, bond angles, torsion angles, and van der Waals contacts. Restraints are "observations" because a penalty is included for disagreement with a restraint. This allows an easy movement of small parts of the structure, but it is difficult to move large parts, for instance, an entire molecule or domain.

Since protein structures are very complicated, their refinement is computationally a large size project. It is, therefore, fair to say that only through the availability of fast computers, especially vector processing machines, has the thorough and at the same time rapid refinement of protein structures become possible. It is perhaps superfluous to state that it makes sense to refine a structure only if careful attention has been paid to the determination of the cell dimensions and the measurement and correction of the X-ray intensity data. It is preferable to measure intensities of the reflections more than once, for instance, in another

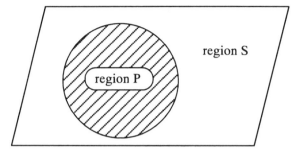

Figure 13.1. In this two-dimensional unit cell the region P represents the size and shape of a protein molecule. Region S contains the disordered solvent.

symmetry-related position. Low order reflections, having Bragg spacings longer than about 7 Å, are usually omitted in the refinement process because their intensities are rather seriously affected by the diffraction of the continuous solvent. Moreover, many data collection techniques for proteins are not designed to measure the low order reflections accurately. That the diffraction by the solvent affects only the low order reflections can be understood in the following way. Suppose the unit cell in Figure 13.1 is homogeneously filled with the same electron density ρ in P and S. P has the size and shape of the protein molecule. The diffraction intensities will be zero; all structure factors are zero. For a structure in which P only is filled with homogeneous density ρ and region S is empty, the structure factors are $\mathbf{G_P}$ where $\mathbf{G_P}$ is the transform of the homogeneously filled particle P. If only region S is filled with homogeneous density ρ, and P is empty, the structure factor must be $-\mathbf{G_P}$, because the sum of the two separate structures has diffraction zero. For the diffraction by a particle of the type P, which is normally close to spherical, we can refer to Figure 10.3, where the shape of the diffraction curve by a homogeneously filled, perfect sphere is given. For such particles, with the size of a protein molecule, the diffraction fades away rapidly with increasing diffraction angle. It has an appreciable contribution to lower order reflections only, and, therefore, this is also true for the disordered solvent in the crystal. As mentioned above low order reflections are usually omitted in the refinement because they are strongly affected by the disordered solvent. Inclusion of this disordered solvent effect improves the refinement and the R-factor drops a few percent. This can be done in the following way (Moews and Kretsinger, 1975; Badger, 1997).

$$\mathbf{F}(\text{disordered solvent}) = -\mathbf{KG_P}$$

K is a scale factor equal to $\dfrac{\text{average solvent density}}{\text{average protein density}}$.

G$_P$ is approximated by **F**(protein), the structure factor calculated for the protein model. This is Babinet's principle: The structure factors of a mask have the same amplitudes, but opposite phase angles as the structure factors of the rest of the unit cell.

An artificial temperature factor is applied to restrict the disordered solvent effect to low order reflections:

$$\mathbf{F}(\text{disordered solvent}) = -K\mathbf{F}(\text{protein}) \times \exp\left[-B(\text{solvent})\frac{\sin^2\theta}{\lambda^2}\right]$$

The calculated structure factor for the low order reflections is then,

$$\mathbf{F}(\text{calc}) = \left\{1-K\exp\left[-B(\text{solvent})\frac{\sin^2\theta}{\lambda^2}\right]\right\} \times \mathbf{F}(\text{protein})$$

The inclusion of the disordered solvent is no more than a correction of the structure factors calculated for the protein model.

Ordered solvent molecules do contribute to the scattering up to high resolution. They are located as electron density peak positions obeying H-bond geometry, and during refinement they are usually introduced into the molecular model at a resolution better than 2.5Å.

13.2. The Mathematics of Refinement

13.2.1. The Method of Least Squares

The refinement techniques in protein X-ray crystallography are based on the principle of least squares or maximum likelihood. An introduction to least squares has been given in Section 7.11, but it will now be treated more extensively. Before we do so it must be stated that the method of least squares is a special case of the more general refinement method: "maximum likelihood." In least squares the observations have fixed values and the parameters are varied such that the calculated values approach the observations as closely as possible in the refinement. It can easily be shown that maximum likelihood gives the same result as least squares for the special case of a Gaussian distribution of the observations. However, for other probability distribution the results are different.

We shall first discuss the least-squares method for the case where the X-ray data are the only observations. Later, in the treatment of more specific refinement methods, constraints and restraints also play a role. Refinement by least squares is an iterative process. In each step (or cycle) the parameters to be refined change, or rather should change, toward their final value without reaching them right away. Usually a great many refinement cycles are carried out before the changes in the parameters become small enough. The refinement has then converged to the final parameter set. The range of convergence is the maximum distance for the atoms to move to their final position. If they are further away there is a good chance that the function to be minimized will be trapped in a local minimum that is not the true minimum. The progress of a refinement can be monitored by calculating after each cycle the crystallographic R-factor [Eq. (13.1)], or better, the more reliable R_{free} (Section 15.2).

For a structure determination with data to 2.0 Å resolution the final R-factor should be close to 20% with R_{free} a few percent higher. In the least-squares refinement $\Sigma_{hkl}(|F_{obs}| - |F_{calc}|)^2$ is minimized. More accurately the function to be minimized is

$$Q = \sum_{hkl} w(h\ k\ l)(|F_{obs}(h\ k\ l)| - |F_{calc}(h\ k\ l)|)^2 \qquad (13.2)$$

The summation is over all crystallographically independent reflections and w is the weight given to an observation. The usual weighting factor in small molecule crystallography is $w(h\ k\ l) = 1/\sigma^2(h\ k\ l)$; σ is the standard deviation. For proteins some $(\sin\theta)/\lambda$ − dependent function is more satis-factory. It is assumed that the $|F_{obs}(h\ k\ l)|$ values are on absolute scale. The minimum of Q is found by varying the atomic parameters u_j ($j = 1 \ldots n$) that determine $|F_{calc}(h\ k\ l)|$, and this is done by putting the differentials of Q with respect to all u_j equal to zero: $\partial Q/\partial u_j = 0$ or

$$\sum_{hkl} w(h\ k\ l)(|F_{obs}(h\ k\ l)| - |F_{calc}(h\ k\ l)|)\frac{\partial|F_{calc}(h\ k\ l)|}{\partial u_j} = 0 \qquad (13.3)$$

Each $|F_{obs}(h\ k\ l)|$ is a constant and each $|F_{calc}(h\ k\ l)|$ depends on the variables u_j. For the solution of these equations $|F_{calc}(h\ k\ l)|$ is expressed in a Taylor expansion:

$$|F_{calc}(h\ k\ l;\ u)| = |F_{calc}(h\ k\ l;\ u_s)| + \sum_i \varepsilon_i\left[\frac{\partial|F_{calc}(h\ k\ l;\ u)|}{\partial u_i}\right]_{u_s}$$

$$+ \left|\frac{1}{2}\sum_j\sum_i \varepsilon_j\varepsilon_i \frac{\partial^2|F_{calc}(h\ k\ l;\ u)|}{\partial u_j\partial u_i} + \cdots\right.$$

$$\left.\rightarrow \text{neglected} \qquad (13.4)\right.$$

$|F_{calc}(h\ k\ l;\ u)|$ indicates that the $|F_{calc}|$ depend on the parameters u. The starting values of u are u_s and are changed by a small amount ε. For parameter u_i we have $\varepsilon_i = u_i - u_{i,s}$. $[\partial|F_{calc}(h\ k\ l;\ u)|/\partial u_i]_{u_s}$ means that the differential of $|F_{calc}(h\ k\ l;\ u)|$ with respect to u_i is calculated at the starting value u_s. Because the ε-values are small, higher order terms in ε can be neglected. Combination of Eqs. (13.4) and (13.3) gives the so called normal equations:

$$\sum_{hkl} w(h\ k\ l)(|F_{obs}(h\ k\ l)| - |F_{calc}(h\ k\ l; u_s)|) \times \left[\frac{\partial |F_{calc}(h\ k\ l; u)|}{\partial u_j}\right]_{u_s}$$

$$- \sum_i \varepsilon_i \sum_{hkl} w(h\ k\ l) \left[\frac{\partial |F_{calc}(h\ k\ l; u)|}{\partial u_i}\right]_{u_s} \times \left[\frac{\partial |F_{calc}(h\ k\ l; u)|}{\partial u_j}\right]_{u_s} = 0$$

With these n equations ($j = 1 \rightarrow n$), the ε-values must be found and then applied to the variables u. Because of the truncation of higher order terms in the Taylor series the final values are approached by iteration. In other words in the next cycle of refinement the process is repeated until convergence is reached. For each cycle new values of $|F_{calc}(h\ k\ l; u)|$ and its derivatives with respect to u_j must be calculated. This requires much computing power.

The principle of solving the normal equations is as follows. In abbreviated form they can be written as $\Sigma_i (\varepsilon_i a_{ij}) - b_j = 0$ or $\Sigma_i \varepsilon_i a_{ij} = b_j$.

$$\text{For} \quad j = 1: \quad a_{11}\varepsilon_1 + a_{21}\varepsilon_2 + a_{31}\varepsilon_3 + \ldots = b_1$$
$$j = 2: \quad a_{12}\varepsilon_1 + a_{22}\varepsilon_2 + a_{32}\varepsilon_3 + \ldots = b_2$$

These equations can be expressed in matrix form:

$$\begin{bmatrix} a_{11}\ a_{21}\ a_{31}\ ----\ \\ \\ a_{12}\ a_{22}\ a_{32}\ ----\ \\ \\ ----------\ \\ ----------\ \end{bmatrix} \times \begin{bmatrix} \varepsilon_1 \\ \varepsilon_2 \\ \varepsilon_3 \\ - \\ - \\ - \end{bmatrix} = \begin{bmatrix} b_1 \\ b_2 \\ b_3 \\ - \\ - \\ - \end{bmatrix}$$

or $[\mathbf{A}] \times [\boldsymbol{\varepsilon}] = [\mathbf{b}]$. $[\mathbf{A}]$ is called the normal matrix. Since $j = 1 \ldots n$ and $i = 1 \ldots n$, it is a square matrix with n rows and columns; moreover the matrix is symmetric. Its elements are

$$\sum_{hkl} w(h\ k\ l) \left[\frac{\partial |F_{calc}(h\ k\ l; u)|}{\partial u_i}\right]_{u_s} \times \left[\frac{\partial |F_{calc}(h\ k\ l; u)|}{\partial u_j}\right]_{u_s}$$

$[\boldsymbol{\varepsilon}]$ is the unknown vector and $[\mathbf{b}]$ the known gradient vector containing the elements

$$\sum_{hkl} w(h\ k\ l)(|F_{obs}(h\ k\ l)| - |F_{calc}(h\ k\ l; u_s)|) \times \left[\frac{\partial |F_{calc}(hkl; u)|}{\partial u_j}\right]_{u_s}$$

Since matrix $[\mathbf{A}]$ is a square matrix another matrix can be derived, $[\mathbf{A}^{-1}]$, which is the inverse or reciprocal of $[\mathbf{A}]$. Its property is $[\mathbf{A}^{-1}] \times [\mathbf{A}] = [\mathbf{E}]$, where $[\mathbf{E}]$ is the unit matrix:

$$[\mathbf{E}] = \begin{bmatrix} 1\ 0\ 0\ ------ \\ 0\ 1\ 0\ ------ \\ 0\ 0\ 1\ ------ \\ -------- \\ -------- \end{bmatrix}$$

$[A^{-1}] \times [A] \times [\varepsilon] = [A^{-1}] \times [b]$ or $[E] \times [\varepsilon] = [A^{-1}] \times [b]$. Since the property of a unit matrix is $[E] \times [\varepsilon] = [\varepsilon]$, the solution of the normal equations is $[\varepsilon] = [A^{-1}] \times [b]$. If $[A]$ is not a very large a matrix its inverse $[A^{-1}]$ can be calculated without too much effort and the problem of finding the parameter shifts $[\varepsilon]$ can easily be solved. However, in protein X-ray crystallography $[A]$ is a huge matrix. Its number of rows and columns is equal to the number of parameters to be refined and this can easily be of the order of 10,000 or more. This prohibits the calculation of $[A^{-1}]$. Fortunately, many elements of matrix $[A]$ are zero or close to zero for the following reason. The elements of matrix $[A]$ are

$$a_{ij} = \sum_{hkl} w(h\ k\ l) \left[\frac{\partial |F_{calc}(h\ k\ l;\ u)|}{\partial u_i} \right]_{u_s} \times \left[\frac{\partial |F_{calc}(h\ k\ l; u)|}{\partial u_j} \right]_{u_s} \quad (13.5)$$

If u_i and u_j are positional parameters x or y or z of the same atom, then for an orthogonal system of axes, the derivatives are not correlated and the elements a_{ij} will have a low value. This is not true for $i = j$, when both derivatives are with respect to the same parameter for the same atom. If u_i and u_j concern different atoms, and these atoms are well resolved, small changes in a parameter of one atom do not affect the other atom and the elements a_{ij} are again small. Correlation between positional parameters and temperature factors is also neglected, although some correlation usually does exist. It follows that the diagonal elements of matrix $[A]$ are in general larger than the off-diagonal elements and the simplification is that all off-diagonal elements are set to zero.

With a diagonal matrix (all off-diagonal elements zero) the calculation of $[A^{-1}]$ is trivial and a set of parameter shifts $[\varepsilon]$ can easily be calculated. However, because the diagonal matrix is not the ideal matrix and because of the neglect of higher terms in the Taylor expansion this set of shifts is not the final one and the procedure is continued as an iterative process: recalculate $[A]$ with the new parameters, find a new set of shifts, etc. until convergence is reached. To increase the rate of convergence the conjugate gradient method is applied, which is discussed below.

If besides X-ray information, geometric or energy information is also incorporated in the refinement, the elements of matrix $[A]$ are

$$a_{ij} = \sum_{hkl} w(h\ k\ l) \left[\frac{\partial |F_{calc}(hkl;\ u)|}{\partial u_i} \right]_{u_s} \times \left[\frac{\partial |F_{calc}(hkl;\ u)|}{\partial u_j} \right]_{u_s}$$

+ terms to account for the extra observations. The elements of vector matrix $[b]$ also contain extra terms:

$$\sum_{hkl} w(h\ k\ l)(|F_{obs}(h\ k\ l)| - |F_{calc}(h\ k\ l;\ u_s)|) \times \left[\frac{\partial |F_{calc}(h\ k\ l;\ u)|}{\partial u_j} \right]_{u_s}$$

+ extra terms. The extra terms contribute to off-diagonal elements and these elements can no longer be set to zero. As a consequence $[A^{-1}]$ is not easy to calculate and the refinement must be carried out by some numerical method, approaching the final parameter set in a cyclic way.

The most popular numerical method in protein X-ray crystallography is the conjugate gradient technique (Tronrud, 1992).

The method depends on matrix $[\mathbf{A}]$ to be symmetric. One starts with an initial estimate of the parameter shifts $[\varepsilon_0]$, which can be taken as zero if no other choice is available. Next a residual vector matrix $[\mathbf{r}_0] = [\mathbf{b}] - [\mathbf{A}] \times [\varepsilon_0]$ is calculated. A second new vector matrix required is the search direction vector $[\mathbf{z}]$ along which the function Q [Eq. (13.2)] is minimized. Choose the first $[\mathbf{z}]$, that is $[\mathbf{z}_0]$, equal to $[\mathbf{r}_0]$. In a number of iterative steps $n = 0, 1, 2, \ldots$, the vectors $[\varepsilon_{n+1}]$, $[\mathbf{r}_{n+1}]$, and $[\mathbf{z}_{n+1}]$ are calculated with the following recipe:

$$[\varepsilon_{n+1}] = [\varepsilon_n] + \alpha_n[\mathbf{z}_n] \quad \text{with} \quad \alpha_n = \frac{[\mathbf{r}_n]\cdot[\mathbf{r}_n]}{[\mathbf{z}_n]\cdot[\mathbf{A}] \times [\mathbf{z}_n]}$$

$$[\mathbf{r}_{n+1}] = [\mathbf{r}_n] - \alpha_n[\mathbf{A}] \times [\mathbf{z}_n]$$

$$[\mathbf{z}_{n+1}] = [\mathbf{r}_{n+1}] + \beta_n[\mathbf{z}_n] \quad \text{with} \quad \beta_n = \frac{[\mathbf{r}_{n+1}]\cdot[\mathbf{r}_{n+1}]}{[\mathbf{r}_n]\cdot[\mathbf{r}_n]}$$

With α_n and β_n chosen as indicated, it can be proven that the search directions $[\mathbf{z}]$ are all conjugate to each other: $[\mathbf{z}_{n+1}]\cdot[\mathbf{A}] \times [\mathbf{z}_j] = 0$ for all $j = 0, 1, 2, \ldots, n$. This guarantees that an efficient path is followed in parameter space toward the minimum of function Q [Eq. (13.2)]. It can also be shown that the residuals $[\mathbf{r}]$ are independent of each other: $[\mathbf{r}_{n+1}]\cdot[\mathbf{r}_j] = 0$, for all $j = 0, 1, 2, \ldots, n$.

The iteration process can be further accelerated by "preconditioning." A matrix $[\mathbf{K}]$ is chosen that resembles $[\mathbf{A}]$, but that can easily be inverted to $[\mathbf{K}^{-1}]$. The expression $[\mathbf{A}] \times [\varepsilon] = [\mathbf{b}]$ is now replaced by $[\mathbf{K}^{-1}] \times [\mathbf{A}] \times [\varepsilon] = [\mathbf{K}^{-1}] \times [\mathbf{b}]$. The more $[\mathbf{K}]$ resembles $[\mathbf{A}]$, the better $[\mathbf{K}^{-1}] \times [\mathbf{A}]$ resembles the unit matrix $[\mathbf{E}]$ and the faster the convergence is. During the procedure matrix $[\mathbf{A}]$ is retained unchanged. This is a great advantage because the nonzero elements can then be stored for the matrix multiplications. After convergence has been reached $[\mathbf{A}]$ is recalculated and another cycle of refinement can be started. In the beginning of the refinement only the X-ray data to moderate resolution are incorporated and during the refinement process the resolution is gradually extended. Structure factors and their derivatives can be conveniently calculated with a method proposed by Agarwal (1978) that uses the fast Fourier transform algorithm (see Section 13.3).

The inverse matrix $[\mathbf{A}^{-1}]$ also serves another purpose. From its diagonal elements a_{jj}, the standard deviation σ of the parameter u_j at the end of the refinement can be estimated:

$$\sigma^2(u_j) = a_{jj}\left(\frac{\displaystyle\sum_{h=1}^{p} w_h(\Delta F_h)^2}{p - n}\right)$$

where p is the number of independent reflections, n is the number of parameters, and $(\Delta F_h) = |F_{obs}(h\,k\,l)| - |F_{calc}(h\,k\,l)|$ calculated with the present parameter value (Cruickshank, 1965). However, this can only be done if the protein is small and the resolution of the diffraction pattern is extremely high. Stec et al. (1995) applied it to crambin for which the data were collected to a resolution of 0.83 Å at 130 °K and Tickle et al. (1998) to γB- and βB2-crystallin. The protein was refined by a full matrix least squares refinement with limited use of restraints. For the majority of macromolecular structure determinations this is impossible and rms. deviations from dictionary values (Engh and Huber, 1991) for bond lengths and angles are listed.

13.2.2. The Formalism of Maximum Likelihood

> "Maximum likelihood changes a world where measurements
> are values to a new world full of statistics"
>
> *Eric de La Fortelle*

As mentioned before, the conditions for least squares refinement are not always fulfilled. These conditions are (de La Fortelle and Bricogne, 1994; Pannu and Read, 1996):

1. The probability distribution of $|F(\text{calc})|$ is a Gaussian centered on $|F(\text{obs})|$.
2. The standard deviation of the Gaussian is an observed quantity and independent of the parameters of the model.
3. Phase angles should be either known or treated as model parameters.

These limiting conditions are also discussed by Tickle et al. (1998). If these conditions are not fulfilled, least squares results are not necessarily the best estimates of the model parameters. One should apply the more general method of maximum likelihood (Fisher, 1912). The principle of this method is rather simple and is based on Bayesian statistics.

Let $(x_1, \ldots, x_n)$ be a set of observations and $(r_1, \ldots, r_m)$ a set of model parameters with a known relationship between the two sets. If there would be no errors at all, one could exactly calculate the observations $(x_1, \ldots, x_n)$ that corresponds to a set of model parameters $(r_1, \ldots, r_m)$. Unfortunately, because of errors in the model and in the observations, only the **probability** that a set of observations corresponds to a set of given model parameters can be calculated. It should be realized that the probability of any set of (not necessarily existing) observations can be calculated if a particular set of model parameters is given. This is the very first step in setting up a maximum likelihood function.

$$p \left[(x_1, \ldots, x_n); (r_1, \ldots, r_m) \right] = p \, [\text{data; model}]$$

The semicolon ";" means here "given the known values of." The crystallographer is not so much interested in the probability of the data, given the model, p [data; model], but in the probability of the model given the data, p [model; data], because the data are the fixed observations. These two probabilities are related through Bayes' Theorem:

$$p[\text{model; data}] = \frac{p[\text{model}] \cdot p[\text{data; model}]}{p[\text{data}]} \qquad (13.6)$$

Proof of Bayes' Theorem

For convenience we introduce A for model and B for data. Let $p[A,B]$ be the joint probability of A and B occuring both and $p[B;A]$ the conditional probability of B given A. By means of Figure 13.2 it can be shown that

$$p[A, B] = p[A] \cdot p[B; A], \text{ and also } p[A, B] = p[B] \cdot p[A; B]$$

It follows that $p[A; B] = \dfrac{p[A] \cdot p[B; A]}{p[B]}$ and this proves Bayes' theorem.

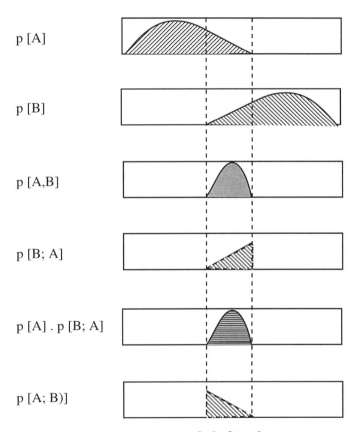

Figure 13.2. Bayes' theorem $p[A; B] = \dfrac{p[A] \cdot p[B; A]}{p[B]}$ explained by means of simple figures. $p[A,B]$ is the joint probability of A and B occuring both. $p[B; A]$ is the conditional probability of B given A, and $p[A; B]$ the conditional probability of A given B.

Bricogne pioneered the introduction of Bayesian statistics and maximum likelihood into crystallography (for a review, see Bricogne, 1997). In crystallography, the data are the structure factor amplitudes $|F(\text{obs})|$ and the parameters are the parameters of the model. However, for convenience, the model parameters are replaced by $\mathbf{F}(\text{calc})$ because the $\mathbf{F}(\text{calc})$'s are directly related to the model parameters[1]. Eq. 13.6. transforms into:

$$p[\mathbf{F}(\text{calc}); |F(\text{obs})|] = p[\mathbf{F}(\text{calc})] \times \frac{p[|F(\text{obs})|; \mathbf{F}(\text{calc})]}{p[|F(\text{obs})|]} \qquad (13.7)$$

The term in the denominator, $p[|F(\text{obs})|]$, can be regarded as a normalization constant and can, for our purposes, be deleted. $p[\mathbf{F}(\text{calc})]$ expresses prior information of the model, for instance restraints. If this is absent, in other words, if it is a constant function, it can also be disregarded. What remains is

$$p[\mathbf{F}(\text{calc}); |F(\text{obs})|] = p[|F(\text{obs})|; \mathbf{F}(\text{calc})] \qquad (13.8)$$

The important result is that we have "inverted" the probability. We started with the probability of finding any set of observations, given the model:

$$p[|F(\text{obs})|; \mathbf{F}(\text{calc})]$$

and have inverted it into the probability of the model with the introduction of the actual observations:

$$p[\mathbf{F}(\text{calc}); |F(\text{obs})|].$$

Because we have disregarded normalization, $p[\mathbf{F}(\text{calc}); |F(\text{obs})|]$ does not exactly obey the requirements for "probability" and is, therefore, given a different notation: $\mathcal{L}$ (likelihood) instead of p.

$$\mathcal{L} = [\mathbf{F}(\text{calc}); |F(\text{obs})|] = p[|F(\text{obs})|; \mathbf{F}(\text{calc})]$$

The problem now is to find the correct expression for $p[|F(\text{obs})|; \mathbf{F}(\text{calc})]$. This can be done for each reflection, and these probabilities must be combined to a joint probability for the entire set of reflections. To simplify this problem, we assume that the observations $|F(\text{obs})|$ are independent of each other. The joint probability for all reflections is then $\prod_{\mathbf{h}} p[|F_{\mathbf{h}}(\text{obs})|; \mathbf{F}_{\mathbf{h}}(\text{calc})]$ and the required likelihood:

$$\mathcal{L}_{\text{total}} = \prod_{\mathbf{h}} p[|F_{\mathbf{h}}(\text{obs})|; \mathbf{F}_{\mathbf{h}}(\text{calc})] \qquad (13.9)$$

---

[1] If the problem concerns the refinement of heavy atom parameters, $\mathbf{F}(\text{calc})$ does not only depend on the model parameters but also on the not exactly known protein phase angles. In that case integration over all possible phase angles (and preferably amplitudes) is the solution.

Note: The mutual independence of the observations $|F(\text{obs})|$ is a matter of debate. Because of their relationship via the structure, they are not independent. However, the issue is that the errors in the $|F(\text{obs})|$s are to a great extent independent, and multiplication of their probabilities is perfectly valid. Weak correlations between errors do exist as a consequence of, for instance, detector inhomogeneities and poor absorption correction. Stronger correlations exist between the partners in a Bijvoet pair of reflections.

In the concept of maximum likelihood, the most reliable set of parameters is obtained if the likelihood $\mathcal{L}_{\text{total}}$ in eq. (13.9) is maximized. Because it is easier to find the maximum of a sum than the maximum of a product, the log of the likelihood, $(\log \mathcal{L}_{\text{total}})$, is taken. Its maximum is at the same position as the maximum of $\mathcal{L}_{\text{total}}$. And it is still easier to minimize $-\log \mathcal{L}_{\text{total}}$:

$$-\sum \log\{p|F_{\mathbf{h}}(\text{obs})|;\ p[\mathbf{F}_{\mathbf{h}}(\text{calc})]\}$$

The minimization is with respect to the parameters of the model structure. In Section 13.2.3, maximum likelihood will be illustrated as applied in the programs MLPHARE and SHARP for the refinement of the parameters of heavy-atoms and anomalous scatterers in MIR and MAD.

13.2.3. The Refinement of the Parameters of Heavy-Atoms and Anomalous Scatterers in MIR and MAD by MLPHARE and SHARP

MLPHARE

Refinement of heavy-atom parameters was discussed in Section 7.11, where the program MLPHARE (Otwinowski, 1991) was mentioned. In this program, the single value of the protein phase angle, as used in the least squares lack of closure error refinement, is replaced by an integrated value of all phase angles, weighted by their likelihood. The likelihood $\mathcal{L}$ in MLPHARE is a very simple one, assuming that errors in the measurements of the native amplitudes are smaller than the combined errors of the measurements of heavy atom derivative amplitudes and nonisomorphism errors.

$$\mathcal{L}(\mathbf{F}_{\text{H}}) = \exp\left[-\frac{\{k_j|F_{\text{PH}}|_{\text{obs}} - |F_{\text{PH}}|_{\text{calc}}\}^2}{2E^2}\right] \qquad (13.10)$$

Where $\mathcal{L}(\mathbf{F}_{\text{H}})$ is a Gaussian function and E^2 is the average value of $\{k_j|F_{\text{PH}}|_{\text{obs}} - |F_{\text{PH}}|_{\text{calc}}\}^2$. The likelihood for one reflection is obtained by integrating over all phase angles. This likelihood for the individual reflections must then be multiplied to generate the global likelihood function for one derivative. Multiplying these global likelihood functions for all derivatives gives the total likelihood, and this must be maximized with respect to the adjustable parameters. As a general procedure, the protein phases resulting from MLPHARE are used as input for solvent flattening or other density modification procedures (Section 8.3). MLPHARE can be applied again after this procedure to give improved protein phase angles (Rould et al., 1992).

MLPHARE has no likelihood for the structure factor amplitudes incorporated, only for the phase angles. A more extensive implementation of the formalism of maximum likelihood is found in the program SHARP (de La Fortelle and Bricogne, 1997), also intended to refine heavy atom parameters. It is presented below.

SHARP

Another program, SHARP, for the refinement of heavy-atom parameters was written by de La Fortelle and Bricogne (1997). This program is entirely based on the maximum likelihood formalism (Bricogne, 1988 and 1991). Therefore, the likelihood of the heavy atom parameters must be derived given the values of the observations and their distribution functions. The required likelihood is equal to the combined probabilities of

- The measured values of $|F_p|$ and $|F_{PH}|$.
- The scaling factors between native and derivative data sets.
- Nonisomorphism.
- Errors in the heavy-atom model.

A central role in their derivation of the likelihood is played by the unknown error-free native structure factor $\mathbf{F}_p^{corr}$.[2] For a reflection $\mathbf{h}$, which is in general acentric, the calculated value for $\mathbf{F}_{PH,j}$ is, in the ideal case, equal to $k_j(\mathbf{F}_P^{corr} + \mathbf{F}_{H,j})$, where k_j is a scaling factor. Because of errors in the native structure, this is not necessarily true and $\mathbf{F}_{PH,j}$ is determined by a Gaussian distribution function, centered at

$$\langle \mathbf{F}_{PH,j} \rangle = k_j \left(D_j \mathbf{F}_P^{corr} + \mathbf{F}_{H,j} \right) \tag{13.11}$$

with a variance V_j^{PH}.

D_j is related to coordinate errors in the native structure (Section 7.7 and eq. 15.2), (Luzzati, 1952; Read, 1990b). In SHARP it also includes deviations in atomic positions due to nonisomorphism. D_j has a value between 0 and 1. Because the phase angle of $\mathbf{F}_{PH,j}$ is not yet known, the Gaussian distribution must be integrated over the phase angle; the result is

---

[2] The actual value of $\mathbf{F}_P^{corr}$ is unknown and at the start of the refinement an $\mathbf{F}_P^{trial}$ is derived from the current native structure. It is removed from the likelihood by integrating over all possible values of $\mathbf{F}_P^{trial}$, that is over its amplitude and phase angle with suitable weighting factors. These weighting factors are derived from the values of the likelihood function at all points of the Harker diagram.

$$p\left[|F_{PH,j}|; \mathbf{F}_P^{corr}, k_j, V_j^{PH}\right] =$$

$$\mathcal{R}_j = \frac{|F_{PH,j}|}{V_j^{PH}} \exp\left[-\frac{|F_{PH,j}|^2 + |\langle F_{PH,j}\rangle|^2}{2V_j^{PH}}\right] \times I_o\left[\frac{|F_{PH,j}| + |\langle F_{PH,j}\rangle|}{V_j^{PH}}\right]$$

$$(13.12)$$

I_0 is a modified Bessel function of order zero (see Section 8.2).

$\mathcal{R}_j$ is called a Rice distribution. Assuming that there are M derivatives and the M data sets are independent, the joint probability distribution of the M structure factor amplitudes is for one reflection $\mathbf{h}$:

$$p\left[|F_{PH,j=1...M}|; \mathbf{F}_P^{corr}, k_{j=1...M}, V_{j=1...M}^{PH}\right] = \prod_{j=1}^{M} \mathcal{R}_j \qquad (13.13)$$

Next, errors in the measurement of the data must be considered. Instead of being true values, they are assigned a Gaussian distribution function $\mathcal{P}^{obs}$ with a mean value $\langle|F_{PH,j}^{obs}|\rangle$ and a standard deviation $\sigma_j(\mathbf{h})$. The probability in eq.13.13 is now multiplied by $\mathcal{P}^{obs}[|F_{PH,j}^{obs}|; \langle|F_{PH,j}^{obs}|\rangle, \sigma_j(\mathbf{h})]$ and integrated over all possible values of $|F_{PH,j}^{obs}|$ yielding the likelihood:

$$\prod_{j=1}^{M} \int_{|F_{PH,j}^{obs}|=0}^{\infty} \mathcal{R}_j \times \mathcal{P}^{obs} d|F_{PH,j}^{obs}|$$

$$\cong \prod_{j=1}^{M} \mathcal{R}_j\left[|F_{PH,j}|, |\langle F_{PH,j}\rangle|, V_j^{total}\right] \qquad (13.14)$$

with $V_j^{total} = V_j^{PH} + \{\sigma_j(\mathbf{h})\}^2$

$|\langle F_{PH,j}\rangle|$ contains $\mathbf{F}_P^{corr}$, $\mathbf{F}_{H,j}$ and k_j. Therefore, $\mathcal{R}_j$ is a function of $\mathbf{F}_P^{corr}$, $\mathbf{F}_{H,j}$, k_j, V_j^{PH} and $\sigma_j(\mathbf{h})$. The likelihood of the model parameters derived from one acentric reflection is then:

$$\mathcal{L}_{acentr.(\mathbf{h})}^{iso} = \int_{\alpha_P=0}^{2\pi} \int_{|F_P|=0}^{\infty} \prod_{j=1}^{M} \mathcal{R}_j\left[|F_{PH,j}|, |\langle F_{PH,j}\rangle|, V_j^{total}\right] d\alpha_P d|F_P| \quad (13.15)$$

In a similar way as for the isomorphous data, a likelihood can be derived from the anomalous differences: $\Delta_j = |F_{PH,j,+}| - |F_{PH,j,-}|$ where + stands for +$\mathbf{h}$ and − for −$\mathbf{h}$. Because of errors, Δ_j is distributed as a one-dimensional Gaussian function $\mathcal{G}$ around a center δ_j with variance V_j^{δ}. Eq. 13.11 for the isomorphous differences is here replaced by

$$\delta_j = k_j\left(|\mathbf{F}_P^{corr} + \mathbf{F}_{H,j,+}|\right) - \left(|\mathbf{F}_P^{corr} + \mathbf{F}_{H,j,-}|\right)$$

$$\mathcal{G}_j\left(\delta_j, \Delta_j, V_j^{\delta}\right) = \frac{1}{\sqrt{2\pi V_j^{\delta}}} \exp\left[-\frac{(\Delta_j - \delta_j)^2}{2V_j^{\delta}}\right] \qquad (13.16)$$

In a procedure similar to the one leading to eq. (13.15), the likelihood of the model parameters derived from anomalous diffraction is

$$\mathscr{L}^{\text{anom.}}_{\text{acentr.}(\mathbf{h})} = \int\limits_{\alpha_P=0}^{2\pi} \int\limits_{|F_P|=0}^{\infty} \prod_{j=1}^{M} \mathscr{G}_j \, d\alpha_P d|F_P| \tag{13.17}$$

The combined likelihood is

$$\mathscr{L}^{\text{iso+anom.}}_{\text{acentr.}(\mathbf{h})} = \int\limits_{\alpha_P=0}^{2\pi} \int\limits_{|F_P|=0}^{\infty} \prod_{j=1}^{M} \mathscr{R}_j \times \mathscr{G}_j \, d\alpha_P d|F_P| \tag{13.18}$$

It depends exclusively on the parameters to be refined: D_j, the heavy-atom parameters incorporated in $\mathbf{F}_{H,j}$, scale factors and variances.

A similar likelihood function can be derived for centric reflections. At the start of the refinement, the current values of the heavy-atom parameters, the scale factors k_j, the variances, and $\mathbf{F}_P^{\text{trial}}$ are introduced. The probability distribution according to eq. (13.12) is calculated for each reflection and each derivative. In the next step, measurement errors are taken into account, and the information from the various derivatives is combined. Finally the integration over $\mathbf{F}_P^{\text{trial}}$ is carried out.

So far, the likelihood of the model parameters was derived from the information of one reflection $\mathbf{h}$. Assuming independent nonisomorphism errors, the likelihood of all reflections can be multiplied, or easier, their logarithmic values are added:

$$\log \mathscr{L}_{\text{total}} = \sum_{\substack{\text{acentr.} \\ \text{reflect.}}} \log \mathscr{L}_{\text{acentr.}} + \sum_{\substack{\text{centr.} \\ \text{reflect.}}} \log \mathscr{L}_{\text{centr.}} \tag{13.19}$$

$\log \mathscr{L}_{\text{total}}$ must be maximized, or $-\log \mathscr{L}_{\text{total}}$ minimized, with respect to the parameters of the heavy atom model, scale factors and nonisomorphism until convergence is reached. The heavy-atom parameters are

- The heavy-atom coordinates.
- Occupancies.
- Temperature factors.

The scale factors k_j include, besides a constant K_j^{sc}, a relative temperature factor with parameter B_j^{sc}:

$$k_j(\mathbf{h}) = K_j^{sc} \exp\left[-\frac{1}{4} B_j^{sc} \left(\frac{2\sin\theta}{\lambda} \right)^2 \right] \tag{13.20}$$

An anisotropic term can also be included.

As mentioned before, in SHARP, D_j is related to coordinate errors in the native structure and to nonisomorphism. Because the errors are assumed to be isotropic, D_j is expressed as an isotropic temperature factor

$$D_j(\mathbf{h}) = \exp\left[-\frac{1}{4} B_j^{non-iso}\left(\frac{2\sin\theta}{\lambda}\right)^2\right] \qquad (13.21)$$

In SHARP the variance for nonisomorphism, V_j^{PH}, consists of two parts: A component V_j^{glob} that increases with resolution and a component that decreases with resolution, V_j^{loc}.

$$V_j^{glob}(\mathbf{h}) = \left\langle\left|F_P^{trial}\right|^2\right\rangle \times \left\{1 - D_j^2(\mathbf{h})\right\} \qquad (13.22)$$

This is based on arguments by Read (1990b).

$\left\langle|F_P^{trial}|^2\right\rangle$ is calculated in resolution bins at the start of the refinement. V_j^{loc} is related to specific local errors and is called the localized component. It decreases with resolution because of a temperature factor with parameter B_j^{loc}:

$$V_j^{loc}(\mathbf{h}) = C_j^{loc}\left\langle\left(\left|F_{j=1}^{PH}\right| - \left|F_{j\neq1}^{PH}\right|\right)^2\right\rangle \left\{\exp\left[-\frac{1}{2}B_j^{loc}\left(\frac{2\sin\theta}{\lambda}\right)^2\right]\right\}$$

C_j^{loc} is a constant < 1 and the mean square isomorphous differences $\left\langle(|F_{j=1}^{PH}| - |F_{j\neq1}^{PH}|)^2\right\rangle$ are estimated in resolution bins at the start of the refinement. The total nonisomorphous variance, V_j^{PH}, is a suitable combination of V_j^{glob} and V_j^{loc}.

13.3. The Principle of the Fast Fourier Transform (FFT) Method

A very time consuming step in the refinement procedure is the calculation of all structure factors and their derivatives from the new parameter set at the end of each cycle in the refinement. To speed up these calculations fast Fourier algorithms are used. A detailed discussion of the application of the fast Fourier technique in crystallography is given by Ten Eyck (1973); see also Agarwal (1978). They enormously reduce the time needed for calculating structure factors from an electron density distribution or the other way around: the calculation of the electron density map from a series of structure factors. The electron density is sampled at grid points, usually at distances equal to one-third of the maximum resolution. The number of grid points is then

$$\frac{V}{\left(\frac{1}{3}d\right)^3} = 27\frac{V}{d^3}$$

where V is the unit cell volume and d is the maximum resolution. At this maximum resolution the number of structure factors in the volume of a reciprocal space sphere is

$$\frac{\frac{4}{3}\pi(1/d)^3}{V^*} = \frac{4}{3}\pi\frac{V}{d^3}$$

because the volume of the unit cell in reciprocal space, V^*, is equal to $1/V$ (Section 4.8).

The FFT technique requires the number of structure factors and the number of grid points to be the same. The former is usually smaller than the latter, but dummy structure factors can be added up to the number of grid points. For simplicity we shall deal with a one-dimensional case only. Let us write the Fourier inversion as

$$X(k) = \sum_{j=0}^{N-1} x(j) W_N^{jk}$$

Suppose $\mathbf{F}(h)$ is calculated from the $\rho(x)$ map; j indicates the grid points: $x = j/N$; k is the index for the structure factors and W_N^{jk} is the exponential term: $W_N^{jk} = \exp[2\pi i(j \cdot k/N)]$. To calculate the $\rho(x)$ map from the structure factors the form of the equation remains the same, but then $W_N^{jk} = \exp[-2\pi i(j \cdot k/N)]$.

The principle of the FFT method is to change the linear series of j terms into a two-dimensional series. To do this N is written as the product of two numbers, N_1 and N_2: $N = N_1 \times N_2$. Also j and k are split up: $j = j_2 + j_1 \times N_2$ and $k = k_1 + k_2 \times N_1$. If $j_2 = 0 \ldots (N_2 - 1)$ and $j_1 = 0 \ldots (N_1 - 1)$ all the integers in the range $0 \le j < N$ are generated once and only once. The same is true for k, if $k_1 = 0 \ldots (N_1 - 1)$ and $k_2 = 0 \ldots (N_2 - 1)$.

$$x(j) = x(j_2, j_1) = x(j_2 + j_1 N_2)$$

and

$$X(k_1, k_2) = \sum_{j_2=0}^{N_2-1} \sum_{j_1=0}^{N_1-1} x(j_2, j_1) \times W_N^{jk}$$

$$W_N^{jk} = W_N^{(j_2+j_1N_2)(k_1+k_2N_1)}$$

$$= W_N^{j_1k_1N_2} \times W_N^{j_2k_1} \times W_N^{j_2k_2N_1} \times W_N^{j_1k_2N}$$

$$W_N^{j_1k_2N} = W_N^N = \exp\left[2\pi i \frac{N}{N}\right] = \cos 2\pi + i \sin 2\pi = 1$$

$$W_N^{N_2} = \exp\left[2\pi i \frac{N_2}{N}\right] = \exp\left[2\pi i \frac{1}{N_1}\right] = W_{N_1}^1$$

$$W_N^{jk} = W_{N_1}^{j_1k_1} \times W_N^{j_2k_1} \times W_{N_2}^{j_2k_2}$$

$$X(k_1, k_2) = \sum_{j_1=0}^{N_1-1} \left(W_{N_1}^{j_1k_1} \times \sum_{j_2=0}^{N_2-1} \{x(j_2, j_1) \times W_N^{j_2k_1} \times W_{N_2}^{j_2k_2}\} \right)$$

Grouping the terms depending on j_1 together:

$$X(k_1, k_2) = \sum_{j_2=0}^{N_2-1} \left(W_N^{j_2k_1} \times W_{N_2}^{j_2k_2} \times \sum_{j_1=0}^{N_1-1} \{x(j_2, j_1) \times W_{N_1}^{j_1k_1}\} \right)$$

$\sum_{j_1=0}^{N_1-1} x(j_2, j_1) \times W_{N_1}^{j_1k_1}$ is a Fourier transform of length N_1 and must be done for each j_2, so N_2 times. The outer summation over j_2 is a Fourier transform of length N_2. To calculate all $X(k)$ it must be done N_1 times.

The time needed for evaluating a Fourier summation is proportional to the number of terms (N) and the number of grid points (N). Therefore, the time needed for calculating a normal Fourier transform of length N would be proportional to N^2 and with FFT:

$$N_2 \times N_1^2 + N_1 \times N_2^2 = N \times (N_1 + N_2)$$

Suppose $N = 2500$ and $N_1 = N_2 = 50$, then we must compare $N^2 = (2500)^2 = 625 \times 10^4$ with $2500 \times 100 = 25 \times 10^4$, an appreciable reduction in time.

13.4. Specific Refinement Methods

13.4.1. Rigid Body Refinement

Sussman and coworkers developed a *co*nstraint/*re*straint *le*ast-*s*quares refinement program, CORELS (Sussman, 1985). In this program a rigid geometry is assigned to parts of the structure and the parameters of these constrained parts are refined rather than individual atomic parameters. Optionally specified dihedral angles within a group can also be refined, which is then no longer completely rigid. It is possible to regard an entire molecule as a rigid entity and to refine its position and orientation in the unit cell. This is often done as a first step in a refinement procedure, for instance, after the molecular replacement procedure has given starting values for the position and orientation of the molecule. With CORELS the molecule can then be more properly positioned. The rigid entity can also be of smaller size, for instance, a folding unit consisting of β-strands and α-helices, or a prosthetic group, or, in nucleic acid structures, a nucleotide. The method increases the data/parameter ratio appreciably and is, therefore, applicable if only moderate resolution data are available.

Because CORELS is based on the conventional method of structure factor calculation and, therefore, is slow, rigid body refinement is performed much faster with a number of other rigid body refinement programs in which fast Fourier transform techniques are employed, for instance, Brünger's XPLOR package, Tronrud's TNT program, or Navaza's AMoRe (Section 10.3.2). However, the principle is the same. In

pure crystallographic refinement without adding stereochemical information, the quantity to be minimized is

$$Q = \sum_{hkl} w(h\ k\ l)\{|F_{\text{obs}}(h\ k\ l)| - |F_{\text{calc}}(h\ k\ l)|\}^2$$

In the CORELS program the constrained entities contribute as such to
the calculation of the $|F_{\text{calc}}(h\ k\ l)|$ values. Restraints modify the least-
squares criterion to the minimization of

$$Q = \sum_{hkl} w(h\ k\ l)\{|F_{\text{obs}}(h\ k\ l)| - |F_{\text{calc}}(h\ k\ l)|\}^2$$
$$+ w_{\text{D}} \sum_d w_d \{D_{\text{obs}}(d) - D_{\text{calc}}(d)\}^2$$
$$+ w_{\text{T}} \sum_i w_{\text{x}} \sum_{\text{x}} \{X_{\text{T}}(i, \mathbf{x}) - X(i, \mathbf{x})\}^2$$

$D_{\text{obs}}(d)$ is the standard distance, corresponding to bond lengths, but also
to angles and van der Waals distances, because in CORELS all restraints
are introduced as distances: a bond length as the distance between nearest
neighbors, a bond angle as the distance between next nearest neighbors,
and a dihedral angle as the distance between a first and a fourth atom

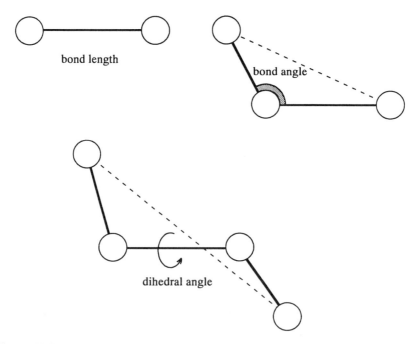

Figure 13.3. In the CORELS refinement method all restraints are introduced as
distances: a bond length as the distance between nearest neighbors, a bond angle
as the distance between next nearest neighbors, and a dihedral angle as the
distance between a first and a fourth atom.

(Figure 13.3). $D_{calc}(d)$ is the distance calculated from the current model. $w_d = 1/\sigma_d$ where σ_d is the standard deviation for distances of type d bonds as observed in small molecule structures. The third term restrains the coordinate vectors $\mathbf{x}$ ($= x, y, z$) of a model atom to a corresponding target position. For crystallographic refinement the third term is omitted by putting $w_T = 0$. It plays a role in model building for which $w(h\ k\ l) = 0$. By choosing the weighting factors $w(h\ k\ l)$ and w_D properly, a relative weight can be given to the X-ray and the stereochemical restraints in the refinement process. The quantity Q is minimized with respect to all positional coordinates and thermal parameters in the usual way with the least-squares method.

13.4.2. Stereochemically Restrained Least-Squares Refinement

Stereochemical restraints are fully incorporated in this method. The occasional calculation of an electron density map, a difference electron density map, or OMIT map (Section 8.2) is useful for manually correcting major imperfections in the model. The Konnert–Hendrickson program, PROLSQ (Hendrickson, 1985), has found wide application in the accurate structure determination of protein molecules, but now a more efficient and flexible program (TNT), written by Tronrud et al. (1987) using the fast Fourier transform algorithm, is mainly used for stereochemically restrained least-squares refinement.

The function to be minimized consists of a crystallographic term (a) and several stereochemical terms (b–f):

$$Q = \sum_{hkl} w(h\ k\ l)\{|F_{obs}(h\ k\ l)| - |F_{calc}(h\ k\ l)|\}^2 \tag{a}$$

$$+ \sum_{dist.j} w_D(j)(d_j^{ideal} - d_j^{model})^2 \tag{b}$$

$$+ \sum_{\substack{planes\ k \\ coplanar\ atoms\ i}} w_P(i, k)(\mathbf{m}_k \cdot \mathbf{r}_{i,k} - d_k)^2 \tag{c}$$

$$+ \sum_{\substack{chiral \\ centers\ l}} w_C(l)(V_l^{ideal} - V_l^{model})^2 \tag{d}$$

$$+ \sum_{\substack{nonbonded \\ contacts\ m}} w_N(m)(d_m^{min} - d_m^{model})^4 \tag{e}$$

$$+ \sum_{\substack{torsion \\ angles\ t}} w_T(t)(X_t^{ideal} - X_t^{model})^2 \tag{f}$$

(a) is the usual X-ray restraint. All other terms are stereochemical restraints.

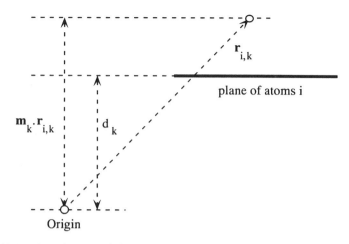

Figure 13.4. Planarity restraining in the Konnert–Hendrickson refinement method is on the deviation of the atoms from the least-squares plane of the group. d_k is the distance from the origin to the current least-squares plane. For an atom at position $\mathbf{r}_{i,k}$, the distance from the least-squares plane is expressed by $\mathbf{m}_k \cdot \mathbf{r}_{i,k} - d_k$. Here $\mathbf{m}_k$ is the unit vector normal to the plane.

(b) restrains the distance between atoms, defining bond lengths, bond angles, or dihedral angles.

(c) imposes the planarity of aromatic rings. The same applies to the guanidyl part of arginine and to peptide planes. The actual restraining is on the deviation of the atoms from the least-squares plane of the group (see Figure 13.4). d_k are the parameters of the current least-square plane. For an atom at position $\mathbf{r}_{i,k}$ the distance from the least-squares plane is expressed by $\mathbf{m}_k \cdot \mathbf{r}_{i,k} - d_k$. Here $\mathbf{m}_k$ is the unit vector normal to the plane and d_k is the distance from the origin to the plane.

(d) restrains the configuration to the correct enantiomer. A protein structure has many chiral centers: at the C_α atoms (except for glycine) and at the $C\beta$ of threonine and isoleucine. Chirality can be expressed by a chiral volume, which is calculated as the scalar triple product of the vectors from a central atom O to three attached atoms 1, 2, and 3; $\{(O \rightarrow 1). [(O \rightarrow 2) \times (O \rightarrow 3)]\}$ or shorter: $\{\mathbf{1} \cdot [\mathbf{2} \times \mathbf{3}]\}$. This is the scalar product of vector $\mathbf{1}$ with a new vector $\mathbf{4}$, which is the vector product of vector $\mathbf{2}$ and vector $\mathbf{3}$ (Figure 13.5). The new vector $\mathbf{4}$ is perpendicular to both vectors $\mathbf{2}$ and $\mathbf{3}$; its length equals the surface area of the parallelogram formed by $\mathbf{2}$ and $\mathbf{3}$; multiplied by $\mathbf{1}$, a scalar results that equals six times the volume of the pyramid that has O as top and triangle 1, 2, 3 as its base. For the enantiomeric configuration two vectors are interchanged and because $\{\mathbf{1} \cdot [\mathbf{2} \times \mathbf{3}]\} = -\{\mathbf{3} \cdot [\mathbf{2} \times \mathbf{1}]\} = -\{\mathbf{1} \cdot [\mathbf{3} \times \mathbf{2}]\} = -\{\mathbf{2} \cdot [\mathbf{1} \times \mathbf{3}]\}$, the triple product will have the opposite sign.

(e) introduces restraints for nonbonded or van der Waals contacts.

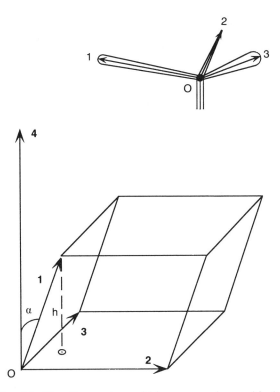

Figure 13.5. The chirality of an amino acid is expressed as a chiral volume, which is calculated as the scalar triple product of the vectors from a central atom O to three attached atoms 1, 2, and 3:{$\mathbf{1} \cdot [\mathbf{2} \times \mathbf{3}]$}. Vector **4** is the vector product of **2** and **3**. The area of the parallelogram formed by vectors **2** and **3** is $|\mathbf{2} \times \mathbf{3}|$. The volume of the parallelepid formed by vectors **1**, **2**, and **3** is $V = h \cdot |\mathbf{2} \times \mathbf{3}| = h \cdot |\mathbf{4}|$, where h is the altitude; h is parallel to **4** and equals $|\mathbf{1}| \cdot |\cos \alpha|$ (because $0 < \alpha < 180°$, $\cos \alpha$ may be negative and, therefore, the absolute value of $\cos \alpha$ is taken). $V = |\mathbf{1}| \cdot |\mathbf{4}| \cdot |\cos \alpha| = |\mathbf{1} \cdot \mathbf{4}| = |\mathbf{1} \cdot [\mathbf{2} \times \mathbf{3}]|$. The volume of the pyramid with O as the top and triangle 1, 2, and 3 as the base is $\frac{1}{3} \cdot h \cdot \frac{1}{2}|\mathbf{2} \times \mathbf{3}| = \frac{1}{6} \cdot V$.

They prevent the close approach of atoms not connected by a chemical bond and are stop signs for those atoms. As such they play an important role in protein structures. Only the repulsive term of the van der Waals potential is taken into acount:

$$E_{\text{repulsive}} = A \times r^{-12}$$

For contacts between C, N, O, and S atoms, and for d-values near and shorter than the equilibrium distance d^{min}, the repulsive term can be approximated by

$$E(d) - E(d^{\text{min}}) = \frac{1}{\sigma^4}(d - d^{\text{min}})^4$$

and the function to be minimized becomes

$$\sum_{\substack{\text{nonbonded} \\ \text{contacts } m}} w_N(m)(d_m^{\min} - d_m^{\text{model}})^4$$

(f) restrains torsion angles. Although a free rotation is in principle possible around single bonds, there are certain restrictions imposed by nonbonded repulsion. For instance the rotations around the $C_\alpha \rightarrow N$ and the $C_\alpha \rightarrow C{=}O$ bond in the peptide main chain are limited to certain combinations. Also in aliphatic side chains the staggered conformation is more stable than the eclipsed one.

Table 13.1. Statistics of data collection and refinement for hevamin at pH 2.0. The structure was refined with TNT against data from 15.0 to 1.9 Å resolution.*

Data processing	
Number of observations	109372
Number of unique reflections	19169
R_{merge} (1.93–1.90 Å)	0.065 (0.213)**
Completeness (1.93–1.90 Å)	0.941 (0.414)

Refinement	
Resolution range (Å)	15.0–1.90
Completeness of working set	0.83
Completeness of test set	0.09
R-factor	0.157
R_{free}	0.199
Number of protein atoms	2087
Number of solvent atoms	140

RMS deviations from ideality	
Bond lengths (Å)	0.010
Bond angles (°)	1.48
Dihedrals (°)	23.0
B-value correlations for bonded atoms (Å^2)	2.1

Average B-values (Å^2)	
All protein atoms	16.5
Main chain atoms	13.2
Side chain atoms	20.1
Solvent atoms	33.2

*From Terwisscha van Scheltinga, thesis 1997, University of Groningen, with permission
**Values in parentheses are for the high resolution shell

All weighting factors w can be chosen with some freedom, although the normal choice is $w = 1/\sigma^2$, except for w_N, which is taken as $1/\sigma^4$; σ is the standard deviation of the expected distribution. Besides the restraints (a)–(f) several others can be added. For instance, it can be expected that the B-parameters of the temperature factors of neighboring atoms are related because large temperature movements of an atom will cause its neighbors to move also over more than average distances. On the other hand, rigid parts restrict the movement of their neighbors. This correlation can be introduced by restraining the B-value of an atom to those of its neighbors.

In structural papers the statistics of the refinement results are usually given in table form, for instance, as in Table 13.1.

13.4.3. SHELXL

Originally this program was devised for the refinement of small molecules, but has been adapted for macromolecule refinement (Sheldrick and Schneider, 1997). The condition is that the X-ray data have been collected to at least 2.5 Å resolution. The program applies a least-squares strategy and refines against F^2 values. It incorporates many features similar to those found in CORELS, PROLSQ, and TNT, such as constraints and geometric restraints with the exception of torsion angle and H-bonding restraints. If noncrystallographic symmetry (NCS) is present, it increases the refinement efficiency considerably. NCS is applied as a restraint in two ways:

1. Distances between atoms 1 and 4 in a connectivity chain $1, 2, 3, 4, \ldots$ are restrained to be equal in the NCS-related molecules.
2. Isotropic temperature factors of related atoms are restrained to be equal.

These restraints reduce the number of parameters appreciably. The power of SHELXL has been attributed to:

- A conventional structure factor calculation that is more precise than an FFT calculation. This causes the program to be relatively slow, but it improves the convergence properties.

- The inclusion of important off-diagonal terms in the least-squares matrix.

Further advantages are the flexible treatment of disorder (e.g., more than one position for a sidechain) and the feature of anisotropic refinement. Moreover, it can refine twinned crystals by fitting the sum of the calculated intensities for the individual components to the observed intensities. This is better than trying to "detwin" the data, which introduces systematic errors.

13.4.4. Energy Refinement (EREF)

In the Konnert–Hendrickson refinement the least squares function

$$\sum_h w(h)(|F_{obs}(h)| - |F_{calc}(h)|)^2$$

is minimized simultaneously with a number of geometric terms related to bond lengths, angles, etc. In another refinement method—proposed by Jack and Levitt (1978)—the X-ray term is minimized together with a potential energy function including terms for bond stretching, bond angle bending, torsion potentials, and van der Waals interactions. Electrostatic interactions are usually ignored, because they act over rather long distances and are not sensitive to small changes in atomic position. Moreover the calculations are performed assuming the molecule to be in a vacuum and the electrostatic energy would be extremely high, unless artificial dielectric constants are introduced. The function to be minimized is

$$Q = (1 - w_X) \times E + w_X \times \sum_h w(h\ k\ l)(|F_{obs}(h\ k\ l)| - |F_{calc}(h\ k\ l)|)^2$$

E is the energy term and w_X controls the relative contribution of the energy and the X-ray term. Its choice, between $w_X = 0$ (pure energy minimization) and $w_X = 1$ (pure X-ray minimization) is rather arbitrary and depends on experience; however, a more objective way for determining w_X has been presented by Brünger (1993) by optimizing the free R-factor (Section 15.2). For the potential energy of a bond the harmonic approximation is used:

$$E_{bond} = \tfrac{1}{2}K_{bond}(b - b_0)^2$$

with b_0 the minimum energy distance and b the actual distance between the atoms. K_{bond} and b_0 can be derived from the vibration spectra of small molecules. The same assumption is made for the bond angles:

$$E_{bond\ angle} = \tfrac{1}{2}K_\tau(\tau - \tau_0)^2$$

The torsion energy around a bond is expressed as

$$E_{torsion} = \tfrac{1}{2}K_\xi(\xi - \xi_0)^2$$

The energy for a dihedral angle is

$$E_{dihedral} = K_\theta\{1 + \cos(m\theta + \delta)\}$$

(see Figure 13.6). θ is the rotation angle and δ a phase angle determining the zero point of rotation. m is the rotation frequency (3 for the C—C bond in ethane). This is a very simple presentation of the dihedral energy and it is only approximately true for bonds that have a large group on either side. For the van der Waals interaction energy both the attractive

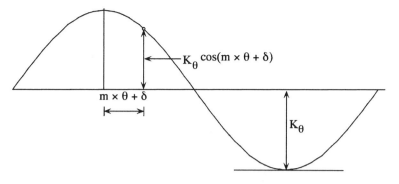

Figure 13.6. The energy for a dihedral angle. θ is the rotation angle and δ a phase angle determining the zero point of rotation. m is the rotation frequency (for instance 3 for the C—C bond in ethane).

term $(B \times r^{-6})$ and the repulsive term $(A \times r^{-12})$ are taken into account:

$$E_{\text{van der Waals}} = A \times r^{-12} + B \times r^{-6}$$

Although EREF is somewhat less popular than stereochemically re-strained refinement, it can play a useful role in minimizing the energy of a system before refining it with the molecular dynamics technique, which will be discussed in the next section.

13.4.5. Molecular Dynamics or Simulated Annealing

The refinement procedures described so far are based on the least-squares method. The function to be minimized follows a downhill path toward its minimum value and if the starting model is not too different from the real structure the refinement easily converges to the correct solution. However, if the distance between the atoms in the model and in the real structure is rather large, the refinement may be trapped in a local minimum instead of reaching the global or true minimum. To avoid this situation a refinement technique is required that allows uphill as well as downhill search directions to overcome barriers in the Q-function. This technique is supplied by molecular dynamics and was introduced by Brünger and incorporated in the X-PLOR package (Brünger, 1987; Brünger and Nilges, 1993). In molecular dynamics the dynamic behavior of a system of particles is simulated. This simulation yields an ensemble of structures that is energetically allowed for a given temperature and pres-sure: the energy distribution of the structures in the ensemble follows Boltzmann's law, which states that the number of structures with a poten-tial energy ε_{pot} is proportional to $\exp[-\varepsilon_{\text{pot}}/kT]$; k is Boltzmann's constant and T is the absolute temperature. The potential energy depends on the relative positions of the atoms and is calculated on the basis of known potential energy functions, which are always only approximately accurate. The potential energy terms are similar to those in EREF (Section 13.4.4). However, in the molecular dynamics refinement an electrostatic term is added. It is calculated as

$$\sum \frac{q_i \times q_j}{4\pi\varepsilon_0\varepsilon_r r_{ij}}$$

ε_0 is the permittivity constant: $4\pi\varepsilon_0 = 1.11264 \times 10^{-10} C^2 N^{-1} m^{-2}$. ε_r is the local dielectric constant, which is given an estimated value between 1 and 80, depending on the position of the atoms i and j with interatomic distance r_{ij} or, alternatively, instead of $\varepsilon_r > 1$ the electric charges can be reduced.

A molecular dynamics calculation on a molecule starts with assigning to the atoms velocities derived from a Maxwellian distribution at an appropriate temperature. At time $t = 0$ the atoms are in a starting configuration that has a potential energy E_{pot} for the entire molecule. On each atom i at position $\mathbf{r}_i$, a force is acting that is the derivative of the potential energy: force $(i) = -\partial E_{pot}/\partial \mathbf{r}_i$. With the equation from Newtonian mechanics, force $(i) = m_i(d^2\mathbf{r}_i/dt^2)$, the acceleration $d^2\mathbf{r}_i/dt^2$ for atom i with mass m_i can be calculated and then applied. After a short time step Δt, in the femtosecond range (1 fsec $= 10^{-15}$ sec), the process is repeated with the atoms in the new positions. If the number of steps is sufficient (10^3–10^4), the minimum of E_{pot} is reached and information about the dynamic behavior of the atoms in the molecule is obtained. If the temperature of the system is raised to a higher value, more atoms have a higher speed and a higher kinetic energy and can overcome higher energy barriers. The basic idea of molecular dynamics as a superior refinement technique is to raise the temperature sufficiently high for the atoms to overcome energy barriers and then to cool slowly to approach the energy minimum. The method is called simulated annealing (SA) in comparison with, for instance, removing internal strain from glass by heating it first and then slowly cooling it down (Brünger, 1991b). In an alternative method the temperature is kept constant but all potential energy terms are scaled by an overall scale factor c, with $0 \leqslant c \leqslant 1$. This is based on Boltzmann's law. An increase in T with constant ε_{pot} has the same effect as diminishing ε_{pot} at constant T.

A typical example of a high energy barrier occurs in the flipping of a peptide plane. It is impossible for the other refinement techniques to overcome this barrier, but not so for simulated annealing. In the application of MD (or SA) to crystallographic refinement, the calculated structure factors of the system are restrained to the observed structure factors as target values, by adding a maximum likelihood target function

$$E_x = \sum_{hk\ell\in W} \left(\frac{1}{\sigma_{ML}^2}\right)(|F_{obs}(hk\ell)| - \langle|F_{obs}(hk\ell)|\rangle)^2 \qquad (13.23)$$

as a pseudoenergy term to the potential energy E_{pot} of the system. W is the working set of reflections, containing 95–100% of the total number of reflections; it is explained in Section 15.2. The expected value $\langle |F_{obs}(hk\ell)| \rangle$ as well as the variance (σ_{ML}^2), can be derived (Pannu and Read, 1996) from $|F_{obs}(hk\ell)|$, $|F_{calc}(hk\ell)|$, and σ_A, which is an error-related parameter (Section 15.6). Eq. (13.23) replaces the original target function E_x based on least squares.

The total energy, $E_x + E_{pot}$, is minimized during the refinement. In principle $\mathbf{F}_{calc}$ must be calculated after each time step Δt. The same is true for the derivatives of $|F_{calc}|$ with respect to atomic parameters. These derivatives are required for calculating the "force." However, the atomic parameters change only very little during Δt and, therefore, it is sufficient to calculate $\mathbf{F}_{calc}$ and its derivatives only after a preset change in the coordinates has been reached, e.g., 1/10 of the resolution. MD refinement is usually preceded by energy refinement (EREF) with or without X-ray restraints, to reduce possibly high energies in the system.

In Section 8.2, OMIT maps were discussed for improving or checking troublesome parts of the structure. An OMIT map is calculated with the observed structure factor amplitudes and phase angles calculated for the reliable part of the structure. The "missing" parts should then show up in the electron density map at half of their actual height. To reduce model bias Sim weighting factors are applied.

OMIT maps are sometimes used to systematically check a molecular model by deleting parts of the structure, one after the other. The omitted part comprises 2–5% of the structure. Model bias can also be reduced by applying simulated annealing to the "known" parts. The atoms in the "known" part at the border of the omitted region are restrained to keep them from moving into the omitted region. The maps calculated in this way are called simulated annealing OMIT maps or composite annealed OMIT maps (Shah et al., 1997).

13.4.5.1. Advantage of MD Refinement

An advantage of MD refinement is the large radius of convergence, which can be several Å units long. In other words, MD draws groups of atoms, originally several Ås away from their final position, corresponding with the potential energy minimum, toward those final positions without any manual intervention. Large errors in the starting model can be corrected. This speeds up the refinement appreciably. The method is very demanding in computer time but with modern high speed computers this is not a problem.

13.4.5.2. The Time-Averaging MD Technique

In MD refinement the dynamic system of moving atoms is restrained by the X-ray data of a static model including isotropic temperature factors.

This is, in principle, not correct because in fact the atomic movements are not necessarily isotropic, but can just as well be anisotropic and anharmonic. Gros et al. (1990) proposed to improve the MD refinement technique by adding as a pseudo-energy term:

$$E_x = k_x \times \sum_{hkl}[|F_{obs}(h\ k\ l)| - k\ |\{F_{calc}(h\ k\ l)\}_{average}|]^2$$

$\{F_{calc}(h\ k\ l)\}_{average}$ is not the structure factor of a single model (based on the atomic coordinates x, y, z and the temperature factor parameter B), but the ensemble average of calculated structure factors without temperature factor:

$$\{\mathbf{F}_{calc}(h\ k\ l)\}_{average} = \frac{\int\limits_{t=0}^{t'} P_t\mathbf{F}_{calc}^t(h\ k\ l)\,dt}{\int\limits_{t=0}^{t'} P_t\,dt}$$

t' is the total trajectory time (for instance, 1.000 time steps). t is any time in the trajectory.

P_t is the weight given to the structure factor $\mathbf{F}_{calc}^t\ (h\ k\ l)$, calculated for the model at time t. P_t is chosen as $\exp[-(t' - t)/\tau_x]$. The effect of this choice is that an exponentially decreasing weight is given to "older" structures (small t) with a relaxation time τ_x, which regulates the contribution from an "old" structure: the larger τ_x, the more it contributes.

$$\int\limits_{t=0}^{t'} P_t\,dt = \tau_x \left\{1 - \exp\left[-\frac{t'}{\tau_x}\right]\right\} \approx \tau_x$$

If $t' \gg \tau_x$ the factor $\left\{1 - \exp\left[-\frac{t'}{\tau_x}\right]\right\}$ can be ignored.

$$\{\mathbf{F}_{calc}(h\ k\ l)\}_{average} = \frac{1}{\tau_x} \int\limits_{t=0}^{t'} \exp\left[-\frac{t' - t}{\tau_x}\right]\mathbf{F}_{calc}^t(h\ k\ l)\,dt$$

The structure factor $\mathbf{F}_{calc}^t(h\ k\ l)$ of an individual structure at time t in the trajectory depends on positional parameters only. No individual thermal parameters are assigned to the atoms but instead their movement is now represented by their spatial distribution during the complete trajectory. This is not necessarily isotropic but can also be anisotropic and anharmonic.

In the execution of the method $\{\mathbf{F}_{calc}(h\ k\ l)\}_{average}$ at time $t = 0$ is taken equal to the classical $\mathbf{F}_{calc}(h\ k\ l)$, including the temperature factor parameter B. In the course of the process these B-parameters are gradually lowered to $1\,\text{Å}^2$ and the relaxation time τ increased from 0 to, for instance, $10\,\text{psec}$ ($1\,\text{psec} = 10^{-12}\,\text{sec}$). In this way a gradual change from classical to time-averaging MD is accomplished. There is no need to calculate $\mathbf{F}_{calc}(h\ k\ l)$ of the structure after each time step Δt. It is sufficient to do it for an ensemble of, for instance, 100 structures. The derivatives of $|F_{calc}|$, as obtained by a fast Fourier technique according to Agarwal (1978), must be recalculated more often because of the small but rapidly changing fluctuations in the atomic positions. It should be remarked that each individual structural model, that is a model calculated after a time step Δt, is a bad representation of the actual structure: the atoms are not in their equilibrium position and no real B-parameters are incorporated. It is the complete collection of the individual models that approaches the ideal one. Therefore, time-averaging MD is not a refinement technique in the pure sense, but a sampling technique that provides us with an excellent model of the structure, including a more complete description of the atomic fluctuations around the equilibrium positions, as the conventional X-ray structure does.

With this time-averaging MD technique remarkably low R-factors—near 10%—can be reached. It will be clear that even more computer time is needed than for the molecular dynamics technique without time averaging, and still more if solvent molecules would be included in the refinement.

Although the convergence range is large for MD refinement and manual intervention is less frequently required than with the previous refinement techniques, nevertheless the electron density map should be checked occasionally for large errors, such as side chains placed in the density of solvent molecules that are not included in the refinement.

13.4.5.3. Torsion Angle Refinement

In 1971, Diamond introduced a refinement technique for protein structures in which only the torsion angles were refined. Peptide planes were kept planar, and bond distances and bond angles were kept constant. It was based on least-squares refinement and turned out not to be a very effective refinement method. In the 1980s, it was replaced by more common refinement protocols. Rice and Brünger (1994) have reintroduced torsion angle refinement with more success by applying it in the context of simulated annealing. Restriction to torsion angles greatly increases the observation/ parameter ratio. The method shows an increased convergence compared with conventional methods, but it does not replace them. It can precede them if the preliminary model is too far from the final solution for a conventional technique.

13.4.6. REFMAC

Murshudov et al. (1997) have written the refinement program REFMAC, which is entirely based on the maximum likelihood formalism. According to Eq. 13.7:

$$p[\mathbf{F}(calc); |F(obs)|] = p[\mathbf{F}(calc)] \times \frac{p[|F(obs)|; \mathbf{F}(calc)]}{p[|F(obs)|]}$$

Disregarding the normalization constant $p[|F(obs)|]$:

$$p[\mathbf{F}(calc); |F(obs)|] = p[\mathbf{F}(calc)] \times p[|F(obs)|; \mathbf{F}(calc)]$$

$p[\mathbf{F}(calc)]$ expresses the prior information as stereochemical restraints. This information is known and will not be discussed here. We restrict ourselves to

$$p[|F(obs)|; \mathbf{F}(calc)]$$

The required likelihood is (Eq. 13.9):

$$\mathscr{L}_{total} = \prod_{\mathbf{h}} p[|F_{\mathbf{h}}(obs)|; \mathbf{F}_{\mathbf{h}}(calc)]$$

We start from eq. 8.6a and Figure 8.1, if vector $\mathbf{F}_K$ is regarded as $\mathbf{F}(calc)$ and vector $\mathbf{F}$ as $\mathbf{F}(obs)$. The conditional probability $p[\xi;|F|]$ is equal to the conditional probability

$$p[\mathbf{F}; \mathbf{F}_K] = p[\mathbf{F}(obs); \mathbf{F}(calc)]$$

Because only the amplitudes of $\mathbf{F}(obs)$ are known, eq. 8.6.a must be integrated over ξ to obtain:

$$p[|F(obs)|; \mathbf{F}(calc)] =$$

$$\frac{|F(obs)|}{\pi \times \sum\limits_{j=1}^{n} f_j^2} \exp\left[-\frac{|F(obs)|^2 + |F(calc)|^2}{\sum\limits_{j=1}^{n} f_j^2} \right]$$

$$\int\limits_{\xi=0}^{2\pi} \exp\left[\frac{2|F(obs)| \times |F(calc)|\cos\xi}{\sum\limits_{j=1}^{n} f_j^2} \right] d\xi$$

$$= 2 \times \frac{|F(obs)|}{\sum\limits_{j=1}^{n} f_j^2} \exp\left[-\frac{|F(obs)|^2 + |F(calc)|^2}{\sum\limits_{j=1}^{n} f_j^2} \right] \times I_0\left(\frac{2|F(obs)| \times |F(calc)|}{\sum\limits_{j=1}^{n} f_j^2} \right)$$

I_0 is a zero order Bessel function. (Compare with eq. 8.7).

This probability function for $p[\mathbf{F}(obs); \mathbf{F}(calc)]$ was derived assuming no errors in atomic positions in the model. The effect of these errors is, as pointed out by Luzzati (1952) and Read (1990b), that $|F(\text{calc})|$ must be replaced by

$$D \times |F(\text{calc})|, \text{ and } \sum_{j=1}^{n} f_j^2 \text{ by } \left(1 - D^2\right) \times \sum_{j=1}^{n} f_j^2 .$$

D is the Fourier transform of the probability distribution $p(\Delta\mathbf{r})$ of $\Delta\mathbf{r}$ (see eq. 15.2).

Experimental errors in the $|F(\text{obs})|$ values can be incorporated by increasing the variance $\sum_{j=1}^{n} f_j^2$ with the variance σ_e^2 of the experimental error. We finally obtain:

$$p[\mathbf{F}(\text{obs}); \mathbf{F}(\text{calc})] =$$

$$2 \times \frac{|F(\text{obs})|}{2\sigma_e^2 + \sum_{j=1}^{n} f_j^2} \exp\left[-\frac{|F(\text{obs})|^2 + D^2 |F(\text{calc})|^2}{2\sigma_e^2 + \sum_{j=1}^{n} f_j^2} \right]$$

$$\times I_0 \left(\frac{2|F(\text{obs})| + D|F(\text{calc})|}{2\sigma_e^2 + \sum_{j=1}^{n} f_j^2} \right)$$

Summary

From a first, non-refined, model of a protein structure, for instance, obtained with isomorphous replacement, important biological information can already be derived. However, for more reliable information and finer details, the structure must be refined. The poor overdetermination limits the refinement, in general, to the positional parameters and an isotropic, but not anisotropic, temperature factor for each atom. Existing stereochemical information from small molecules adds additional "observations" and changes the observation/parameter ratio favorably. The refinement methods use the least-squares or maximum likelihood formalism for approaching the final solution, often after a great many cycles in which the fast Fourier transform method is essential.

Molecular dynamics or simulated annealing is one of the most popular refinement techniques. It combines real energy terms with restraints on the structure factor amplitudes as pseudoenergy terms. With the simulated annealing concept the system can jump over local energy minima and, therefore, has a long radius of convergence.

In many refinement programs the least-squares method is replaced by the maximum likelihood formalism.

Chapter 14
The Combination of Phase Information

14.1. Introduction

In the multiple isomorphous replacement method the phase information from the various heavy atom derivatives and from anomalous scattering is combined by multiplying the individual phase probability curves. If the electron density map, which results from isomorphous replacement, can be fully interpreted, the crystallographer immediately starts with model refinement and the isomorphous phase information is left behind. However, if the electron density map is inadequate for complete interpretation, map improvement (= phase refinement) should precede model refinement. Solvent flattening and the inclusion of molecular averaging are examples of map improvement techniques (Chapter 8). Another way to improve the existing model is by combining the isomorphous replacement phase information with phase information from the known part of the structure. It is clear that a general and convenient way of combining phase information from these various sources would be most useful. Such a method has been proposed by Hendrickson and Lattman (1970) and has been based on previous studies by Rossmann and Blow (1961). Hendrickson and Lattman propose an exponential form for each individual probability curve of the following type:

$$P_s(\alpha) = N_s \exp[K_s + A_s \cos\alpha + B_s \sin\alpha + C_s \cos 2\alpha + D_s \sin 2\alpha]$$

$P_s(\alpha)$ is the probability for phase angle α derived from source s. K_s and the coefficients A_s, B_s, C_s, and D_s contain, e.g., structure factor amplitudes but not the protein phase angles α. N_s is a normalization factor. The multiplication of the available $P_s(\alpha)$ functions to the overall probability function $P(\alpha)$ is now simplified to an addition of all K_s and of the coefficients A_s–D_s in the exponential term.

$$P(\alpha) = \prod_s P_s(\alpha) = N' \exp\left[\sum_s K_s + \left(\sum_s A_s\right) \cos\alpha + \left(\sum_s B_s\right) \sin\alpha \right.$$
$$\left. + \left(\sum_s C_s\right) \cos 2\alpha + \left(\sum_s D_s\right) \sin 2\alpha \right]$$

or, combining N' and $\exp[\Sigma_s K_s]$:

$$P(\alpha) = N \exp[A \cos\alpha + B \sin\alpha + C \cos 2\alpha + D \sin 2\alpha] \quad (14.1)$$

The value of N is not important. Moreover it disappears if the "best" Fourier map is calculated with Eq. (7.36):

$$\mathbf{F}_{hkl}(\text{best}) = \frac{\displaystyle\int_\alpha P_{hkl}(\alpha)\mathbf{F}_{hkl}(\alpha)\,d\alpha}{\displaystyle\int_\alpha P_{hkl}(\alpha)\,d\alpha}$$

We shall now derive the form of the coefficients A_s–D_s for

- isomorphous replacement
- anomalous scattering
- partial structures
- solvent flattening
- molecular averaging

14.2. Phase Information from Isomorphous Replacement

In Section 7.12 the probability function $[P(\alpha)]_j$ for one reflection and derivative j in the isomorphous replacement has been presented as

$$[P(\alpha)]_j = N \exp\left[-\frac{\varepsilon_j^2(\alpha)}{2E_j^2}\right]$$

$\varepsilon_j(\alpha)$ is the "lack of closure error" for the structure factor amplitude $|F_{PH}(\text{calc})|$ of derivative j.

$$\varepsilon_j(\alpha) = \{|F_{PH}(\text{obs})|_j - k_j|F_{PH}(\text{calc})|_j\}$$

The derivation of a suitable form of the phase probability curve is easier if the error is redefined as an error in the derivative intensity instead of the structure factor amplitude. Assuming that $|F_{PH}(\text{calc})|^2$ and $|F_{PH}(\text{obs})|^2$ are on the same scale, we have

$$\varepsilon_j'(\alpha_P) = |F_{PH,j}(calc)|^2 - |F_{PH,j}(obs)|^2$$

$$|F_{PH,j}(calc)|^2 = |\mathbf{F}_P + \mathbf{F}_{H,j}|^2$$
$$= |F_P|^2 + |F_{H,j}|^2 + 2|F_P| \times |F_{H,j}| \times \cos(\alpha_{H,j} - \alpha_P)$$

(see Figure 7.15b)

$$\varepsilon_j'(\alpha_P) = |F_P|^2 + |F_{H,j}|^2 - |F_{PH,j}(obs)|^2$$
$$+ 2|F_P| \times |F_{H,j}| \times \cos(\alpha_{H,j} - \alpha_P)$$

A Gaussian distribution for $\varepsilon_j'(\alpha_P)$ is assumed:

$$[P_{iso}(\alpha_P)]_j = N_j \exp\left[-\frac{\{\varepsilon_j'(\alpha_P)\}^2}{2(E_j')^2}\right] \qquad (14.2)$$

N_j is a normalizing factor and E_j' is the estimated standard deviation of the errors in the derivative intensity.

$$\{\varepsilon_j'(\alpha_P)\}^2 = (|F_P|^2 + |F_{H,j}|^2 - |F_{PH,j}(obs)|^2)^2$$
$$+ 4 \times (|F_P|^2 + |F_{H,j}|^2 - |F_{PH,j}(obs)|^2)$$
$$\times |F_P| \times |F_{H,j}| \times \cos(\alpha_{H,j} - \alpha_P)$$
$$+ 4 \times |F_P|^2 + |F_{H,j}|^2 \times \cos^2(\alpha_{H,j} - \alpha_P)$$

We want to separate functions with α only from the rest. This can be done by writing for

$$\cos^2(\alpha_{H,j} - \alpha_P) \rightarrow \tfrac{1}{2}\{1 + \cos 2(\alpha_{H,j} - \alpha_P)\}$$
$$\cos(\alpha_{H,j} - \alpha_P) \rightarrow \cos\alpha_{H,j}\cos\alpha_P + \sin\alpha_{H,j}\sin\alpha_P$$

With the separation of $\mathbf{F}_{H,j}$ into its components (Figure 14.1), this results in

$$\{\varepsilon_j'(\alpha_P)\}^2 = (|F_P|^2 + |F_{H,j}|^2 - |F_{PH,j}(obs)|^2)^2 + 2 \times |F_P|^2 \times |F_{H,j}|^2$$
$$+ 4 \times (|F_P|^2 + |F_{H,j}|^2 - |F_{PH,j}(obs)|^2) \times |F_P| \times A_{H,j} \times \cos\alpha_P$$
$$+ 4 \times (|F_P|^2 + |F_{H,j}|^2 - |F_{PH,j}(obs)|^2) \times |F_P| \times B_{H,j} \times \sin\alpha_P$$
$$+ 2 \times |F_P|^2 \times (A_{H,j}^2 - B_{H,j}^2) \cos 2\alpha_P$$
$$+ 4 \times |F_P|^2 \times A_{H,j} \times B_{H,j} \times \sin 2\alpha_P$$

Comparison of this equation with Eqs. (14.1) and (14.2) gives

$$A_{iso,j} = -\frac{2(|F_P|^2 + |F_{H,j}|^2 - |F_{PH,j}(obs)|^2) \times |F_P|}{(E_j')^2} \times A_{H,j}$$

$$B_{iso,j} = -\frac{2(|F_P|^2 + |F_{H,j}|^2 - |F_{PH,j}(obs)|^2) \times |F_P|}{(E_j')^2} \times B_{H,j}$$

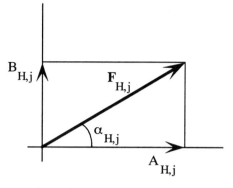

Figure 14.1. The separation of $\mathbf{F}_{H,j}$ into its components: $A_{H,j} = |F_{H,j}| \cos \alpha_{H,j}$ and $B_{H,j} = |F_{H,j}| \sin \alpha_{H,j}$.

$$C_{iso,j} = -\frac{|F_P|^2}{(E_j')^2} \times (A_{H,j}^2 - B_{H,j}^2)$$

$$D_{iso,j} = -\frac{2|F_P|^2}{(E_j')^2} \times A_{H,j} \times B_{H,j}$$

14.3. Phase Information from Anomalous Scattering

The improvement of protein phase angles by incorporating information from anomalous scattering has already been discussed in Section 9.3.

$$P_{ano}(\alpha_P) = N' \exp\left[-\frac{\varepsilon_{ano}^2(\alpha_P)}{2(E')^2}\right]$$

[see Eq. (9.2)] with $\varepsilon_{ano}(\alpha_P) = \Delta PH(obs) - \Delta PH(calc)$ and

$$\Delta PH(obs) = |F_{PH}(+)| - |F_{PH}(-)|$$

$$\Delta PH(calc) = -\frac{2 \times |F_P| \times |F_H|}{k \times |F_{PH}|} \sin(\alpha_H - \alpha_P)$$

(Section 9.3)

$$\varepsilon_{ano}(\alpha_P) = |F_{PH}(+)| - |F_{PH}(-)| + \frac{2|F_P| \times |F_H|}{k \times |F_{PH}|} \sin(\alpha_H - \alpha_P)$$

$$= \underline{\hspace{1.5cm}} \Delta \underline{\hspace{1.5cm}} + \kappa \sin(\alpha_H - \alpha_P)$$

with

$$\Delta = |F_{PH}(+)| - |F_{PH}(-)| \quad \text{and} \quad \kappa = \frac{2|F_P| \times |F_H|}{k \times |F_{PH}|}$$

$$\varepsilon_{ano}^2(\alpha_P) = \Delta^2 + 2 \times \Delta \times \kappa \sin(\alpha_H - \alpha_P) + \kappa^2 \sin^2(\alpha_H - \alpha_P)$$

As before, terms with α_P only, should be separated.

$$\sin(\alpha_H - \alpha_P) = \sin\alpha_H \cos\alpha_P - \cos\alpha_H \sin\alpha_P$$

$$\sin^2(\alpha_H - \alpha_P) = \tfrac{1}{2}\{1 - \cos 2(\alpha_H - \alpha_P)\}$$
$$= \tfrac{1}{2} - \tfrac{1}{2}\cos 2\alpha_H \cos 2\alpha_P - \tfrac{1}{2}\sin 2\alpha_H \sin 2\alpha_P$$

$$\varepsilon_{ano}^2(\alpha_P) = \Delta^2 + \tfrac{1}{2}\kappa^2 + (2 \times \Delta \times \kappa \sin\alpha_H)\cos\alpha_P$$
$$- 2 \times \Delta \times \kappa \cos\alpha_H \sin\alpha_P$$
$$- \tfrac{1}{2} \times \kappa^2 \cos 2\alpha_H \cos 2\alpha_P$$
$$- \tfrac{1}{2} \times \kappa^2 \sin 2\alpha_H \sin 2\alpha_P$$

$$-\frac{\varepsilon_{ano}^2(\alpha_P)}{2(E')^2} = -\frac{\Delta^2 + \tfrac{1}{2}\kappa^2}{2(E')^2} \qquad\qquad (= K_{ano})$$

$$-\frac{\Delta \times \kappa \sin\alpha_H}{(E')^2}\cos\alpha_P \qquad (= A_{ano}\cos\alpha_P)$$

$$+\frac{\Delta \times \kappa \cos\alpha_H}{(E')^2}\sin\alpha_P \qquad (= B_{ano}\sin\alpha_P)$$

$$+\frac{\kappa^2 \cos 2\alpha_H}{4(E')^2}\cos 2\alpha_P \qquad (= C_{ano}\cos 2\alpha_P)$$

$$+\frac{\kappa^2 \sin 2\alpha_H}{4(E')^2}\sin 2\alpha_P \qquad (= D_{ano}\sin 2\alpha_P)$$

14.4. Phase Information from MAD

Pähler et al. (1990) derived the phase probability function from measurements with MAD. For a specific observation j, made at a particular λ, they find for A_j, B_j, C_j and D_j:

$$A_j = Q\frac{R\cos\alpha_A - S\sin\alpha_A}{E^2}$$

$$B_j = Q\frac{R\sin\alpha_A - S\cos\alpha_A}{E^2}$$

$$C_j = \frac{(S^2 - R^2)\cos 2\alpha_A + 2RS\sin 2\alpha_A}{4E^2}$$

$$D_j = \frac{(S^2 - R^2)\sin 2\alpha_A - 2RS\cos 2\alpha_A}{4E^2}$$

with $Q = |^\lambda F(\mathbf{h})|^2 - |F_{BA}|^2 - p(\lambda)|F_A|^2$

$$R = q(\lambda)|F_{\text{BA}}||F_{\text{A}}| \text{ and } S = s(\mathbf{h})r(\lambda)|F_{\text{BA}}||F_{\text{A}}|$$

F_{BA}, F_{A}, p, q and r have the same meaning as in Section 9.4. E is the expectation value for errors at the true phase angles. $s(\mathbf{h}) = +1$ for $h\,k\,\ell$ and -1 for $\bar{h}\,\bar{k}\,\bar{\ell}$.

Coefficients for the complete MAD phase probability function are obtained by summation: $\sum_j A_j$, $\sum_j B_j$, $\sum_j C_j$ and $\sum_j D_j$.

14.5. Phase Information from Partial Structure Data, Solvent Flattening, and Molecular Averaging

For partial structure information as well as for solvent flattening and molecular averaging, we encountered the same form of the phase probability function:

$$P_{\text{SF}}(\alpha_{\text{P}}) = P_{\text{average}}(\alpha) = N \exp[X' \cos(\alpha_{\text{P}} - \alpha_{\text{calc}})]$$

[see Eq. (8.14)]

$$P_{\text{partial}}(\alpha_{\text{P}}) = N \exp[X \cos(\alpha_{\text{P}} - \alpha_{\text{partial}})]$$

see Eq. (8.9) in which $1/2\pi I_0(X)$ is now replaced by the normalizing constant N.

The exponential term for phase combination is simply

$$\underset{= A_{\text{partial}}}{\underline{X \cos \alpha_{\text{partial}} \; \cos \alpha_{\text{P}}}} + \underset{= B_{\text{partial}}}{\underline{X \sin \alpha_{\text{partial}} \; \sin \alpha_{\text{P}}}}$$

or

$$\underset{= A_{\text{SF}}}{\underline{X' \cos \alpha_{\text{SF}} \; \cos \alpha_{\text{P}}}} + \underset{= B_{\text{SF}}}{\underline{X' \sin \alpha_{\text{SF}} \; \sin \alpha_{\text{P}}}}$$

or

$$\underset{= A_{\text{average}}}{\underline{X' \cos \alpha_{\text{average}} \; \cos \alpha_{\text{P}}}} + \underset{= B_{\text{average}}}{\underline{X' \sin \alpha_{\text{average}} \; \sin \alpha_{\text{P}}}}$$

Summary

The major advantage of the Hendrickson and Lattman formalism is that one need not calculate phase probability distributions afresh every time some new information is added to the protein phases. The new information can easily be combined with previous information by simple addition to the coefficients A, B, C, and D in the general phase probability distribution $P(\alpha) = N \exp[A \cos \alpha + B \sin \alpha + C \cos 2\alpha + D \sin 2\alpha]$.

Chapter 15
Checking for Gross Errors and Estimating the Accuracy of the Structural Model

15.1. Introduction

After the molecular model of the protein structure has been refined, it may still contain errors that have creeped into the model during the interpretation of the electron density map, particularly of the regions where the electron density is weak. Some of the errors are obvious and should cause immediate suspicion, for instance, the presence of left-handed helices can almost always be ruled out. Most of the available modeling programs allow regularization of geometry, but do not guarantee overall good quality of the final model. A very qualitative impression of the accuracy of the structural model can be obtained by inspection of the electron density map:

- the connectivity of the main chain and the side chains
- the bulging out of the carbonyl oxygen atoms from the main chain
- the interpretation of the side chain electron density.

15.2. R-Factors

A more quantitative impression of the accuracy of the structure is obtained from the various residual indices (see Appendix 2). The common crystallographic R-factor is

$$
R = \frac{\sum_{hkl} ||F_{obs}| - k|F_{calc}||}{\sum_{hkl} |F_{obs}|}
$$

where k is a scale factor. For acentric model structures with the atoms randomly distributed in the unit cell, $R = 0.59$. For structures refined to high resolution, for instance, 1.6 Å, the R-factor should not be much higher than 0.16. This R-factor is an overall number and does not indicate major local errors. More useful in this respect is $R_{real\ space}$ (Jones et al., 1991). It is obtained in the following way: The final electron density map is plotted on a grid G_1 and a calculated map on an identical grid G_2. This calculated map is obtained by a Gaussian distribution of electron density around the average position for each atom in the model, with the same temperature factor for all atoms. The two density sets are scaled together. Now the electron densities of separate residues, or groups of atoms in a residue, are selected on both grids and built on a grid $G_3(obs)$ and a grid $G_3(calc)$, respectively. For nonzero elements in the two G_3 grids R is calculated as

$$R_{real\ space} = \frac{\sum |\rho_{obs} - \rho_{calc}|}{\sum |\rho_{obs} + \rho_{calc}|}$$

It is plotted as a function of the residues along the polypeptide chain (Figure 15.1). The fitting of the main chain alone can be obtained by incorporating just the N, C(α), C(β), C, and O atoms in the calculation. The fitting of the side chains in the density can be checked by taking only the side chain atoms.

It has been shown that the normal crystallographic R-factor can reach surprisingly low values in the refinement of protein structural models that

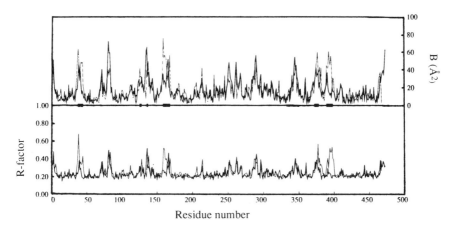

Figure 15.1. Real space R-factor (lower panel) and average B-factor (upper panel) of the *Azotobacter vinelandii* lipoamide dehydrogenase. Misplaced loops have been indicated by a thin line and after their correction by a thick line. Note the correspondence between the R- and the B-factor at the problem sites. (Source: Dr. Andrea Mattevi.)

appear later to be incorrect, for instance, because the number of model parameters is taken too high. Brünger (1992, 1993) suggested improving this situation with the introduction of a free R-factor, which is unbiased by the refinement process. In this method one divides the reflections into a "test set (T)" of unique reflections and a "working set (W)." The test set is a random selection of 5–10% of the observed reflections. Noncrystallographic symmetry causes a correlation between certain reflections. It must be avoided that reflections in the test set are correlated with reflections in the working set. To assure that correlated reflections are either in the working set or in the test set, the reflections in a number of very thin shells in reciprocal space are taken as the test set (Rees, 1983; Kleywegt and Jones, 1995; Kleywegt et al., 1996). The refinement is carried out with the working set only, and the free R-factor is calculated with the test set of reflections only:

$$ R^{\text{free}} = \frac{\displaystyle\sum_{hkl \subset T} ||F_{\text{obs}}| - k|F_{\text{calc}}||}{\displaystyle\sum_{hkl \subset T} |F_{\text{obs}}|} $$

where $hkl \subset T$ means all reflections belonging to test set T. Brünger could show that a high correlation exists between the free R-factor and the accuracy of the atomic model phases. Separating the working set and the test set is called "cross-validation" (Brünger, 1997b). The underlying principle is that if a structure is really improved in a refinement step, both $R_{\text{working set}}$ and R_{free} will decrease. If, however, $R_{\text{working set}}$ decreases as a result of fitting to noise, R_{free} will not decrease but increase. For instance, it can happen that too many peaks in the electron density map are assigned to water molecules, whereas in fact they are noise peaks and should be removed in the refinement. Applications of R_{free} are reviewed by Kleywegt and Brünger (1996).

15.3. The Ramachandran Plot

The stereochemistry of the main chain folding can be investigated with a Ramachandran plot (Ramachandran et al., 1963) in which the dihedral angles ϕ and Ψ for each residue are plotted in a square matrix (Figure 15.2). It is customary to have the conformation of the fully extended chain in the corners of the square. Short contacts between atoms of adjacent residues prevent ϕ and Ψ from taking on all possible angles between $-180°$ and $+180°$. They are clustered in regions in the Ramachandran matrix, with boundaries depending on the choice of the permitted van der Waals distances and tetrahedral angles. Usually a conservative and a more relaxed boundary are given, as in Figure 15.2. For highly refined structures nearly all the ϕ/Ψ values do lie within the allowed regions.

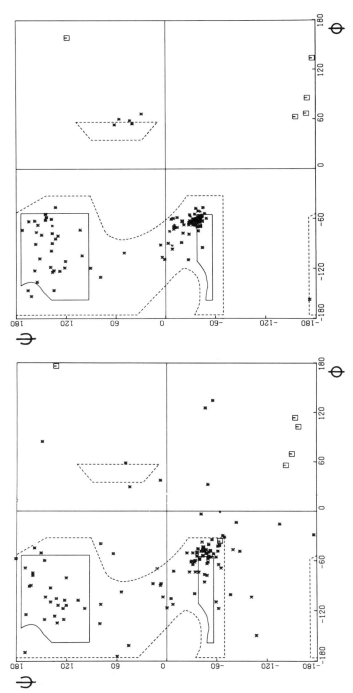

Figure 15.2. Ramachandran plot of the polypeptide chain conformation in phospholipase A_2 at 1.7 Å resolution. Glycine is denoted by open squares, all other residues by asterisks. Continuous lines enclose regions in which the tetrahedral angle $(N, C_\alpha, C) = 110°$ and broken lines enclose regions with the angle $(N, C_\alpha, C) = 115°$. The left figure is for the structural model before the refinement and the right one after the refinement. The convention for the sense of the ϕ and Ψ angle rotation is the following: the angles are positive for a right-handed rotation: when looking along any bond from N to C_α and from C_α to C, the far end rotates clockwise relative to the near end, as is indicated by the arrows in Figure 15.3.

Due to the lack of a side chain, glycyl residues can adopt a larger range of ϕ and Ψ angles. The electron density for nonglycine residues lying outside the allowed region should be carefully checked.

15.4. Stereochemistry Check

Unusual ω angles (Figure 15.3) and eclipsed dihedral angles in side chains should cause suspicion. Another feature to observe is whether the structure shows more than a few unsatisfied H-bonds. If it does, this would be energetically extremely unfavorable. Attention should also be given to residues or parts of residues with conspicuously high B-values as well as to unpaired charged residues in the interior of the molecule and to abnormally close van der Waals contacts. Cooperation among several laboratories resulted in a number of programs for the validation of macromolecular structures: PROCHECK (Laskowski et al., 1993a, 1993b; Mac Arthur et al., 1994), WHAT_CHECK (Hooft et al., 1996), and SURVOL (Wodak et al., 1995). The first two programs are now routinely used in the majority of all X-ray and NMR laboratories. PROCHECK makes a very detailed analysis of all geometric aspects of proteins. Bond lengths, bond angles, planarities, and so on are compared with ideal values as determined from an analysis of peptide structures extracted from the Cambridge Crystallographic Data Centre (http://www.ccdc.cam.ac.uk). Torsion angles, such as φ, ψ, and ω (Figure 15.3) are compared with values observed in high resolution protein structures in the Protein Data Bank (PDB; http://rutgers.rcsb.org).

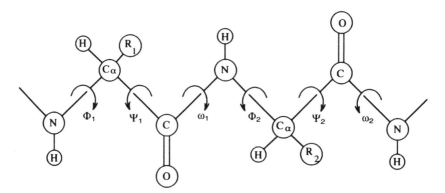

Figure 15.3. A schematic picture of a fully stretched polypeptide chain. The rotations around the C_α–NH bonds are given by the angle ϕ and around the C_α–C=O bonds by the angle Ψ. The peptide planes are usually flat with ω = 180°.

WHAT_CHECK provides a whole battery of checks, ranging from administrative checks, as space groups, V_M (Section 3.9), and so on, via geometric checks similar to the ones in PROCHECK, to a series of normality determinations for rotamers, peptide flip likelihood, and so on. SURVOL compares atomic volumes with a database of average atomic volumes determined from high resolution PDB files.

Another method based on existing knowledge was proposed by Sippl (1993) and is incorporated in the program PROSA. It is a threading program which determines whether the aminoacid sequence is correctly threaded through the density. It is useful in the early stages of density fitting but not for a detailed atomic analysis of the final structure.

15.5. The 3D–1D Profile Method

An interesting method for checking the quality of a protein molecular model has been developed by Eisenberg and co-workers (Bowie et al., 1991; Lüthy et al., 1992; Wilmanns and Eisenberg, 1993). In this method, a so-called 3D profile is compared with the amino acid sequence of the protein. The 3D profile is obtained in the following way. Each residue in the chain is assigned to one of six classes of side chain environment. These classes are determined by two parameters (see Figure 15.4): (1) the area of the residue that is buried and (2) the fraction of side chain area that is covered by polar atoms (O and N). In addition to the six classes of

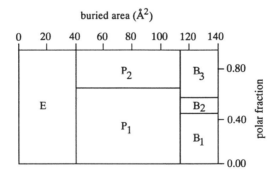

Figure 15.4. The six side chain environment categories in the Eisenberg profile method. A residue was assigned to category E (exposed) if less than $40 \, \text{Å}^2$ of the side chain was buried. For a buried area between 40 and $114 \, \text{Å}^2$, the residue was placed in environment category P_1(polar) if the polar fraction was smaller than 0.67, and in P_2 if the polar fraction was 0.67 or larger, where the polar fraction is the fraction of the side chain area covered by polar atoms. For buried side chains, side chains with a buried area of more than $114 \, \text{Å}^2$, three classes were distinguished: B_1 with a polar fraction smaller than 0.45, B_2 with a polar fraction between 0.45 and 0.58, and B_3 with a polar fraction of 0.58 and larger.

side chain environment, three classes of local secondary structure are distinguished: α-helix, β-sheet, and others. Together there are $6 \times 3 = 18$ classes. On the basis of a number of well-refined three-dimensional structures a standard matrix has been constructed giving the 3D–1D score for every type of amino acid residue in each of the 18 classes. For each residue in the polypeptide chain, the 3D–1D score is read from the matrix. The overall 3D score S for the compatibility of the model with the sequence is the sum of the 3D–1D scores for all residues in the chain (Figure 15.5). A high score is found for a correct structure, a low one for an incorrect structure. A low one means that the residue is in a class where it is not frequently found. Since S depends on the length of the polypeptide chain, it should be compared for proteins with the same length. Besides the overall 3D–1D score S, a profile score for each position in the polypeptide chain can be determined to locate improperly built segments in the 3D model. This score is calculated as the average of the 3D–1D scores for 21 residues in a window with the particular residue in the center of that window (Figure 15.6).

The weakness of the method is that the division into discrete classes means that, because of sharp boundaries, the residue preferences can change dramatically with even extremely small changes in buried area (b) or in polarity (p) of the residue. Therefore, a continuous representation of the residue preferences as a function of the buried area and the polar fraction was introduced (Kam and Eisenberg, 1994).

residue number	residue type	class	3D–1D score
....	....	....	....
46	T	Eα	-0.17
47	D	E	0.22
48	Y	P1	-1.31
49	L	B1	1.10
....	....	....	....

$$+ \underline{\hspace{4cm}}$$

$$S = \text{total score}$$

Figure 15.5. This figure lists four residues of an arbitrary polypeptide chain. In column 3 the class to which each residue belongs is given and in column 4 the 3D–1D score that is derived from the standard matrix. The total score S is obtained as the sum over all residues of the 3D–1D scores.

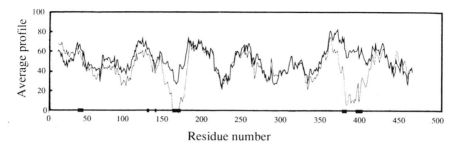

Figure 15.6. Plot of the average 3D–1D score of the first subunit in *Azotobacter vinelandii* lipoamide dehydrogenase before (thin line) and after (thick line) corrections to the model versus residue number (profile score). Because a 21 residue sliding window was used, the scores for the first and last 10 residues have no meaning and were omitted. A score below 0.20 indicates a bad part of the model. Note the improvement in the model after application of the corrections. (Source: Dr. Andrea Mattevi.)

$P(i; j,b,p)$ is the conditional probability that a residue i in a secondary structure state j occurs in one of the 6 environment classes determined by b and p. They define an information value $S_{i,j}^{o}(b,p)$ as

$$S_{i,j}^{o}(b, p) = \ln\left[\frac{P(i; j, b, p);}{P(i)} \right]$$

$P(i)$ is the a priori probability of the occurrence of residue i. $P(i)$ is known for each type of amino acid and $P(i; j,b,p)$ can be read from the standard matrix. Therefore, $S_{i,j}^{o}(b,p)$ is a function of the discrete values of b and p. This function is smoothened to a two-dimensional function $S_{i,j}(b,p)$ that best represents the observed values $S_{i,j}^{o}(b,p)$. A 3D profile can be constructed from the continuous representation of residue preferences instead of the original discrete representation.

The advantage of the profile method is that it is completely independent of any assumption introduced into the model construction, because it depends exclusively on the compatibility of the model with its own amino acid sequence.

15.6. Quantitative Estimation of the Coordinate Error in the Final Model

The indicators so far mentioned do not give a value for the error in the atomic coordinates of the molecular model. An estimation of the average value of this error, $\overline{|\Delta r|}$, can be obtained by methods proposed by, Luzzati (1952) and Read (1986, 1990b). Luzzati has derived a relationship between the average error $\overline{|\Delta r|}$ in the atomic coordinates and the difference between $|F_{obs}|$ and $|F_{calc}|$, as expressed in the crystallographic reliability factor R. The R-factor is plotted as a function of $(\sin\theta)/\lambda$ and this curve is compared with a family of calculated lines, which are functions of $\overline{|\Delta r|} \times (\sin\theta)/\lambda$. The members of the family are calculated for different values of $\overline{|\Delta r|}$ and from the line that is closest to the experimental curve, $\overline{|\Delta r|}$ for the crystal structure is derived. The assumption is that the difference between $|F_{obs}|$ and $|F_{calc}|$ is due exclusively to errors in the positional coordinates of the atoms. The mathematics of the Luzzati method is rather complicated and we shall not go into any details. For an example see Figure 15.7. Because the crystallographic R-factor is a less reliable indicator for the accuracy of the model than R_{free} of the test set, it has been suggested that R_{free}–values be used instead of R-values in a Luzzati plot (Kleywegt et al., 1994; Brünger, 1997b). In the method proposed by Read (1986), which is based on previous work by Luzzati and others, for instance Srinivasan and Ramachandran (1965), the root mean square value of the coordinate error $\sqrt{(\Delta|r|)^2}$ is obtained in a graphic way; the difference is that now a single plot is sufficient, the σ_A plot, based on Eq. (15.1):

$$\ln\sigma_A = \frac{1}{2}\ln\left(\frac{\Sigma_P}{\Sigma_N}\right) - \frac{8}{3}\pi^2\overline{(|\Delta r|)}^2\left(\frac{\sin\theta}{\lambda}\right)^2 \qquad (15.1)$$

σ_A is defined as

$$\sigma_A = D\left(\frac{\Sigma_P}{\Sigma_N}\right)^{1/2}$$

where $\Sigma_N = \Sigma_{j=1}^{N}f_j^2$. The summation is over all N atoms in the structure for Σ_N and over all atoms in the partially known structure for Σ_P. Σ_P/Σ_N

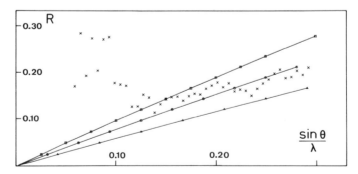

Figure 15.7. Plot of the crystallographic R-factor as a function of $(\sin\theta)/\lambda$ (crosses). Superimposed are calculated Luzzati lines for a coordinate error $\overline{|\Delta r|}$ of 0.08 Å (triangles), 0.12 Å (circles), and 0.16 Å (squares). (Reproduced with permission from Dijkstra et al. (1981).)

should be 1 if the structure is completely identified. But this is never true because of rather disordered solvent atoms in the crystal. These atoms contribute appreciably to low order reflections that should, therefore, be ignored. D is the Fourier transform of the probability distribution $p(\Delta \mathbf{r})$ of $\Delta \mathbf{r}$:

$$D(\mathbf{S}) = \int p(\Delta \mathbf{r}) \, \exp[2\pi i(\Delta \mathbf{r}) \cdot \mathbf{S}] \, d\Delta \mathbf{r}$$

D is in general complex but a centrosymmetric distribution for $p(\Delta \mathbf{r})$ is assumed and, therefore, D can be written as

$$D(\mathbf{S}) = \int p(\Delta \mathbf{r}) \, \cos[2\pi(\Delta \mathbf{r}) \cdot \mathbf{S}] \, d\Delta \mathbf{r} \qquad (15.2)$$

If $\ln \sigma_A$ is plotted vs. $\{(\sin \theta)/\lambda\}^2$ a straight line is obtained with slope $\{-\frac{8}{3}\pi^2 (|\Delta r|)^2\}$ and intercept of $\frac{1}{2}\ln(\Sigma_P/\Sigma_N)$ [see Eq. 15.1]. σ_A is obtained from an independent estimate. Different methods are available. They are discussed by Read (1986). It can, for example, be done with a method suggested by Hauptman (1982). He regarded σ_A as the square root of the correlation coefficient between the observed and calculated normalized structure factors:

$$\sigma_A = \left[\frac{\sum(|E(\text{obs})|^2 - \overline{|E(\text{obs})|^2}) \, (|E(\text{calc})|^2 - \overline{|E(\text{calc})|^2})}{\{\sum(|E(\text{obs})|^2 - \overline{|E(\text{obs})|^2})^2 \, \sum(|E(\text{calc})|^2 - \overline{|E(\text{calc})|^2})^2\}^{1/2}} \right]^{1/2}$$

with $\overline{|E(\text{obs})|^2} = \overline{|E(\text{calc})|^2} = 1$. It was shown by Read that this estimated value of σ_A can be refined by finding the zero of the residual function:

$$R = \sum w(\sigma_A - m|E_N| |E_P^c|)$$

w is 1 for centric and 2 for noncentric reflections; m is the figure of merit. An alternative method is given by Srinivasan and Chandrasekaran (1966). They give the following expression for σ_A:

$$\sigma_A = \overline{|E(\text{obs})| \, |E(\text{calc})| \, \cos[\alpha(\text{obs}) - \alpha(\text{calc})]}$$

Because D, and therefore also σ_A, is a function of resolution, σ_A must be estimated in ranges of resolution. As the example in Figure 15.8 shows, the data points at lower resolution do not fit the line. The same is true for the points at high resolution, which is probably due to measurement errors.

For well-refined structures the Luzzati method and the σ_A plot give nearly invariably estimated errors in the coordinates between 0.2 and 0.3 Å.

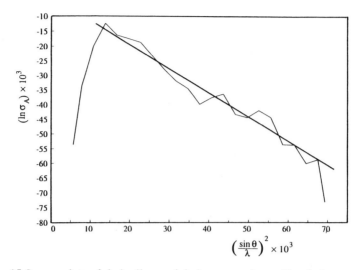

Figure 15.8. σ_A plot of haloalkane dehalogenase from *Xanthobacter autotrophicus*. The structure was determined to a resolution of 2.5 Å. (Source: Dr. K.H.G. Verschueren.)

The Derivation of the σ_A Plot

We start with Eq. (15.2) and must first of all find $p(\Delta r)$ as a function of $|\Delta r|$. If the assumption is made that $P(\Delta r)$ is spherically symmetric (does not depend on the direction of Δr) and moreover that $|\Delta r|$ is distributed as a Gaussian function:

$$p(\Delta \mathbf{r}) = N \exp\left[-\frac{|\Delta r|^2}{2\sigma^2}\right]$$

in which N is the normalization constant and σ the standard deviation. N can be obtained from

$$\int_0^\infty p(\Delta \mathbf{r})\, d(\Delta \mathbf{r}) = 1$$

$$N \int_0^\infty \exp\left[-\frac{|\Delta r|^2}{2\sigma^2}\right] 4\pi |\Delta r|^2\, d|\Delta r| = (2\pi\sigma^2)^{3/2} \times N = 1$$

$$N = (2\pi\sigma^2)^{-3/2} \quad \text{and} \quad p(\Delta \mathbf{r}) = (2\pi\sigma^2)^{-3/2} \exp\left[-\frac{|\Delta r|^2}{2\sigma^2}\right] \quad (15.3)$$

$$D(\mathbf{S}) = \int p(\Delta \mathbf{r}) \cos[2\pi(\Delta \mathbf{r}) \cdot \mathbf{S}]\, d\Delta \mathbf{r}$$

$$= (2\pi\sigma^2)^{-3/2} \int \exp\left[-\frac{|\Delta r|^2}{2\sigma^2}\right] \cos[2\pi(\Delta \mathbf{r} \cdot \mathbf{S})]\, d\Delta \mathbf{r}$$

Replace $\Delta \mathbf{r}$ and $\mathbf{S}$ by their orthogonal components:

$$\Delta \mathbf{r} = \Delta \mathbf{x} + \Delta \mathbf{y} + \Delta \mathbf{z} \quad \text{and} \quad \mathbf{S} = \mathbf{h} + \mathbf{k} + \mathbf{l}$$

$$(\Delta \mathbf{r} \cdot \mathbf{S}) = h \times \Delta x + k \times \Delta y + l \times \Delta z$$

$$\cos 2\pi (\Delta \mathbf{r} \cdot \mathbf{S}) = \cos 2\pi (h \times \Delta x + k \times \Delta y + l \times \Delta z)$$
$$= \cos(2\pi h \times \Delta x) \cos(2\pi k \times \Delta y) \cos(2\pi l \times \Delta z)$$

All sin terms have vanished because of the spherical symmetry of $\Delta \mathbf{r}$.

$$\int\limits_{\text{all } \Delta r} = \int\limits_{\Delta x=-\infty}^{+\infty} \int\limits_{\Delta y=-\infty}^{+\infty} \int\limits_{\Delta z=-\infty}^{+\infty} = 8 \times \int\limits_{\Delta x=0}^{+\infty} \int\limits_{\Delta y=0}^{+\infty} \int\limits_{\Delta z=0}^{+\infty}$$

$$D = 8(2\pi\sigma^2)^{-3/2} \int\limits_{\Delta x=0}^{\infty} \int\limits_{\Delta y=0}^{\infty} \int\limits_{\Delta z=0}^{\infty} \exp\left[-\frac{(\Delta x)^2 + (\Delta y)^2 + (\Delta z)^2}{2\sigma^2}\right]$$

$$\times \cos(2\pi h \times \Delta x) \cos(2\pi k \times \Delta y) \cos(2\pi l \times \Delta z)\, d(\Delta x)\, d(\Delta y)\, d(\Delta z)$$

Using the fact that the definite integral

$$\int\limits_0^{\infty} \exp[-a^2 x^2] \cos bx \, dx = \frac{\sqrt{\pi}}{2a} \exp\left[-\frac{b^2}{4a^2}\right]$$

it follows that $D = \exp[-2\pi^2 |S|^2 \sigma^2]$. $\hspace{2cm}$ (15.4)

Next the relation between σ^2 and $\overline{|\Delta r|^2}$ must be found.

$$\overline{|\Delta r|^2} = \int\limits_0^{\infty} p(|\Delta r|)(|\Delta r|)^2 \, d|\Delta r| \hspace{2cm} \text{[See Eq. (15.3)]}$$

$$= (2\pi\sigma^2)^{-3/2} \times 4\pi \int\limits_0^{\infty} (|\Delta r|)^2 \times (|\Delta r|)^2 \times \exp\left[-\frac{|\Delta r|^2}{2\sigma^2}\right] d|\Delta r|$$

$$= (2\pi\sigma^2)^{-3/2} \times 4\pi \int\limits_0^{\infty} (|\Delta r|)^4 \times \exp\left[-\frac{|\Delta r|^2}{2\sigma^2}\right] d|\Delta r|$$

Knowing that $\int\limits_0^{\infty} x^{2n} \exp[-ax^2] dx = \dfrac{1 \times 3 \times 5 \times ... (2n-1)}{2^{n+1} a^n} \sqrt{\dfrac{\pi}{a}}$

we obtain $\sigma^2 = \dfrac{1}{3}\overline{|\Delta r|^2}$ and $D = \exp\left[-\dfrac{8}{3}\pi^2 \overline{|\Delta r|^2}\left(\dfrac{\sin\theta}{\lambda}\right)^2\right]$

$$\sigma_A = D\left(\frac{\sum P}{\sum N}\right)^{\frac{1}{2}} \text{ and } \ln \sigma_A = \frac{1}{2}\ln\left(\frac{\sum P}{\sum N}\right) - \frac{8}{3}\pi^2 \overline{|\Delta r|^2}\left(\frac{\sin\theta}{\lambda}\right)^2$$

In a later paper Read (1990b) showed that his (and Luzzati's) assumption of a Gaussian Δr distribution, equal for all atoms, is in general not allowed. As a consequence, the Luzzati plot and the σ_A plot can, in principle, be used only for comparative, rather than absolute measures of the coordinate errors. However, from the comparison of independently determined structures it seems that, somewhat surprisingly, the σ_A plot gives reasonable results although the actual errors may be somewhat larger. For instance, Daopin et al. (1994) compared two crystal structures of transforming growth factor (TGF-β2) and found that the r.m.s. difference between the two structures was 0.33 Å, whereas 0.21 Å was obtained from the Luzzati plot and 0.18 Å from the σ_A plot.

Ohlendorf (1994) came to the conclusion that the "errors" derived from the comparison of four independently refined models of human interleukin 1β could be partly due to real differences in conformation.

Summary

In the literature a number of protein structures have been presented that afterward turned out to be entirely or partly wrong. Apparently the density map was incorrectly interpreted. It is not quaranteed that interpretation errors will be removed in the refinement process and the best way to follow is to apply the methods presented in this chapter to check the accuracy of the model. If any suspicion is raised against part of the model, it should be carefully checked for alternative conformations. The Luzzati method and the σ_A plot give an estimate of the coordinate error. Because of underlying assumptions, which are not always true, they can, in principle, be used only for comparative, rather than absolute measures of the coordinate errors. However, from the comparison of independently determined structures it seems that the σ_A plot gives reasonable results. It is clear that a better method for the estimation of errors in a protein structure is needed, but as long as that is not available the best one can do is to use the σ_A plot.

Appendix 1
A Compilation of Equations for Calculating Electron Density Maps

Straightforward Electron Density Map

$$\rho(x\ y\ z) = \frac{1}{V}\sum_{hkl}|F(h\ k\ l)|\ \exp[-2\pi i(hx + ky + lz) + i\alpha\ (h\ k\ l)]$$

$$= \frac{1}{V}\sum_{hkl}|F(h\ k\ l)|\ \cos[2\pi(hx + ky + lz) - \alpha(h\ k\ l)]$$

The $|F(h\ k\ l)|$s are the structure factor amplitudes of the reflections $(h\ k\ l)$.

Difference Electron Density Map

$$\Delta\rho\ (x\ y\ z) = \frac{1}{V}\sum_{hkl}\Delta|F(h\ k\ l)|_{iso}\ \exp[-2\pi i(hx + ky + lz) + i\alpha_P(h\ k\ l)]$$

$$= \frac{1}{V}\sum_{hkl}\Delta|F(h\ k\ l)|_{iso}\ \cos[2\pi(hx + ky + lz) - \alpha_P(h\ k\ l)]$$

$\Delta|F(h\ k\ l)|_{iso}$ is the difference between the structure factor amplitudes for the protein and some isomorphous derivative of that protein. The phase angles $\alpha_P(h\ k\ l)$ are those of the native protein. The map shows the electron density, which is extra (or which is missing) in the derivative at half the actual height.

A $2F_{obs} - F_{calc}$ Map

$$\rho(x\ y\ z) = \frac{1}{V}\sum_{hkl}(2|F_{obs}| - |F_{calc}|)\ \exp[-2\pi i(hx + ky + lz) + i\alpha_{calc}]$$

This map can be regarded as the sum of the electron density of a model and of a difference electron density map at double height. It shows, besides the electron density of the model, the difference between the actual structure and the model at normal height. The phase angles are those calculated for the model.

A Residual, or Double Difference, Electron Density Map

$$\Delta\rho(x\ y\ z) = \frac{1}{V}\sum_{hkl}\{|F_{obs}| - |\mathbf{F}_{native} + \mathbf{F}_{attached}|\}$$
$$\times \exp[-2\pi i(hx + ky + lz) + i\alpha_{calc}]$$

$|F_{obs}|$ is the structure factor amplitude for the derivative. $\mathbf{F}_{attached}$ is the structure factor contribution by those attached atoms or groups of atoms for which the parameters are already known. The phase angles α_{calc} are for the native protein and the attached heavy atoms. This is a useful Fourier summation for the detection of extra attached atoms or groups of atoms.

An OMIT Map

$$\Delta\rho(x\ y\ z) = \frac{1}{V}\sum_{hkl}(|F_{obs}| - |F_{calc}|)\ \exp[-2\pi i(hx + ky + lz) + i\alpha_{calc}]$$

F_{calc} is the structure factor of a partial model, that is a model from which a fragment has been deleted. The phase angles α_{calc} are for the model with fragment deleted. It is a difference Fourier summation that is often used if part of the electron density map cannot be interpreted satisfactorily. This part is then deleted in the model and does not contribute to the phase angle calculation. The map should show the density corresponding to the fragment, at half the height. Alternatively one can calculate

$$\rho(x\ y\ z) = \frac{1}{V}\sum_{hkl}|F_{obs}|\ \exp[-2\pi i(hx + ky + lz) + i\alpha_{calc}]$$

This map should show the entire model with the deleted fragment at half height. Or with coefficients $2|F_{obs}| - |F_{calc}|$, which also shows the entire model but the deleted fragment at full height.

A Simulated Annealing OMIT Map

This is an OMIT map with simulated annealing applied to the "known" part of the structure. It is also called: Composite Annealed OMIT Map.

An OMIT Map with Sim Weighting

$$\Delta\rho(x \, y \, z) = \frac{1}{V}\sum_{hkl}m(|F_{obs}| - |F_{calc}|) \exp[-2\pi i(hx + ky + lz) + i\alpha_{calc}]$$

$m = I_1(X)/I_0(X)$ for noncentric reflections and $m = \tanh (X/2)$ for centric reflections, where

$$X = \frac{2|F_{obs}| \times |F_K|}{\sum\limits_{1}^{n}f_i^2}$$

I_0 are I_1 are modified Bessel functions of order zero and one, respectively, $|F_{obs}|$ is the observed structure factor amplitude, and $|F_K|$ is the amplitude for the known part of the structure. The f_is are the atomic scattering factors for the n missing atoms. It is assumed that the partial structure is error-free. In practice this will not be true and then X must be taken as

$$X = \frac{2\sigma_A|E_{obs}| \times |E_K|}{1 - \sigma_A^2}$$

$|E|$ is the normalized structure factor amplitude. σ_A is defined in Section 15.6.

A Weighted Electron Density Map Calculated with Phase Angles α_{calc} from the Partial Structure

$$\rho(x \, y \, z) = \frac{1}{V}\sum_{hkl}(2m|F_{obs}| - |F_{calc}|) \exp[- 2\pi i(hx + ky + lz) + i\alpha_{calc}]$$

for noncentric reflections and

$$\rho(x \, y \, z) = \frac{1}{V}\sum_{hkl}m|F_{obs}| \exp [-2\pi i(hx + ky + lz) + i\alpha_{calc}]$$

for centric reflections. This map is an improvement over the $(2F_{obs} - F_{calc})$ map because it applies Sim weighting to the observed structure factor amplitudes (see above). Possible missing parts in the structure will show up more clearly in the electron density map than without Sim weighting. Sim assumed the partial structure to be error-free, but in practice this is never true. The effect of these errors is taken care of by defining X differently:

$$X = \frac{2\sigma_A|E_{obs}| \times |E_K|}{1 - \sigma_A^2}$$

and by weighting down $|F_{calc}|$:

$$\rho(x\ y\ z) = \frac{1}{V}\sum_{hkl}(2m|F_{obs}| - D|F_{calc}|)\ \exp[-\ 2\pi i(hx + ky + lz) + i\alpha_{calc}]$$

with $D = \sigma_A(\Sigma_P/\Sigma_N)^{-1/2}$ (Section 15.6). $|E|$ is the normalized structure factor amplitude.

Appendix 2
A Compilation of Reliability Indices

Common Crystallographic R-Factor for Indicating the Correctness of a Model Structure

$$R = \frac{\sum_{hkl} ||F_{\text{obs}}| - k|F_{\text{calc}}||}{\sum_{hkl} |F_{\text{obs}}|}$$

The Free R-Factor

$$R_{\text{free}} = \frac{\sum_{hkl \subset T} ||F_{\text{obs}}| - k|F_{\text{calc}}||}{\sum_{hkl \subset T} |F_{\text{obs}}|}$$

where $hkl \subset T$ means all reflections belonging to test set T of unique reflections. The refinement is carried out with the remaining reflections, the working set W. The advantage of using this R-factor over the regular crystallographic R-factor is that it is unbiased by the refinement process.

R-Factor for Comparing the Intensity of Symmetry-Related Reflections

$$R_{\text{sym}}(I) = \frac{\sum_{hkl} \sum_{i} |I_i(h\ k\ l) - \overline{I(h\ k\ l)}|}{\sum_{hkl} \sum_{i} I_i(h\ k\ l)}$$

for n independent reflections and i observations of a given reflection. $\overline{I(h\ k\ l)}$ is the average intensity of the i observations.

R-Factor for Comparing the Structure Factor Amplitude for Symmetry-Related Reflections

$$R_{\text{sym}}(F) = \frac{\sum\limits_{hkl}\sum\limits_{i} ||F_i(h\ k\ l)| - |\overline{F(h\ k\ l)}||}{\sum\limits_{hkl}\sum\limits_{i} |F_i(h\ k\ l)|}$$

for i observations each of n independent reflections. $|\overline{F(h\ k\ l)}|$ is the average value for the structure factor amplitude of the i observations of a given reflection.

R-Factor for the Comparison of N Data Sets after Merging

On $|F_{hkl}|$:

$$R_{\text{merge}} = \frac{\sum\limits_{hkl}\sum\limits_{j=1}^{N} ||F_{hkl}| - |F_{hkl}(j)||}{\sum\limits_{hkl} N \times |F_{hkl}|}$$

$|F_{hkl}|$ is the final value of the structure factor amplitude.

On I_{hkl}:

$$R_{\text{merge}} = \frac{\sum\limits_{hkl}\sum\limits_{j=1}^{N} |I_{hkl} - I_{hkl}(j)|}{\sum\limits_{hkl} N \times I_{hkl}}$$

Real Space R-Factor

$$R_{\text{real space}} = \frac{\sum |\rho_{\text{obs}} - \rho_{\text{calc}}|}{\sum |\rho_{\text{obs}} + \rho_{\text{calc}}|}$$

The function is calculated per residue for either all atoms, or the main chain atoms only, or the side chain atoms. The summation is over all grid points for which ρ_{calc} has a nonzero value for a particular residue. The function shows how good the fit is between the model and the electron density map.

$R_{Cullis}(iso)$

$$R_{Cullis}(iso) = \frac{\sum_{hkl} \|F_{PH} \pm F_P| - F_H(calc)\|}{\sum_{hkl} |F_{PH} \pm F_P|}$$

for centric reflections only. F_P, F_{PH}, and F_H include their sign ($+$ or $-$): $F_{PH} + F_P$ if the signs of F_{PH} and F_P are opposite and $F_{PH} - F_P$ if they are equal.

$R_{Cullis}(ano)$

$$R_{Cullis}(ano) = \frac{\sum_{hk\ell} \|\Delta F_{PH}^{\pm}(obs)| - |\Delta F_{PH}^{\pm}(calc)\|}{\sum_{hk\ell} \Delta F_{PH}^{\pm}(obs)}$$

where $\Delta F_{PH}^{\pm}$ (obs) is the structure factor amplitude difference between Bijvoet pairs and $\Delta F_{PH}^{\pm}(calc) = 2\dfrac{f''}{\Delta f}|F_H|\sin(\alpha_{PH} - \alpha_H)$ (Note that an alternative nomenclature for Δf is f').

$R_{Cullis}(\lambda)$

$$R_{Cullis}(\lambda) = \frac{\sum_{hk\ell} (|\mathbf{F}_P(\lambda_i) - \mathbf{F}_P(\lambda_o)| - \mathbf{F}(\lambda_i)(calc))}{\sum_{hk\ell} |\mathbf{F}_P(\lambda_i) - \mathbf{F}_P(\lambda_o)|}$$

where $\mathbf{F}_P(\lambda_i)$ is the structure factor of the protein at λ_i, and $\mathbf{F}_P(\lambda_0)$ at the parent λ; $\mathbf{F}(\lambda_i)(calc)$ is the calculated contribution by the anomalous scatterer with respect to the parent.

$R_{Kraut}(iso)$

$$R_{Kraut}(iso) = \frac{\sum_{hkl} \||F_{PH}| - |\mathbf{F}_P + \mathbf{F}_H(calc)\|}{\sum_{hkl} |\mathbf{F}_{PH}|}$$

This R-factor is used in isomorphous replacement methods to check the heavy atom refinement.

R_{Kraut}(ano)

$$R_{\text{kraut}}(\text{ano}) = \frac{\sum\limits_{hk\ell} \left| \left|F_{\text{PH}}^+(\text{obs})\right| - \left|F_{\text{PH}}^+(\text{calc})\right| \right| + \left| \left|F_{\text{PH}}^-(\text{obs})\right| - \left|F_{\text{PH}}^-(\text{calc})\right| \right|}{\sum\limits_{hk\ell} \left(\left|F_{\text{PH}}^+(\text{obs})\right| + \left|F_{\text{PH}}^-(\text{obs})\right| \right)}$$

Derivative R-Factor

$$R_{\text{deriv}} = R_{\text{iso}} = \frac{\sum\limits_{hkl} \left| \left|F_{\text{deriv}}(h\,k\,l)\right| - \left|F_{\text{native}}(h\,k\,l)\right| \right|}{\sum\limits_{hkl} \left|F_{\text{native}}\right|}$$

This R-factor is used for checking the quality of an isomorphous derivative.

$R_{\text{anomalous}}$

$$R_{\text{anomalous}} = \frac{\sum\limits_{hk\ell} \left| \left|F_{\text{PH}}^+\right| - \left|F_{\text{PH}}^-\right| \right|}{\sum\limits_{hk\ell} \dfrac{\left|F_{\text{PH}}^+\right| + \left|F_{\text{PH}}^-\right|}{2}}$$

R_{lambda}

$$R_{\text{lambda}} = \frac{\sum\limits_{hk\ell} \sum\limits_{i} \left| \left|F_P(\lambda_i)\right| - \left|F_P(\lambda_0)\right| \right|}{\sum\limits_{hk\ell} \left|F_P(\lambda_0)\right|}$$

where $|F_P(\lambda_i)|$ is the structure factor amplitude of the protein at λ_i, and $|F_P(\lambda_0)|$ at the parent λ.

Standard Linear Correlation Coefficient Between Observed and Calculated Structure Factor Amplitudes

$$C = \frac{\sum\limits_{h} (|F_h(\text{obs})| - \overline{|F_h(\text{obs})|}) \times (|F_h(\text{calc})| - \overline{|F_h(\text{calc})|})}{\left[\sum\limits_{h} (|F_h(\text{obs})| - \overline{|F_h(\text{obs})|})^2 \times \sum\limits_{h} (|F_h(\text{calc})| - \overline{|F_h(\text{calc})|})^2\right]^{1/2}}$$

The same but in a different form:

$$C = \frac{\overline{|F_h(\text{obs})| \times |F_h(\text{calc})|} - \overline{|F_h(\text{obs})|} \times \overline{|F_h(\text{calc})|}}{[\{\overline{|F_h(\text{obs})|^2} - (\overline{|F_h(\text{obs})|})^2\} \times \{\overline{|F_h(\text{calc})|^2} - (\overline{|F_h(\text{calc})|})^2\}]^{1/2}}$$

Standard Linear Correlation Coefficient Between two Electron Density Maps, $\rho_1(xyz)$ and $\rho_2(xyz)$

$$C = \frac{\sum \left(\rho_1(xyz) - \overline{\rho_1(xyz)}\right) \times \left(\rho_2(xyz) - \overline{\rho_2(xyz)}\right)}{\left[\sum \left(\rho_1(xyz) - \overline{\rho_1(xyz)}\right)^2 \times \sum \left(\rho_2(xyz) - \overline{\rho_2(xyz)}\right)^2\right]^{1/2}}$$

The same but in a different form:

$$C = \frac{\overline{\rho_1(xyz) \times \rho_2(xyz)} - \overline{\rho_1(xyz)} \times \overline{\rho_2(xyz)}}{\left[\left\{\overline{(\rho_1(xyz))^2} - \left(\overline{\rho_1(xyz)}\right)^2\right\} \times \left\{\overline{(\rho_2(xyz))^2} - \left(\overline{\rho_2(xyz)}\right)^2\right\}\right]^{1/2}}$$

The Phasing Power of Heavy Atoms in an Isomorphous derivative

Isomorphous phasing power: $\left[\dfrac{\sum\limits_{hk\ell} |F_H(\text{calc})|^2}{\sum\limits_{hk\ell} |E|^2}\right]^{1/2}$ with

$$\sum_{hk\ell} |E|^2 = \sum_{hk\ell} \{|F_{PH}(\text{obs})| - |F_{PH}(\text{calc})|\}^2$$

An alternative expression is: $\dfrac{\sum\limits_{hk\ell} |F_H(\text{calc})|}{\sum\limits_{hk\ell} |E|}$

The Phasing Power of Anomalously Scattering Atoms

Anomalous phasing power: $\left[\dfrac{\sum\limits_{hk\ell} |F_H(\text{imag.})|^2}{\sum\limits_{hk\ell} \left||\Delta F_{PH}^{\pm}(\text{obs})| - |\Delta F_{PH}^{\pm}(\text{calc})|\right|^2}\right]^{1/2}$

or alternatively: $\dfrac{\sum\limits_{hk\ell} |F_H(\text{imag.})|}{\sum\limits_{hk\ell} \left||\Delta F_{PH}^{\pm}(\text{obs})| - |\Delta F_{PH}^{\pm}(\text{calc})|\right|}$

where $\triangle F^{\pm}$ is the structure factor amplitude difference between Bijvoet pairs, and F_H (imag.) is the imaginary component of the calculated structure factor contribution by the anomalously scattering atoms.

$$\text{Dispersive phasing power:} \left[\frac{\sum\limits_{hk\ell} |F_H(\text{real})|^2}{\sum\limits_{hk\ell} \left| |\triangle F_{PH}^{\pm}(\text{obs})| - |\triangle F_{PH}^{\pm}(\text{calc})| \right|^2} \right]^{1/2}$$

$$\text{or alternatively:} \quad \frac{\sum\limits_{hk\ell} |F_H(\text{real})|}{\sum\limits_{hk\ell} \left| |\triangle F_{PH}^{\pm}(\text{obs})| - |\triangle F_{PH}^{\pm}(\text{calc})| \right|}$$

where $\triangle F^{\pm}$ is the structure factor amplitude difference between Bijvoet pairs, and $F_H(\text{real})$ is the real component of the calculated structure factor contribution by the anomalously scattering atoms.

Figure of Merit

The figure of merit for a given reflection $(h\,k\,l)$ is defined as

$$m = \frac{|F(h\ k\ l)_{\text{best}}|}{|F(h\ k\ l)|}$$

where

$$\mathbf{F}(h\ k\ l)_{\text{best}} = \frac{\sum\limits_{\alpha} P(\alpha)\mathbf{F}_{hkl}(\alpha)}{\sum\limits_{\alpha} P(\alpha)}$$

It can be shown that the figure of merit is the weighted mean of the cosine of the deviation of the phase angle from α_{best}: $m = \overline{\cos\{\alpha - \alpha(\text{best})\}}$. It is also equal to $\dfrac{I_1(X)}{I_0(X)}$ for acentric reflections and to $\tanh\left(\dfrac{X}{2}\right)$ for centric reflections. For the definition of X see Section 8.2.

Appendix 3
The Variation in the Intensity of X-ray Radiation

When the anode of an X-ray tube is bombarded by electrons their deceleration causes the emission of photons. One electron impact gives rise to a photon with a certain amount of energy. There is no relation between the photons, either in time or in energy. Therefore, the number of emitted photons with the same energy, if measured during a time t, is not a fixed number (Figure App. 3.1). If that number is measured n times (where n is very large) with an average value of N_0, the probability of measuring N photons is

$$P(N) = \frac{1}{N!} N_0^N \exp[-N_0] \text{ (Poisson distribution)} \quad \text{(App. 3.1)}$$

For sufficiently large N_0 ($N_0 \geqslant 9$) this distribution can be replaced by the Gauss distribution:

$$P(N) = \frac{1}{\sqrt{2\pi N_0}} \exp\left[-\frac{(N - N_0)^2}{2N_0}\right] \quad \text{(App. 3.2)}$$

The general form of the Gauss distribution is

$$f(x) = \frac{1}{\sigma\sqrt{2\pi}} \exp\left[-\frac{(x - \bar{x})^2}{2\sigma^2}\right] \quad \text{(App. 3.3)}$$

The spread of the curve is usually expressed in the variance of x, which is defined as

$$\sigma^2 = \int_{x=-\infty}^{+\infty} (x - \bar{x})^2 f(x)\, dx \quad \text{(App. 3.4)}$$

Figure App. 3.1. Photons with the same energy leaving the anode as a function of time.

By comparing Eqs. (App. 3.2) and (App. 3.3) it is found that the standard deviation σ for the X-ray photon emission is $\sqrt{N_0}$.

In practice the number of photons is usually measured only once and the value of N found in that single measurement is taken as the best value with a standard deviation $\sigma = \sqrt{N}$. This is also true for synchrotron radiation, because we do not have perfect control over the physical state of the charged particles.

References

Abrahams, J.P. Acta Crystallogr. (1997) **D53**, 371–376.

Abrahams, J.P., Leslie, A.G.W., Lutter, R., and Walker, J.E. Nature (1994) **370**, 621–628.

Agarwal, R.C. Acta Crystallogr. (1978) **A34**, 791–809.

Amemiya, Y. Methods in Enzymol. (1997) **276**, 233–243.

Antson, A.A., Otridge, J., Brzozowski, A.M., Dodson, E.J., Dodson, G.G., Wilson, K.S., Smith, T.M., Yang, M., Kurecki, T., and Gollnick, P. Nature (1995) **374**, 693–700.

Arndt, U.W. and Wonacott, A.J., eds. *The Rotation Method in Crystallography.* North-Holland, Amsterdam, 1977.

Asselt, E.J. van, Perrakis, A., Kalk, K.H., Lamzin, V.S., and Dijkstra, B.W. Acta Crystallogr. (1998) **D54**, 58–73.

Azaroff, L. Acta Crystallogr. (1955) **8**, 701–704.

Badger, J. Methods in Enzymol. (1997) **277**, 344–352.

Bella, J. and Rossmann, M.G. Acta Crystallogr. (1998) **D54**, 159–174.

Bentley, G.A. and Houdusse, A. Acta Crystallogr. (1992) **A48**, 312–322.

Bhat, T.N. and Cohen, G.H. J. Appl. Cryst. (1984) **17**, 244–248.

Bhat, T.N. J. Appl. Cryst. (1988) **21**, 279–281.

Blessing, R.H., Guo, D.Y., and Langs, D.A. Acta Crystallogr. (1996) **D52**, 257–266.

Blow, D.M. and Crick, F.H.C. Acta Crystallogr. (1959) **12**, 794–802.

Blow, D.M. and Matthews, B.W. Acta Crystallogr. (1973) **A29**, 56–62.

Bowie, J.U., Lüthy, R., and Eisenberg, D. Science (1991) **253**, 164–170.

Bricogne, G. Acta Crystallogr. (1974) **A30**, 395–405.

Bricogne, G. Acta Crystallogr. (1976) **A32**, 832–847.

Bricogne, G. Acta Crystallogr. (1988) **A44**, 517–545.

Bricogne, G. in: *Isomorphous Replacement and Anomalous Scattering*, Proc. Daresbury Study Weekend, pp. 60–68; Wolf, W., Evans, P.R., and Leslie, A.G.W., eds. SERC Daresbury Laboratory, Warrington, UK, 1991.

Bricogne, G. Acta Crystallogr. (1993) **D49**, 37–60.

Bricogne, G. Methods in Enzymol. (1997) **276**, 361–423.

Brünger, A.T. Acta Crystallogr. (1990) **A46**, 46–57.

Brünger, A.T. Acta Crystallogr. (1991a) **A47**, 195–204.

Brünger, A.T. Ann. Rev. Phys. Chem. (1991b) **42**, 197–223.

Brünger, A.T. Nature (London) (1992) **355**, 472–475.

Brünger, A.T. Acta Crystallogr. (1993) **D49**, 24–36.

Brünger, A.T. Methods in Enzymol. (1997a) **276**, 558–580.

Brünger, A.T. Methods in Enzymol. (1997b) **277**, 366–396.

Brünger, A.T., Kuriyan, J., and Karplus, M. Science (1987) **235**, 458–460.

Brünger, A.T. and Nilges, M. Quarterly Rev. Biophys. (1993) **26**, 49–125.

Brzozowski, A., Derewenda, Z.S., Dodson, E.J., Dodson, G.G., and Turkenburg, J.P. Acta Crystallogr. (1992) **B48**, 307–319.

Carter, C.W. Methods: Companion to Methods Enzymol. (1990) **1**, 12–24.

Castellano, E.E., Oliva, G., and Navaza, J. J. Appl. Cryst. (1992) **25**, 281–284.

CCP4, Collaborative Computational Project, Number 4 Acta Crystallogr. (1994) **D50**, 760–763. Web page: http://www.dl.ac.uk/ccp/ccp4/main.html

Chayen, N.E. Structure (1997) **5**, 1269–1274.

Chayen, N.E. Acta Crystallogr. (1998) **D54**, 8–15.

Chayen, N.E., Shaw Stewart, P.D., Maeder, D.L., and Blow, D.M. J. Appl. Crystallogr. (1990) **23**, 297–302.

Chayen, N.E., Show Stewart, P.D., and Blow, D.M. J. Crystal Growth (1992) **122**, 176–180.

Cohen, S.L. Structure (1996) **4**, 1013–1016.

Cowtan, K.D. and Main, P. Acta Crystallogr. (1996) **D52**, 43–48.

Crick, F.H.C. and Magdoff, B.S. Acta Crystallogr. (1956) **9**, 901–908.

Crowther, R.A. Acta Crystallogr. (1969) **B25**, 2571–2580.

Crowther, R.A. in: *The Molecular Replacement Method*, pp. 173–178; Rossmann, M.G., ed. Gordon & Breach, New York, 1972.

Crowther, R.A. and Blow, D.M. Acta Crystallogr. (1967) **23**, 544–548.

Cruickshank, D.W.J. in Computing Methods in Crystallography, p. 113. J.S. Rollett, ed, Pergamon Press, Oxford, 1965.

Cruickshank, D.W.J., Helliwell, J.R., and Moffat, K. Acta Crystallogr. (1987) **A43**, 656–647.

Cruickshank, D.W.J., Helliwell, J.R., and Johnson, L.N.: "Time resolved macromolecular crystallography," Oxford Science Publications (Oxford University Press) 1992.

Cudney, B., Patel, S., Weisgraber, K., Newhouse, Y., and McPherson, A. Acta Crystallogr. (1994) **D50**, 414–423.

Cygler, M. and Desrochers, M. Acta Crystallogr. (1989) **A45**, 563–572.

Daopin, S., Davies, D.R., Schlunegger, M.P., and Grütter, M.G. Acta Crystallogr. (1994) **D50**, 85–92.

Dauter, Z. Methods Enzymol. (1997) **276**, 326–344.

Debaerdemaeker, T. and Woolfson, M.M. Acta Crystallogr. (1983) **A39**, 193–196.

Derewenda, U., Swenson, L., Green, R., Wei, Y., Morosoli, R., Sharec, F., Kluepfel, D., and Derewenda, Z.S. J. Biol. Chem. (1994) **269**, 20811–20814.

DeTitta, G.T., Weeks, C.M., Thuman, P., Miller, R., and Hauptman, H.A. Acta Crystallogr. (1994) **A50**, 203–210.

Diamond, R. Acta Crystallogr. (1971) **27A**, 436–452.

Dickerson, R.E., Weinzierl, J.E., and Palmer, R.A. Acta Crystallogr. (1968) **B24**, 997–1003.

Dijkstra, B.W., Kalk, K.H., Hol, W.G.J., and Drenth, J. J. Mol. Biol. (1981) **147**, 97–123.

Dodson, E. and Vijayan, M. Acta Crystallogr. (1971) **B27**, 2402–2411.

Driessen, H.P.C., Bax, B., Slingsby, C., Lindley, P.F., Mahadevan, D., Moss, D.S., and Tickle, I.J. Acta Crystallogr. (1991) **B47**, 987–997.

Ducruix, A. and Giegé, R. *Crystallization of Nucleic Acids and Proteins, a Practical Approach.* IRL Press, Oxford, 1992.

Eagles, P.A.M., Johnson, L.N., Joynson, M.A., McMurray, C.H., and Gutfreund, H. J. Mol. Biol. (1969) **45**, 533–544.

Ealick, S.E. Structure (1997) **5**, 469–472.

Engh, R.A. and Huber, R. Acta Crystallogr. (1991) **A47**, 392–400.

Epp, O., Steigemann, W., Formanek, H., and Huber, R. Eur. J. Biochem. (1971) **20**, 432–437.

Ferré-d'Amaré, A.R. and Burley, S.K. Structure (1994) **2**, 357–359.

Fisher, R.A. Mess. of Math. (1912) **41**, 155–160.

Freund, A.K. Structure (1996) **4**, 121–125.

Fujinaga, M. and Read, R.J. J. Appl. Crystallogr. (1987) **20**, 517–521.

Furey, W. and Swaminathan, S. Methods Enzymol. (1997) **277**, 590–620.

Garcia, A.E., Krumhansl, J.A., and Frauenfelder, H. PROTEINS, Structure, Function and Genetics (1997) **29**, 153–160.

García-Ruiz, J.M. and Moreno, A. Acta Crystallogr. (1994) **D50**, 484–490.

Garman, E.F. and Schneider, T.R. J. Appl. Cryst. (1997) **30**, 211–237.

George, A. and Wilson, W.W. Acta Crystallogr. (1994) **D50**, 361–365.

Gewirth, D., The HKL manual. A description of the programs DENZO, XDISPLAYF, and SCALEPACK: An Oscillation Data Processing Suite for Macromolecular Crystallography (1997).

Gilliland, G.L. and Ladner, J.E. Curr. Opin. Struct. Biol. (1996) **6**, 595–603.

Gilliland, G.L. Methods Enzymol. (1997) **277**, 546–556.

Gonzalez, A., Denny, R., and Nave, C. Acta Crystallogr. (1994) **D50**, 276–282.

Gouaux, E. Structure (1998) **6**, 5–10.

Gros, P., van Gunsteren, W.F., and Hol, W.G.J. Science (1990) **249**, 1149–1153.

Gruner, S.M. Curr. Opin. Struct. Biol. (1994) **4**, 765–769.

Guss, J.M., Merritt, E.A., Phizackerley, R.P., Hedman, B., Murata, M., Hodgson, K.O., and Freeman, H.C. Science (1988) **241**, 806–811.

Hahn, T. ed., *International Tables for Crystallography*, Vol. A. D. Reidel, Dordrecht, 1993.

Hajdu, J. and Johnson, L.N. Biochemistry (1990) **29**, 1669–1678.

Harada, Y., Lifchitz, A., Berthou, J., and Jolles, P. Acta Crystallogr. (1981) **A37**, 398–406.

Harker, D. Acta Crystallogr. (1956) **9**, 1–9.

Harp, J.M., Timm, D.E., and Bunick, G.J. American Crystallographic Association, pp. 66 (WeBo1) and 148 (P205) July 19–25, St. Louis, Missouri, 1997.

Harp, J.M., Timm, D.E., and Bunick, G.J. Acta Crystallogr. (1998) **D54**, 622–628.

Hauptman, H.A. Acta Crystallogr. (1982) **A38**, 289–294.

Hauptman, H.A. Methods Enzymol. (1997a) **277**, 3–13.

Hauptman, H.A. Current Opin. Structural Biology (1997b) **7**, 672–680.

Heitler, W.G. The Quantum theory of radiation, 3rd ed. Oxford University Press, London, 1966.

Helliwell, J.R. *Macromolecular Crystallography with Synchrotron Radiation.* Cambridge University Press, Cambridge, 1992.

Helliwell, J.R., Ealick, S., Doing, P., Irving, T., and Szebenyi, M. Acta Crystallogr. (1993) **D49**, 120–128.

Hendrickson, W.A. and Lattman, E.E. Acta Crystallogr. (1970) **B26**, 136–143.

Hendrickson, W.A. and Ward, K.B. Acta Crystallogr. (1976) **A32**, 778–780.

Hendrickson, W.A. Methods Enzymol. (1985) **115**, 252–270.

Hendrickson, W.A., Love, W.E., and Murray, G.C. J. Mol. Biol. (1968) **33**, 829–842.

Hendrickson, W.A., Smith J.L., Phizackerley, R.P., and Merritt, E.A. Proteins (1988) **4**, 77–88.

Hendrickson, W.A., Horton, J.R., and LeMaster, D.M. EMBO J. (1990) **5**, 1665–1672.

Hiremath, C.N., Munshi, S.K., and Murthy, M.R.N. Acta Crystallogr. (1990) **B46**, 562–567.

Hodel, A., Kim, S.-H., and Brünger, A.T. Acta Crystallogr. (1992) **A48**, 851–858.

Hönl, H. Annalen der Physik, 5. Folge (1933) **18**, 625–655.

Hooft, R.W.W., Vriend, G., Sander. C., and Abola, E.E. Nature (1996) **381**, 272.

Hoppe, W.Z. Elektrochemie (1957) **61**, 1076–1083.

Huber, R. in: *Molecular Replacement. Proceedings of the Daresbury Study Weekend*, 15–16 February, pp. 58–61; Machin, P.A., ed. SERC Daresbury Laboratory, Warrington, 1985.

Huber, R. and Schneider, M.J. Appl. Cryst. (1985) **18**, 165–169.

International Tables for Crystallography, published for the International Union of Crystallography by Kluwer Academic Publishers, Dordrecht 1995.

Jack, A. and Levitt, M. Acta Crystallogr. (1978) **A34**, 931–935.

Jacobson, R.H., Zhang, X-J., DuBose, R.F., and Matthews, B.W. Nature (1994) **369**, 761–766.

James, R.W. *The Optical Principles of the Diffraction of X-Rays*, Chapter 4. London, Bell, 1965.

Jancarik, J. and Kim, S.-H. J. Appl. Cryst. (1991) **24**, 409–411.

Johnson, J.E. Acta Crystallogr. (1978) **B34**, 576–577.

Jones, T.A., Zou, J.-Y., and Cowan, S.W. Acta Crystallogr. (1991) **A47**, 110–119.

Kahn, R., Fourme, R., Gadet, A., Janin, J., Dumas, C., and André, D. J. Appl. Cryst. (1982) **15**, 330–337.

Kam, Y.J.Z. and Eisenberg, D. Protein Science (1994) **3**, 687–695.

Karle, J. Int. J. Quant. Chem. Symp. (1980) **7**, 357–367.

Karle, J. Acta Crystallogr. (1989) **A45**, 765–781.

Kartha, G. and Parthasarathy, R. Acta Crystallogr. (1965) **18**, 749–753.

Kauzmann, W., Quantum Chemistry. Academic Press, New York, 1957.

Klein, O. and Nishina, Y. Z. für Physik (1929) **52**, 853–868.

Kleywegt, G.J., Bergfors, T., Senn, H., Le Motte, P., Gsell, B., Shudo, K., and Jones, T.A. Structure (1994) **2**, 1241–1258.

Kleywegt, G.J. and Jones, T.A. Structure (1995) **3**, 535–540.

Kleywegt, G.J. and Brünger, A.T. Structure (1996) **4**, 897–904.

Kleywegt, G.J., Hoier, H., and Jones, T.A. Acta Crystallogr. (1996) **D52**, 858–863.

Kleywegt, G.J. and Read, R.J. Structure (1997) **5**, 1557–1569.

Krishna Murthy, H.M., Hendrickson, W.A., Orme-Johnson, W.H., Merritt, E.A., and Phizackerley, R.P. J. Biol. Chem. (1988) **263**, 18430–18436.

La Fortelle, E. de and Bricogne, G. Trans. of the ACA Meeting, Likelihood, Bayesian Inference and their Application to the Solution of New Structures, Bricogne, G. and Carter, C., eds., 1994.

La Fortelle, E. de and Bricogne, G. Methods Enzymol. (1997) **276**, 472–494.

Lamzin, V.S. and Wilson, K.S. Acta Crystallogr. (1993) **D49**, 129–147.

Lamzin, V.S. and Wilson, K.S. Methods in Enzymol. (1997) **277**, 269–305.

Landau, E.M. and Rosenbusch, J.P. Proc. Natl. Acad. Sci. (1996) **93**, 14532–14535.

Landau, E.M., Rummel, G., Cowan-Jacob, S.W., and Rosenbusch, J.P. J. Phys. Chem. (1997) **101**, 1935–1937.

Laskowski, R.A., MacArthur, M.W., Moss, D.S., and Thornton, J.M. J. Appl. Crystallogr. (1993a) **26**, 283–291.

Laskowski, R.A., Moss, D.S., and Thornton, J.M. J. Mol. Biol. (1993b) **231**, 1049–1067.

Leahy, D.J., Hendrickson, W.A., Aukhil, I., and Erickson, H.P. Science (1992) **258**, 987–991.

Leslie, A.G.W. Acta Crystallogr. (1987) **A43**, 134–136.

Lindblom, G. and Rilfors, L. Biochim. Biophys. Acta (1989) **988**, 221–256.

Lunin, V.Y. Acta Crystallogr. (1988) **A44**, 144–150.

Lunin, V.Y. and Skovoroda, T.P. Acta Crystallogr. (1991) **A47**, 45–52.

Lunin, V.Y., Urzhumtsev, A.G., and Skovoroda, T.P. Acta Crystallogr. (1990) **A46**, 540–544.

Lunin, V.Y. and Vernoslova, E.A. Acta Crystallogr. (1991) **A47**, 238–243.

Lüthy, R., Bowie, J.U., and Eisenberg, D. Nature (London) (1992) **356**, 83–85.

Lutter, R., Abrahams, J.P., van Raaij, M.J., Todd, R.J., Lundqvist, T., Buchanan, S.K., Leslie, A.G.W., and Walker, J.E. J. Mol. Biol. (1993) **229**, 787–790.

Luzzati, V. Acta Crystallogr. (1952) **5**, 802–810.

MacArthur, M.W., Laskowski, R.A., and Thornton, J.M. Curr. Opin. Struct. Biol. (1994) **4**, 731–737.

Machin, P.A. ed., *Molecular Replacement. Proceedings of the Daresbury Study Weekend*, 15–16 February. SERC Daresbury Laboratory, Warrington, 1985.

Main, P. Acta Crystallogr. (1967) **23**, 50–54.

Matthews, B.W. J. Mol. Biol. (1968) **33**, 491–497.

Matthews, B.W. and Czerwinski, E.W. Acta Crystallogr. (1975) **A31**, 480–487.

McKenna, R., Di Xia, Willingmann, P., Ilag, L.L. Krishnaswamy, S., Rossmann, M.G., Olson, N.H., Baker, T.S., and Incardona, N.L. Nature (London) (1992) **355**, 137–143.

McPherson, A., Jr. Methods Biochem. Anal. (1976) **23**, 249–345.

McPherson, A., Jr. J. Crystal Growth (1992) **122**, 161–167.

McPherson, A., Malkin, A.J., and Kuznetsov, Y.G. Structure (1995) **3**, 759–768.

Michel, H. *Crystallization of Membrane Proteins*. CRC Press, London, 1990.

Miller, R., Gallo, S.M., Khalak, H.G., and Weeks, C.M. J. Appl. Cryst. (1994) **27**, 613–621.

Moews, P.C. and Kretsinger, R.H. J. Mol. Biol. (1975) **91**, 201–228.

Moffat, K. Ann. Rev. Biophys. Biophys. Chem. (1989) **18**, 309–332.

Moffat, K. Methods in Enzymol. (1997) **277**, 433–447.

Morris, A.L., MacArthur, M.W., Hutchinson, E.G., and Thornton, J.M. Proteins (1992) **12**, 345–364.

Moss, D.S. Acta Crystallogr. (1985) **A41**, 470–475.

Murshudov, G.N., Vagin, A.A., and Dodson, E.J., Acta Crystallogr. (1997) **D53**, 240–255.

Nagai, K., Oubridge, C., Jessen, T.-H., Li, J., and Evans, P.R. Nature (London) (1990) **348**, 515–520.

Nagai, K., Evans, P.R., Li, J., and Oubridge, Ch. *Proceedings of the CCP4 Study Weekend*, January, pp. 141–149. Daresbury Laboratory Warrington, 1991.

Narasinga Rao, S., Jyh-Hwang Jih, and Hartsuck, J.A. Acta Crystallogr. (1980) **A36**, 878–884.

Narten, A.H. and Levy, H.A. in: Water, a comprehensive treatise, Vol. I (Frank, F., ed.) p. 311–332; Plenum Press, New York, 1972.

Navaza, J. Acta Crystallogr. (1994) **A50**, 157–163.

Navaza, J. and Vernoslova, E. Acta Crystallogr. (1995) **A51**, 445–449.

Navaza, J. and Soludjian, P. Methods Enzymol. (1997) **276**, 581–594.

Ohlendorf, D.H. Acta Crystallogr. (1994) **D50**, 805–812.

Otwinowski, Z. Proceedings of the CCP4 Study Weekend, 25–26 January, pp. 80–86, Wolf, W., Evans, P.R., and Leslie, A.G.W., eds., 1991.

Pähler, A., Smith, J.L., and Hendrickson, W.A. Acta Crystallogr. (1990) **A46**, 537–540.

Pai, E.F. Curr. Opin. Struct. Biol. (1992) **2**, 821–827.

Pannu, N.S. and Read, R.J. Acta Crystallogr. (1996) **A52**, 659–668.

Perrakis, A., Sixma, T.K., Wilson, K.S., and Lamzin, V.S. Acta Crystallogr. (1997) **D53**, 448–455.

Pflugrath, J.W. Curr. Opin. Struct. Biol. (1992) **2**, 811–815.

Phillips, W.C. Methods Enzymol. (1985) **114**, 300–329.

Podjarny, A.D., Bhat, T.N., and Zwick, M. Ann. Rev. Biophys. Biophys. Chem. (1987) **16**, 351–373.

Polikarpov, I. and Sawyer, L. Joint CCP4 and ESF-EACBM Newsletter on Protein Crystallography, Daresbury Laboratory, (1995) **31**, 5–11.

Polikarpov, I., Teplyakov, A., and Oliva, G. Acta Crystallogr. (1997) **D53**, 734–737.

Rabinovich, D. and Shakked, Z. Acta Crystallogr. (1984) **A40**, 195–200.

Ramachandran, G.N., Ramakrishnan, C., and Sasisekharan, V.J. Mol. Biol. (1963) **7**, 95–99.

Rayment, I. Acta Crystallogr. (1983) **A39**, 102–116.

Read, R.J. Acta Crystallogr. (1986) **A42**, 140–149.

Read, R.J. Crystallographic Computing School, Bischenberg, 1990a.

Read, R.J. Acta Crystallogr. (1990b) **A46**, 900–912.

Read, R.J. Methods in Enzymol. (1997) **277**, 110–128.

Read, R.J. and Schierbeek, A.J. J. Appl. Crystallogr. (1988) **21**, 490–495.

Rees, D.C. Acta Crystallogr. (1983) **A39**, 916–920.

Ren, Z. and Moffat, K. J. Appl. Cryst. (1995) **28**, 482–493.

Rice, L.M. and Brünger, A.T. Proteins (1994) **19**, 277–290.

Rosenbaum, G., Holmes, K.G., and Witz, J. Nature (London) (1971) **230**, 434–437.

Rossmann, M.G. Acta Crystallogr. (1990) **A46**, 73–82.

Rossmann, M.G. Structure (1995) **5**, 650–655.

Rossmann, M.G. and Blow, D.M. Acta Crystallogr. (1961) **14**, 641–647.

Rossmann, M.G. and Blow, D.M. Acta Crystallogr. (1962) **15**, 24–31.

Rossmann, M.G. and Blow, D.M. Acta Crystallogr. (1963) **16**, 39–45.

Rossmann, M.G., McKenna, R., Liang Tong, Di Xia, Jin-Bi Dai, Hao Wu, Hok-Kin Choi, and Lynch, R.E. J. Appl. Crystallogr. (1992) **25**, 166–180.

Rould, M.A., Perona, J.J., and Steitz, T.A. Acta Crystallogr. (1992) **A48**, 751–756.

Sakabe, N. J. Appl. Crystallogr. (1983) **16**, 542–547.

Sakabe, N. Nuclear Instruments and Methods in Physics Research (1991) **A103**, 448–463.

Sauer, O., Schmidt, A., and Kratky, C. J. Appl. Cryst. (1997) **30**, 476–486.

Shah, S.A., Shen, B.W., and Brünger, A.T. Structure (1997) **5**, 1067–1075.

Sheldrick, G.M., Dauter, Z., Wilson, K.S., Hope, H., and Sieker, L.C. Acta Crystallogr. (1993) **D49**, 18–23.

Sheldrick, G.M. and Gould, R.O. Acta Crystallogr. (1995) **B51**, 423–431.

Sheldrick, G.M. and Schneider, T.R. Methods Enzymol. (1997) **277**, 319–343.

Sigler, P.B. Biochemistry (1970) **9**, 3609–3617.

Sim, G.A. Acta Crystallogr. (1959) **12**, 813–815.

Sim, G.A. Acta crystallogr. (1960) **13**, 511–512.

Sippl, M.J. Proteins (1993) **17**, 355–362.

Skilling, J. in: *Maximum Entropy and Bayesian Methods*, pp. 45–52. Skilling, J., ed. Kluwer Academic Publishers, Dordrecht, 1988.

Smith, G.D., Nagar, B., Rini, J.M., Hauptman, H.A., and Blessing, R.H. Acta Crystallogr. (1998) **D54**, 799–804.

Sowadski, J.M. Curr. Opin. Struct. Biol. (1994) **4**, 761–764.

Spurlino, J.C., Smallwood, A.M., Carlton, D.D., Banks, T.M., Vavra, K.J., Johnson, J.S., Cook, E.R., Falvo, J., Wahl, R.C., Pulvino, T.A., Wendoloski, J.J., and Smith, D.L. Proteins (1994) **19**, 98–109.

Srajer, V., Teng, T.-Y., Ursby, T., Pradervand, C., Ren, Z., Adachi, S., Bourgeois, D., Wulff, M., and Moffat, K. Science (1996) **274**, 1726–1729.

Srinivasan, R. Acta Crystallogr. (1966) **20**, 143–144.

Srinivasan, R. and Chandrasekaran, R. Indian J. Pure Appl. Phys. (1966) **4**, 178–186.

Srinivasan, R. and Ramachandran, G.N. Acta Crystallogr. (1965) **19**, 1008–1014.

Stec, B., Zhou, R., and Teeter, M.M. Acta Crystallogr. (1995) **D51**, 663–681.

Stewart, J.M. and Karle, J. Acta Crystallogr. (1976) **A32**, 1005–1007.

Stoddard, B.L., Dean, A., and Bash, P.A. Nature Struct. Biol. (1996) **3**, 590–595.

Strijtveen, B. and Kellogg, R.M. Tetrahedron (1987) **43**(21), 5045.

Stuart, D.I. and Jones, E.Y. Curr Opin. Struct. Biol. (1993) **3**, 737–740.

Stubbs, M.T., Nar, H., Lowe, J., Huber, R., Ladenstein, R., Spangfort, M.D., and Svensson, L.A. Acta Crystallogr. (1996) **D52**, 447–452.

Sussman, J.L. Methods Enzymol. (1985) **115B**, 271–303.

Sweet, R.M., Singer, P.T., and Smalås, A. Acta Crystallogr. (1993) **D49**, 305–307.

Templeton, D.H. and Templeton, L.K. Acta Crystallogr. (1991) **A47**, 414–420.

Ten Eyck, L.F. Acta Crystallogr. (1973) **A29**, 183–191.

Teng, T.Y. J. Appl. Cryst. (1990) **23**, 387–391.

Terwilliger, T.C. Acta Crystallogr. (1994a) **D50**, 11–16.

Terwilliger, T.C. Acta Crystallogr. (1994b) **D50**, 17–23.

Terwilliger, T.C. and Eisenberg, D. Acta Crystallogr. (1983) **A39**, 813–817.

Tickle, I.J., Laskowski, R.A., and Moss, D.S. Acta Crystallogr. (1998) **D54**, 243–252.

Tollin, P. and Rossmann, M.G. Acta Crystallogr. (1966) **21**, 872–876.

Tollin, P., Main, P., and Rossmann, M.G. Acta Crystallogr. (1966) **20**, 404–407.

Tong, L. and Rossmann, M.G. Acta Crystallogr. (1990) **A46**, 783–792.

Tronrud, D.E. Acta Crystallogr. (1992) **A48**, 912–916.

Tronrud, D.E., Ten Eyck, L.F., and Matthews, B.W. Acta Crystallogr. (1987) **A43**, 489–501.

Tucker, A.D., Baty, D., Parker, M.W., Pattus, F., Lazdunski, C., and Tsernoglou, D. Protein Eng. (1989) **2**, 399–405.

Turner, M.A., Yuan, C.-S., Borchardt, R.T., Hershfield, M.S., Smith, G.D., and Howell, P.L. Nature Struct. Biol. (1998) **5**, 369–376.

Vellieux, F.M.D. and Read, R.J. Methods in Enzymol. (1997) **277**, 18–53.

Walter, R.L., Thiel, D.J., Barna, S.L., Tate, M.W., Wall, M.E., Eikenberry, E.F., Gruner, S.M., and Ealick, S.E. Structure (1995) **3**, 835–844.

Wang, B.-C. Methods Enzymol. (1985) **115**, 90–112

Weeks, C.M., DeTitta, G.T., Hauptman, H.A., Thuman, P., and Miller, R. Acta Crystallogr. (1994) **A50**, 210–220.

Westbrook, E.M. Methods Enzymol. (1985) **114**, 187–196.

Wilmanns, M. and Eisenberg, D. Proc. Natl. Acad. Sci. U.S.A. (1993) **90**, 1379–1383.

Wilson, A.J.C. Nature (1942) **150**, 151–152.

Wilson, A.J.C. Acta Crystallogr. (1950) **3**, 397–398.

Wodak, S.J., Pontius, J., Vaguine, A., and Richelle, J. Proc. CCP4 Study Weekend 6–7 January, Hunter, W.N., Thornton, J.M., and Bailey, S., eds., 1995.

Xiang, S., Carter, C.W., Bricogne, G., and Gilmore, C.J. Acta Crystallogr. (1993) **D49**, 193–212.

Zhang, K.Y.J. and Main, P. Acta Crystallogr. (1990a) **A46**, 41–46.

Zhang, K.Y.J. and Main, P. Acta Crystallogr. (1990b) **A46**, 377–381.

Zhang, K.Y.J. Cowtan, K., and Main, P. Methods in Enzymol. (1997) **277**, 53–64.

Zhang, X.-J. and Matthews, B.W. Acta Crystallogr. (1994) **D50**, 675–686.

Index